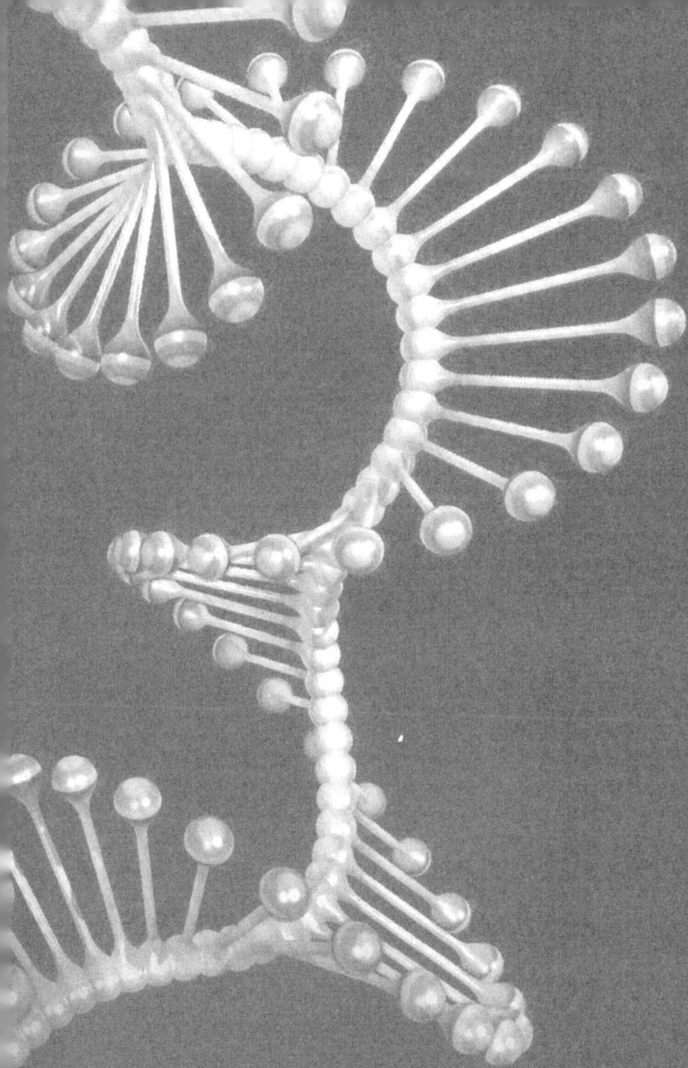

Templated DNA Nanotechnology

Templated DNA Nanotechnology

Functional DNA Nanoarchitectonics

edited by

Thimmaiah Govindaraju

PAN STANFORD PUBLISHING

Published by

Pan Stanford Publishing Pte. Ltd.
Penthouse Level, Suntec Tower 3
8 Temasek Boulevard
Singapore 038988

Email: editorial@panstanford.com
Web: www.panstanford.com

British Library Cataloguing-in-Publication Data
A catalogue record for this book is available from the British Library.

Templated DNA Nanotechnology:
Functional DNA Nanoarchitectonics

ISBN 978-981-4800-21-1 (Hardcover)
ISBN 978-0-429-42866-1 (eBook)

Contents

Preface

Nucleic acids have structurally evolved over billions of years to effectively store and transfer genetic information. One of the simplest, elegant, and information-rich nucleic acid structures is the double-stranded DNA (dsDNA). The remarkable specificity of molecular recognition between complementary nucleobases and structural robustness have made DNA an attractive biomolecule for molecular programming that involve information coding and decoding purposes. Two complementary single-stranded DNA (ssDNA) molecules spontaneously hybridize through canonical Watson–Crick (WC) base pairing to form a right-handed DNA double helix (dsDNA), which is ~2 nm thick and exhibits a persistence length of ~50 nm, making it a perfect nanomaterial at the molecular level. The disruptive thinking of Nadrian Seeman in the 1980s led to the utilization of DNA as a nanomolecular building block (analogy to a brick) to construct emergent molecular systems and nanomaterial objects of varied size and shape. This bottom-up approach to construct nanoscale architecture with DNA marked the beginning of a new and exciting field of research, namely DNA nanotechnology, which is contributing significantly to the broad area of nanoscience and nanotechnology. DNA nanotechnology was further advanced by the concept of DNA origami, which offers complexity and diversity, developed by Paul Rothemund and others. DNA nanotechnology has been an active and celebrated area of research. It has evolved significantly from its original form, although it relies on complex computer-based design, programming, and the use of very long DNA sequences, while concerns related to the ease of adoptability, reproducibility, and cost-effectiveness for practically utility need to be addressed. On the other hand, templated or co-assembly of functional molecules or any designed structural units and short DNA/RNA oligonucleotides (dB_n, where B = nucleobase, $n < 40$–50, which corresponds to persistence length of DNA) offers simple and cost-effective means to construct hybrid DNA nanosystems and architectures with novel properties and guaranteed applications.

Such a solution-processible technology based on controlled molecular assembly of functional organic molecules and nucleic acids is an emerging area of research, and its full potential needs to be harnessed through dedicated research efforts. Specifically, covalent and noncovalent co-assembly of functional units and short oligonucleotides constitutes a minimalistic approach and an attractive alternative to construct functional nanoarchitectures. The "molecular architectonics" (Avinash and Govindaraju, 2018) of designer small molecules and short DNA sequences through complementary binding interaction is certain to generate well-defined functional nanoarchitectures with realistic applications in areas ranging from biology to materials science and is termed "DNA nanoarchitectonics." The concept of nanoarchitectonics has been introduced and pioneered by Masakazu Aono and Katsuhiko Ariga, NIMS, Japan.

Contextually, in line with the proverb famously mentioned by Niels Bohr, "It is very hard to predict, especially about the future," it is impossible to predict the future, especially of DNA, which evolved through rigorous selection of nature and synthetic chemistry innovations. We have to agree on one thing that there is a bright future ahead for DNA, either as a genetic material or as a construction material! All we need to do is simplify the concepts and working models to maximize the applications of DNA/RNA in all spheres of scientific and technological disciplines. Synthetic organic chemistry plays a key role in accessing novel functional molecules, modified ssDNA/RNA and their conjugates, and incorporation of desired chromophores. The DNA/RNA nanoarchitectonics is highly dependent on the conventional or unconventional complementary base pairing, and finding suitable molecular partners is quite challenging but certainly not impossible. For instance, generating 3D objects using templated assemblies of small molecules and DNA with precise control of molecular organization on the nano-, meso-, and macroscale without any dispersion is another challenging task. Structural and functional switching at molecular and material levels, with reasonable control on the time and response to the chosen stimulus, is the key to success. The future directions for the field also include programming of templated assemblies of synthetic

---

Avinash, M. B. and Govindaraju, T. (2018). *Acc. Chem. Res.*, **51**, 414–426.

small molecules and DNA to control biological events. The fields of molecular architectonics and DNA nanoarchitectonics involving small functional molecules and short oligonucleotides are expected to grow quickly and find a wide range of applications in chemistry, biology, engineering, (bio)materials, and biomedicine.

In this book, novel approaches adapted by leading researchers in different parts of the world to create functional nucleic acid molecular systems and nanoarchitectures is covered under the topic "Templated DNA Nanotechnology: Functional DNA Nanoarchitectonics," a major subtopic envisaged under the broad subject area of DNA nanotechnology. Individual chapters contributed by active practitioners provide the readers with fundamental and advanced knowledge emanated from their own work and others in the subject area. Each chapter provides plenty of illustrations, historical perspectives, case studies and practical examples, critical discussions, and future prospects. This book can serve as a practical handbook or advanced graduate textbook for students and advanced-level researchers.

Hans-Achim Wagenknecht and Robert Hofsäß discuss covalent DNA-chromophore architectures, photoinduced processes within the chromophore stacks, and their use as artificial light-harvesting systems in the first chapter. This thorough and detailed discussion of a variety of covalent DNA conjugates and their applications sets the necessary foundation for the book. In Chapter 2, Prolay Das and Seema Singh describe pre- and postsynthetic modifications of DNA, programmed DNA–synthetic molecule hybrids, characterization, and their nanostructures, properties, and applications. From our group, we describe a novel concept of mutually templated functional molecule–DNA architectures supported by the canonical and noncanonical hydrogen-bonding interactions of nucleobases, and their functional properties and applications in Chapter 3. The use of functional molecules and their assemblies as a template for ssDNA to generate hybrid DNA ensembles with emergent properties guarantees efficient and cost-effective practical applications across disciplines. Reji Varghese et al. provide a detailed description of design, challenging synthesis, self-assembly of DNA amphiphiles, and the resultant DNA nanostructure as soft nanoscaffolds for various applications in Chapter 4. The self-assembly of DNA amphiphiles leads to nanoarchitectures with a dense display of ssDNA on the surface, which is extremely hard to achieve using the principles of structural DNA nanotechnology and DNA origami approaches. In Chapter 5, Philippe Barthélémy et al. present hybrid amphiphiles featuring DNA nucleobase–lipid conjugates and their interesting self-assembly properties to design a new generation of functional and responsive materials. The authors discuss the behaviors of these systems at the nanometer scale, translation of the nanosystems into advanced soft materials at the macroscopic level, and new opportunities that these soft materials bring to the field of biomedicine. In Chapter 6, Mahesh Hariharan et al. introduce the reader to thorough and eloquent details of photoexcited state processes in templated DNA–chromophore assemblies. The advanced photophysical and photofunctional properties help in understanding the mechanism of light-induced lesions and electronic communication between the chromophores and DNA nucleobases, which in turn help in deciphering the oxidative damage repair pathways in the biological systems and the development of light-harvesting and optoelectronic devices, presented in this and other chapters. Mitsunobu Nakamura et al. deal with another interesting concept of metal-supported construction of a templated assembly of multichromophore arrays and DNA in Chapter 7. In addition to providing a general introduction to the concept of metal-supported functional chromophore–DNA assemblies, the authors describe multichromophore arrays supported on a Au substrate and their photocurrent responses due to electron transfer through the π-stacked array. Chapter 8 by Alberto Bianco et al. discusses the merger of DNA nanotechnology and carbon nanotechnology, which is very exciting and futuristic. Interactions of DNA and RNA with carbon materials and the utility of their complexes in gene and cancer therapy are described. A description of noncovalent and covalent conjugates of single- and double-stranded DNA/RNA with carbon nanotubes (CNTs), graphene oxide (GO), and their novel functional properties is particularly noteworthy. Joshy Joseph et al. present functional nanoarchitectures fabricated through the self-assembly of fullerene derivatives and DNA, with special focus on optoelectronic and biomedical applications, in Chapter 9. The authors introduce the reader to different modes of DNA–fullerene interactions, fullerene-induced DNA condensation, and higher-order nanostructures, including fullerene cluster–assisted DNA assemblies. In Chapter 10,

Ashwani Sharma et al. discuss nucleic acid–based molecular devices and biosensors. A major portion of this chapter deals with aptamers and their application in designing biosensors for detection, imaging, diagnostics, and therapeutics. The final chapter (Chapter 11) discusses the utilization of highly predictable molecular recognition of nucleobases, unique characteristics of flexible ssDNA, rigid dsDNA, and associated reversible conformational transformations of canonical and noncanonical DNA structures for developing nanoswitches and devices with numerous applications, including sensors of ions, pH, and biomolecules, as well as to develop diagnostic and therapeutic tools. Overall, all the chapters are well organized, interconnected, and futuristic in presentation.

I thank all the authors for their excellent contributions to this unique book through plenty of case studies, attractive illustrations, thought-provoking concepts, and future outlook. Finally, I take this opportunity to thank Prof. C. N. R. Rao, the messiah of materials science and nanotechnology, for his constant support and encouragement and dedicate this book to his 85th birthday.

Thimmaiah Govindaraju
Jawaharlal Nehru Centre for Advanced Scientific Research,
Bengaluru 560064,
Karnataka, India
2018

Chapter 1

Multichromophore Stacks in DNA: Novel Light-Harvesting Systems

Robert Hofsäß and Hans-Achim Wagenknecht
Karlsruhe Institute of Technology (KIT), Institute of Organic Chemistry, Fritz-Haber-Weg 6, 76131 Karlsruhe, Germany
wagenknecht@kit.edu

1.1 Introduction

Depletion of fossil fuel resources and increasing power consumption of the world led to a rise in the demand for new alternative energy sources. Sunlight offers a green, sustainable, and unlimited source of energy. The major tasks for scientists are to harness this energy source and to develop methods to store it. An archetype for potential light-harvesting systems can be found in nature. Plants use photosynthesis to convert light energy into chemical energy. In these systems, a so-called light-harvesting antenna (LHA) plays a key role in the collection of sunlight. Different chromophores absorb the incoming light and pass the energy onto a reaction center, where the chemical part of photosynthesis takes place and the energy is finally stored as sugar [1, 2]. This multichromophore system offers some

Templated DNA Nanotechnology: Functional DNA Nanoarchitectonics
Edited by Thimmaiah Govindaraju

ISBN 978-981-4800-21-1 (Hardcover), 978-0-429-42866-1 (eBook)
www.panstanford.com

great advantages over the use of single monomers. A greater part of the visible light is absorbed and a larger area can be irradiated since not only the reaction center absorbs the light but also the antennas, which subsequently transfer the energy to the reaction center. This way, an overall broader and more effective absorption is guaranteed.

Characterization and elucidation of the behavior of chromophores in large multichromophore stacks require the use of well-structured templates, which are able to prevent solvation and aggregation effects. Besides the biological function, the rigid DNA structure offers a suitable scaffold for the preparation of supramolecular chromophore aggregates. The regular geometry and the canonical base pairing allow excellent programmability of the sequence. Large polymers, networks, and architecture of astonishing complexity can be created this way. These structures possess high physicochemical stabilities and rigidity and show monodispersity, as seen in DNA objects of Seeman [3] and Rothemund [4]. This unique combination of properties can hardly be realized with organic building blocks—covalently, as polymers, as well as noncovalently, through self-assembly. If synthetic chromophore-modified building blocks are used in addition to the natural 2′-deoxyadenosine, thymidine, 2′-deoxyguanosine, and cytidine building blocks, DNA nanostructures with interesting optoelectronic properties may be created. The underlying DNA structure dictates the flow of electrons, photons, and charge-separated states (excitons). The efficiency and rate of these processes are controlled by the DNA-based orientation of the chromophores with respect to each other. In this way, surprisingly straightforward artificial light-harvesting systems can be created.

This chapter focuses on the design and synthesis of DNA-based multichromophore systems and their use in artificial light-harvesting systems. Photoinduced processes within the chromophore stacks will be discussed as well as important lessons learned for the development of novel systems.

1.2 Covalent DNA-Chromophore Architectures: Facile Sequence Control; Challenging Synthesis

One of the major prerequisites for the development of artificial DNA-based light-harvesting systems is the investigation of chromophore–

chromophore interactions in DNA. Especially the detailed understanding of electron transfer (EnT) processes and energy transfer (ET) processes between chromophores yields crucial information for the design of new systems. Therefore, the position of the chromophores in the strand, their distance, and their geometry must be controlled very precisely.

Besides its geometry, the most significant advantage of DNA as a scaffold is the building block chemistry, which was greatly improved over the last 50 years [5]. Today, oligonucleotides are mostly synthesized by automated, solid phase phosphoramidite chemistry, first introduced by Caruthers and Beaucage [6–9]. This method offers a couple of advantages for the organization of multichromophore arrays. The automated synthesis allows exact programming of the desired sequence and mixing of natural with chromophore-modified nucleosides. Furthermore, building blocks of the unmodified nucleosides are commercially available, even in larger scales. Chromophores can be incorporated into DNA via the corresponding artificial building blocks, which can be obtained by means of organic synthesis. This paves the way to DNA-chromophore architectures with precise and easy-to-control chromophore arrangements [10].

There are several structural types of artificial building blocks for the incorporation of chromophores into DNA (Fig. 1.1). In the following section the four most common structures, (a) substitution of the nucleoside with an acyclic or cyclic linker, (b) C-nucleosides as substitution of the nucleobase, (c) modification of the 2′-position hydroxy group, and (d) modification of the nucleobase itself will be covered in detail.

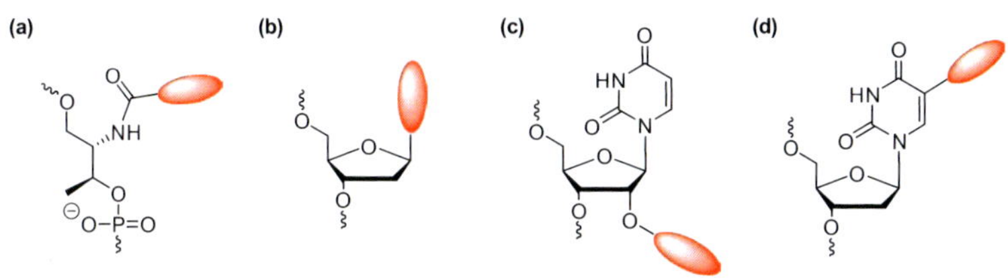

Figure 1.1 Structural types of chromophore-modified DNA building blocks. (a) D-threolinol as representative artificial acyclic linker [11], (b) C-nucleoside, (c) modification of the 2′-position, and (d) chromophore attached to the 5-position of 2′-deoxyuridine (chromophores shown in red).

1.2.1 Chromophores as Nucleoside Surrogates

The rigid and chiral ring structure of the 2′-deoxyribose moiety makes a considerable contribution to the structure of DNA and its three-dimensional, helical duplex. However, the glycosidic C–N bond between the sugar moiety and the purine base is vulnerable to acidic hydrolysis, especially with positively charged dyes as aglycon [12, 13]. In addition, the synthesis of the modified phosphoramidite building blocks requires elaborate protecting group strategies. A synthetically more straightforward and often published approach uses acyclic linkers between the phosphodiester bridges, for instance, D-threolinol [11], (S)-serinol [14], and glycol derivatives [15–17] (Fig. 1.2). In addition, flexible linkers allow flexible intercalation of the chromophore in the DNA base stack, in contrast to the forced intercalation (FIT) probes that apply short linkers [18, 19].

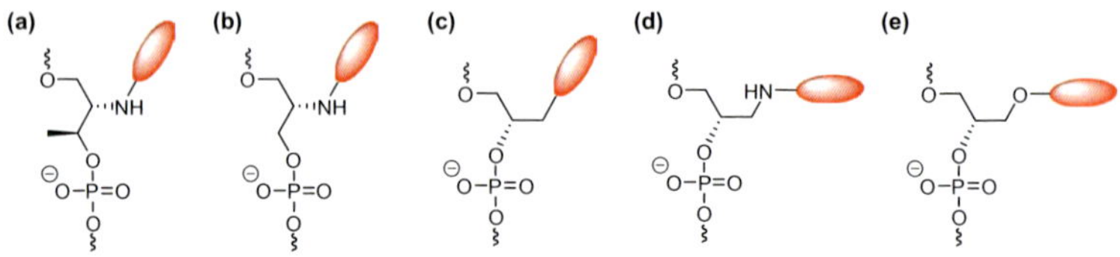

Figure 1.2 Structures of acyclic linkers (a) D-threolinol [11], (b) (S)-serinol [14] and the glycol derivatives, (c) (S)-1,2-propanediol [15], (d) (S)-3-amino-1,2-propanediol [17], and (e) (S)-1,2,3-propanetriol [16] (chromophores shown in red).

By incorporating up to six methyl red chromophores via D- and L-threolinol linkers, Asanuma and coworkers could show the remarkable influence of the DNA double helix on the arrangement of the dyes [20]. Increase of the size of the methyl red stack in the single strand led to a hypsochromic shift, increased absorption, and the absence of fluorescence. This proves the formation of H-aggregates along the single strand. If the modified strand is annealed with the complementary counterstrand, the H-absorption band disappears and a new, bathochromic-shifted band appears. Together with a bathochromic shift of the CD signal this shows that the chromophores are helically twisted in the stack inside the DNA duplex. This process is fully reversible by thermal dehybridization of the duplex. In another study Asanuma et al. investigated homo- and hetero-dye aggregates in double strands, consisting of methyl red and naphthyl red as base surrogates in the middle of the sequence using D-threolinol linkers (Fig. 1.3) [21, 22]. These D-threolinol spacers separated the chromophores but allowed efficient formation of aggregates. The methyl red-modified strand **ML1** only shows the absorption of methyl red due to the absence of dye-dye interactions. If the strand is hybridized with the methyl red–modified counterstrand **ML2**, a hypsochromic shift clearly indicates the formation of H-aggregates in the DNA duplex [21]. Similar observations were made when the methyl red strand **ML1** was hybridized with the naphthyl red–modified counterstrand **NL1**. While the single strands just showed absorption of unstacked dyes, a hypsochromic-shifted absorption band indicated the existence of hetero-H-aggregates in the duplex [23]. Both duplexes were investigated via CD spectroscopy, and a zipper-like chromophore stacking in DNA was proposed.

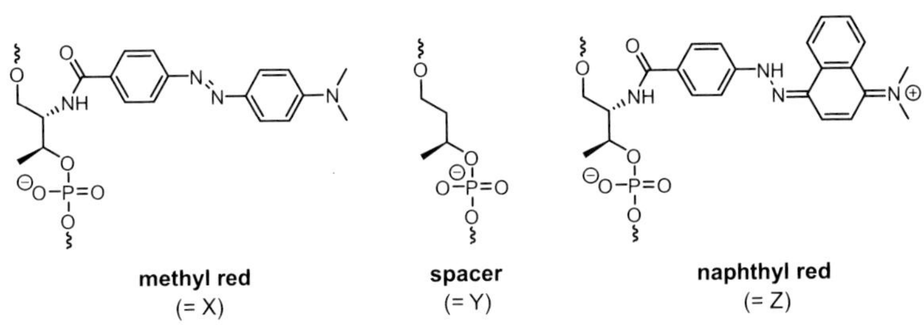

5'—G–G–T–A–T–C–X–Y–X–Y–X–Y–G–C–A–A–T–C–3' **ML1**

3'—C–C–A–T–A–G–Y–X–Y–X–Y–X–C–G–T–T–A–G–5' **ML2**

3'—C–C–A–T–A–G–Y–Z–Y–Z–Y–Z–C–G–T–T–A–G–5' **NL3**

Figure 1.3 Structures of the chromophore-modified D-threolinol linkers and sequences of the modified DNA single strand of Asanuma et al. for the investigation of homo- and hetero-dye aggregation in DNA duplexes [21, 22].

Using the same type of linker, pyrene and perylene were investigated as a potential Förster resonance energy transfer (FRET) pair in the DNA duplex (Fig. 1.4) [24]. Pyrene shows fluorescence between 370 nm and 430 nm and therefore overlaps with the absorption of perylene (390 nm – 470 nm). Selective excitation of pyrene resulted in perylene emission, proving a successful and efficient FRET. In this system, DNA plays a key role as a template, which allows the exact positioning of donor (pyrene) and acceptor (perylene) in close proximity to each other but prevents the formation of undesired excimers or exciplexes. The highest FRET efficiency could be achieved if the dyes were separated by just one base pair and only one donor was close to the acceptor. The introduction of a second donor could not improve the FRET efficiency. In further studies, Asanuma et al. investigated the distance and orientation dependency of the FRET pair by inserting up to 21 A-T pairs between donor and acceptor [25]. As expected, increase of the distance between the dyes results in lower acceptor fluorescence. FRET has a strong distance dependency ($k_{ET} \propto 1/r^6$; r = distance between donor and acceptor) and therefore can be used as a molecular ruler in biological applications [26]. However, every five A-T pairs (i.e., $n = 3$, 8, 13, and 18 base pairs) the FRET efficiency dropped significantly, which corresponds to a half-turn of the B-helix. Similar results were obtained by Wilhelmsson et al. using chromophore nucleosides as the base surrogate [27, 28]. This clearly demonstrates that the FRET efficiency is dependent not only on the distance between the chromophores but also on the orientation and angle of the transition dipole moments and that DNA is a highly suitable scaffold for precisely controlling these two factors [25].

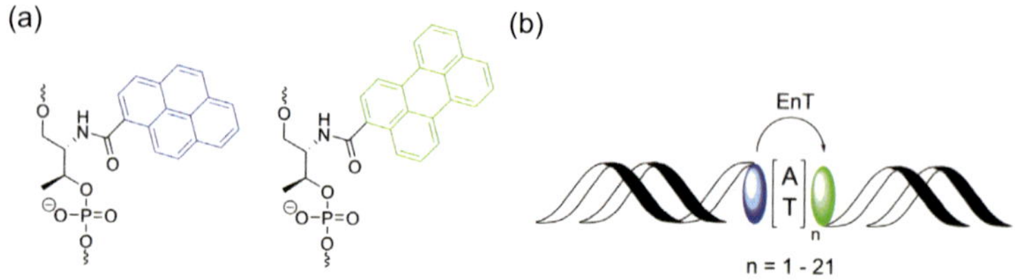

Figure 1.4 Study of distance and orientation dependency of FRET in DNA by Asanuma et al. [24, 25]. (a) Modified D-threolinol linkers carrying pyrene (blue) and perylene (green) chromophores and (b) schematic illustration of insertion of A-T pairs for increased distance between donor and acceptor.

Häner and Langenegger incorporated pyrenes [29], phenanthrolines [30], and phenanthrenes [31] as non-nucleosidic base surrogates in DNA using achiral carboxamide linkers. By hybridizing two strands containing up to 7 pyrenes each, a chromophore stack with an astonishing 14 pyrenes could be achieved inside the duplex [32]. Single strands with more than one pyrene and their respective double strands emitted excimer fluorescence and therefore proved interstrand and intrastrand interactions between the dyes. If more than 10 pyrenes were present in the DNA, they formed a right-handed helix, which was determined by CD spectroscopy. This is an astonishing result because the linker-chromophore building blocks were achiral. On the basis of these results a DNA-based LHA was developed consisting of DNA as scaffold, phenanthrenes as light-collecting antenna, and an exciplex-forming pyrene as the energy-collection center [33]. By using up to eight phenanthrenes, it could be shown that the number of photons emitted is proportional to the number of light-collecting chromophores, although the overall quantum yield of the system stayed the same. Another light-harvesting complex (LHC) was assembled with a DNA three-way junction as the framework [34]. While the phenanthrene light-collecting complex was located in one of the three stems, an exchangeable acceptor was arranged in proximity of the complex through the geometry of the three-way junction. Pyrene, a perylenebisimide, and a cyanine dye were tested as acceptors in the system. While quenching was observed with perylenebisimide, pyrene and the cyanine dye showed acceptor fluorescence and therefore a successful ET.

As a last example for the incorporation of dyes using acyclic linkers, perylene-3,4:9,10-tetracarboxylic acid bisimide (PBI) will be covered. PBIs possess excellent photostability, high fluorescence quantum yields, and strong hydrophobicity and therefore were often used for the assembly of supramolecular systems [35–37]. We used (S)-aminopropandiol linkers (Fig. 1.2) to locate PBIs in the middle and at the end of single strands [36, 37]. The respective double strands emitted excimer fluorescence of PBI dimers, proving the existence of strong π-π interactions between the dyes. In the case of the 5′-terminally modified strands, this means that two double strands are glued together by the hydrophobic interactions of the PBI termini. On the basis of these results, we developed a DNA network with triangular three-way junction building blocks (Fig. 1.5) [38]. The monomers were connected by sticky ends, which were achieved by terminal incorporation of PBI and the hydrophobic interactions between the building blocks. Furthermore, six PBIs were placed in the middle of a double strand and the helicity of the resulting chromophore stack was examined by CD spectroscopy [39]. The two single strands were modified with three PBIs each, which were separated by either thymidine or an abasic site, to allow the formation of a zipper-like structure. Interestingly, the helical arrangement of the chromophore stack depends on the counter base of the dyes. If thymidine is chosen as the counter base, a left-handed helix is formed, whereas abasic sites promote the generation of right-handed aggregates.

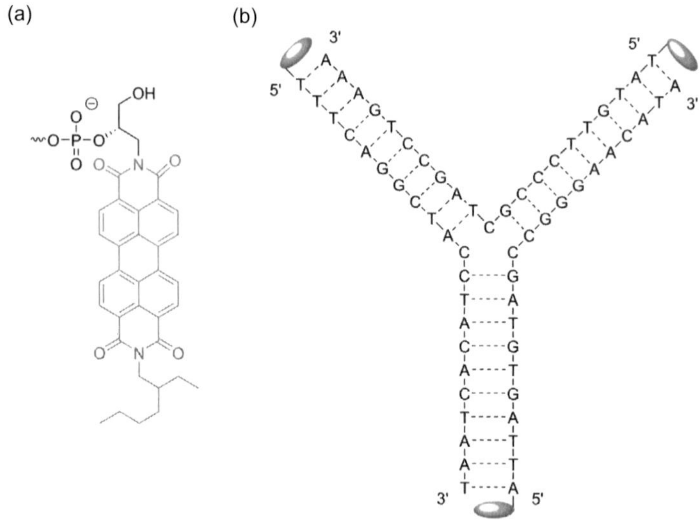

Figure 1.5 Perylenebisimide-DNA conjugates for the creation of DNA networks by Wagenknecht et al. [38]. (a) Perylenebisimide attached to an (S)-aminopropandiol linker and (b) a Y-shaped three-way DNA building block with PBI sticky ends.

1.2.2 Chromophores as Nucleobase Surrogates

The incorporation of chromophores as non-nucleosidic base surrogates is a significant alteration of the natural DNA structure. If this structure, namely the sugar-phosphate backbone, should be preserved, incorporation of chromophores as nucleobase surrogates is a suitable option. In these cases, the purine or pyrimidine bases at the anomeric position of the 2′-deoxyribose are exchanged by aromatic fluorophores.

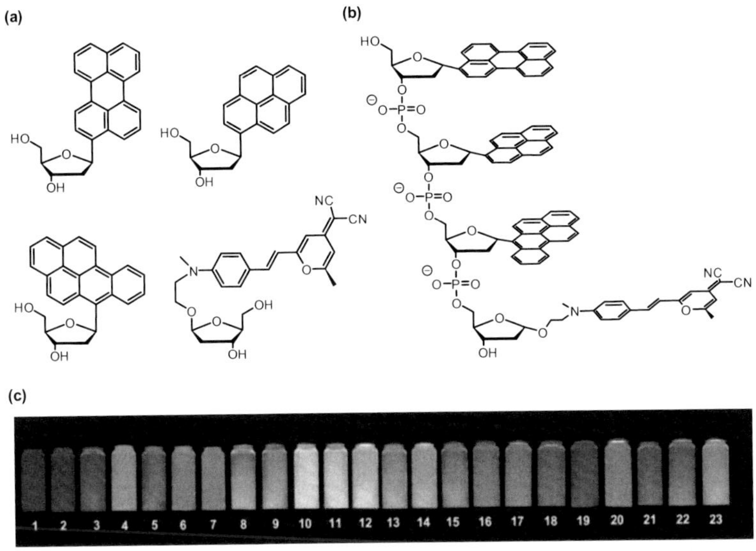

Figure 1.6 Exemplary oligodeoxyfluorosides (ODFs) developed by Kool et al. [40–42]. (a) Some of the monomers used for synthesis of the ODFs, (b) schematic structure of tetrafluor ODFs, and (c) photograph of 23 multispectral ODFs upon irradiation with λ_{ex} = 354 nm [43]. (c) Reprinted with permission from Ref. [43]. Copyright (2009) American Chemical Society.

Kool et al. synthesized an array of C-nucleosides, bearing different planar, aromatic chromophores at the 2′-deoxyribofuranoside (Fig. 1.6) [40, 41]. Four molecules were connected in a single-strand-like structure to form the so-called oligodeoxyfluorosides (ODFs) [41]. The optical characteristics of the resulting ODFs differ strongly from the characteristics of the used monomers, since the DNA scaffold promotes strong photophysical interactions between the chromophores. Therefore, EnT processes, ET processes, and excimer/exciplex formation can take place, resulting in high molecular extinction coefficients and large Stokes shifts. In a first combinatory approach, molecules carrying pyrene, oxoperylene, dimethylaminostilbene, and quinacridone were combined to deoxyfluoroside tetramers. The resulting library possessed 256 items, in which the kind of chromophores and their arrangement controlled the fluorescence of the respective ODFs [41]. Furthermore, a library with over 14,000 ODFs, consisting of 11 different fluorophores [42], and a system with 23 separate ODFs [43] were developed and characterized [44, 45]. The investigated ODFs share the same excitation wavelength but emit light all over the visible part of the spectrum. Although the photophysical interactions were not determined in detail, this astonishing example does not only prove what can be achieved by the controlled arrangement of chromophore but also offers great benefits for modern imaging and also sensor techniques. The demands made on the instruments could be significantly decreased since only one excitation wavelength is needed to detect multiple color emission.

In the previously shown examples, the chromophore aggregates were almost exclusively made up of planar, aromatic molecules. The working groups of Seitz and Leumann investigated the incorporation of twisted biphenyl-base surrogates in DNA [46–49]. However, upon annealing of the single strands stacking energy is gained. This energy is sufficient to overcome the small rotation barrier of roughly 10 kJ mol^{-1} and therefore to planarize the biphenyl groups in the DNA-chromophore stack. If multiple biphenyl-biphenyl pairs were placed in the duplex, they formed a zipper-like chromophore stack, which was able to stabilize the DNA double helix by hydrophobic interactions between the dyes alone. The resulting double helix was investigated by CD spectroscopy, whereby a B-helix with no deformations could be observed. Seitz and coworkers studied binaphthyl as a torsionally flexible, nonplanar base surrogate [50]. Successive introduction of multiple binaphthyl units led to duplex stabilization, resulting from hydrophobic chromophore-chromophore interactions. It is presumed that the binaphthyl system is capable of adjusting the two flexibly linked aromatic units to allow interactions with interhelical and extrahelical partners. These interactions do not include fluorescence quenching, making these base surrogates interesting for use in DNA nanotechnology- or oligonucleotide-based light-harvesting systems.

1.2.3 Modification of the 2′-Position

If RNA instead of DNA is used as an oligonucleotide scaffold, the sugar part can be used to create multichromophore systems. The hydroxyl group at the 2′-position of the ribofuranoside is not part of the covalent sugar-phosphate backbone and therefore is available to covalently incorporate chromophores into RNA. Although modifications of the 2′-position are commonly used to incorporate dyes into RNA, there are only a few studies in which this position was applied to create multichromophore arrays.

Yamana and coworkers created RNA oligonucleotides that carried between two and four pyrene moieties (Fig. 1.7). These chromophores were located at the 2′-position of uridine (**PyU**) [51] or adenosine (**PyA**) [52] via small methoxy linkers. By hybridization with an unmodified counterstrand, RNA duplexes were obtained that emitted strong excimer fluorescence, whose intensity increased with the number of pyrenes in the chromophore stack (**RNA1**). CD spectroscopy and molecular dynamics simulations confirmed that a helical pyrene array is formed along the right-handed RNA duplex in the minor groove. Through hybridization of two pyrene-modified single strands, an RNA duplex (**RNA2**) with an astonishing amount of 10 chromophores could be obtained [53]. This pyrene array consists of a zipper-like array along the minor groove of the RNA duplex and shows strong excimer emission.

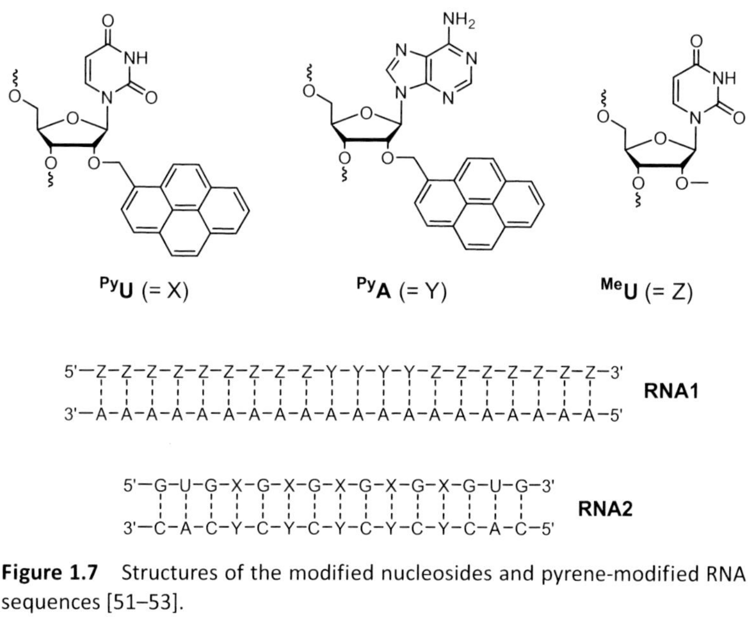

Figure 1.7 Structures of the modified nucleosides and pyrene-modified RNA sequences [51–53].

Korshun et al. could show that by changing the linker between the ribofuranoside and the chromophore moieties from a methoxy structure to a carbamate structure, it is possible to locate the chromophore stack in the major instead of the minor groove [54]. Hybridization of two single strands containing a pyrenemethyl carbamate nucleoside gave duplexes with an excimer located in the major groove if the two chromophores were separated by one base pair. These results provide important insight for the development of not only DNA but also RNA chromophore arrays.

1.2.4 Modification of the DNA Base

One of the major goals in the development of artificial light-harvesting systems is the design of programmable and sequence-specific chromophore arrays. DNA, with its highly specific Watson–Crick (WC) base pairing offers programmable sequences if chromophores can be incorporated without replacing the nucleobase and therefore destroying their binding motives. The position for the modification must be chosen carefully. For example, modification of the C8-position of purines leads to *syn*-conformation of the nucleoside, which can perturb the WC base pairing [55]. By attaching the chromophore to the C5-position of pyrimidines or C7-position of 7-deazapurines, the natural anticonformation is preserved and the WC pairing may stay intact [56, 57]. Furthermore, the linkage between base and chromophore is usually done by simple C–C bonds or an acetylene bridge to guarantee strong electron coupling between the components [58].

Recently, Stulz et al. reported the formation of DNA-based porphyrin arrays. Up to 11 porphyrins were placed in a single strand by linking diphenylporphyrin [59] and tetraphenylporphyrin [60] via acetylene moieties to uridine. Surprisingly, it could be shown by CD spectroscopy and molecular dynamics calculations that even in a single strand, a helical arrangement of H-aggregates is formed and that the chirality of the DNA scaffold is induced into the chromophore stack. Upon duplex formation the porphyrins are located in the major groove, which proves that the WC pairing stays fully intact. Unexpectedly, no duplex stabilization could be observed in these examples. However, if the porphyrin sequence was alternated between the two counterstrands, duplex stabilization was gained by interactions between the zipper-like arranged chromophores [61]. The presence of an EnT within the chromophore array could be proven by fluorescence quenching of Zn-porphyrins by metal-free porphyrins (Fig. 1.8).

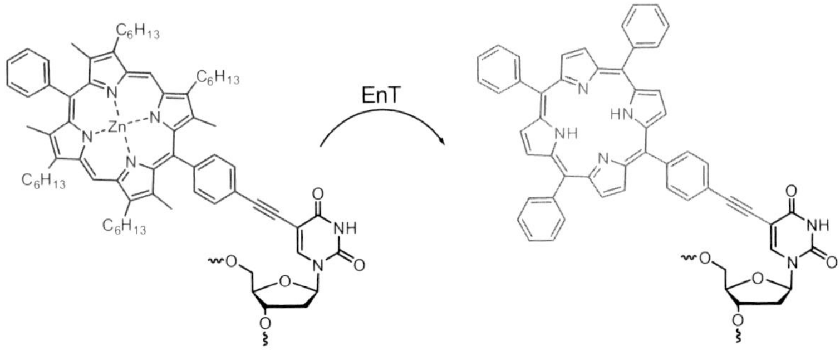

Figure 1.8 Porphyrin-modified 2′deoxyuridines used for the formation of oligoporphyrin assemblies in DNA by Stulz et al. with schematic fluorescence quenching through energy transfer from Zn-porphyrins to metal-free analogs [59–61].

We attached phenothiazin (Pz-dU) [62] and pyrene (Py-dU) [63] at the 5-position of 2′-deoxyuridine using a single C–C bond and created DNA duplex arrays with up to five chromophores. Fivefold incorporation of Pz-dU gave right-handed aggregates in the major groove, showing broad absorption between 300 nm and 400 nm, and strongly red-shifted fluorescence [62]. If five Py-dU were incorporated, again a right-handed array is formed. Astonishingly, the fluorescence intensity of the fivefold-modified single strand is 10 times higher than the intensity of a single-modified strand and even increases from the single strand to the double strand by a factor of 22 [63]. This shows that a highly ordered pyrene assembly is formed along the DNA duplex, in which the pyrenes are in contact through electrostatic and π-π interactions and a homo-EnT is possible. If one of the Py-dU units in the assembly is paired with a false base in the counterstrand, the fluorescence intensity drops significantly. It is expected that the mismatch destroys the high order of the chromophore stack, which disturbs the electrostatic interactions and the EnT between the pyrenes. In a last example, Py-dU and Pz-dU were incorporated side by side in an alternating order. As expected, a strong quenching of the pyrene typical fluorescence at 465 nm was observed, again proving the presence of EnT processes in the right-handed chromophore aggregate [62].

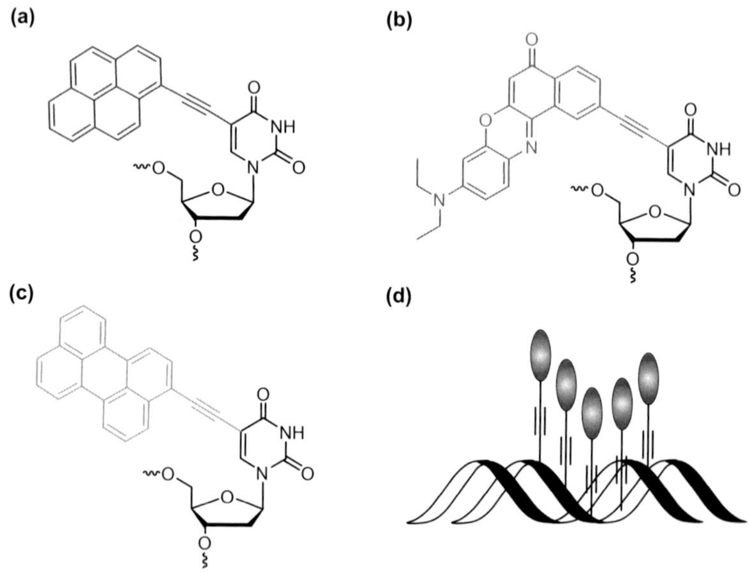

Figure 1.9 Covalent chromophore assemblies based on (a) Py-≡-dU, (b) Nr-≡-dU, and (c) Pe-≡-dU. (d) Schematic representation of the resulting right-handed helix upon fivefold incorporation of Py-≡-dU and Nr-≡-dU [64, 65].

In a similar approach ethynyl pyrene (Py-≡-dU) [64] and ethynyl nile red (Nr-≡-dU) [65] were attached to DNA at the 5-position of 2′-deoxyuridine (Fig. 1.9). The incorporation of two to five adjacent Py-≡-dU gave nonlinear rise to pyrene absorption for each additional building block since excitonic interactions occur between the chromophores [64]. Interestingly, only the highly ordered right-handed array in the major groove, as described in the previous experiments, is formed if more than three Py-≡-dU units are located next to each other in the double strand. If only two chromophores are placed adjacent to each other, no biphasic signal in the pyrene absorption range between 340 nm and 440 nm can be observed in the CD spectra. Furthermore, it could be shown, that the right counter base (dA) is crucial for the assembly of the pyrene-DNA arrays. Contradictory to the porphyrin assemblies, the fivefold pyrene modification leads to an increase in the duplex stability compared with the corresponding single-modified DNA. The higher melting temperature is explained by π-π interactions between the pyrenes. Nr-≡-dU shows, just like nile red itself, strong hydrophobicity, a solvatochromic behavior, and an excellent fluorescence quantum yield [65]. DNA duplexes with three to five Nr-≡-dU next to each other possessed a strong hypsochromic shift, a biphasic signal between 500 nm and 700 nm, as well as complete fluorescence quenching around 660 nm [66]. All three observations demonstrate, that similar to the pyrene assemblies, H-aggregates are formed in the right-handed nile red DNA arrangements. However, in contrast to the pyrene-DNA conjugates, each additional Nr-≡-dU lowers the melting temperature of DNA and therefore destabilizes the duplex.

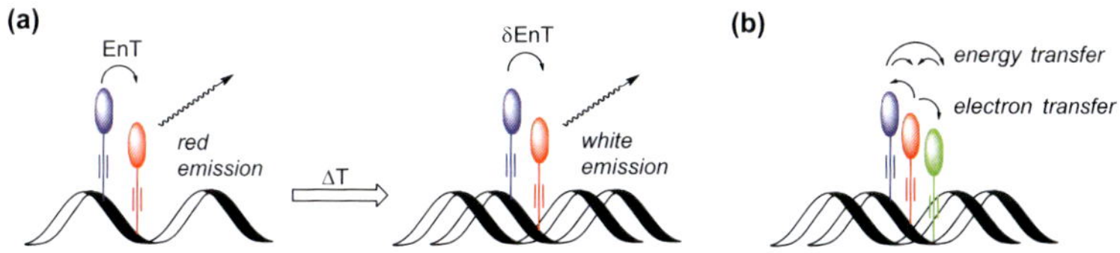

Figure 1.10 Energy transfer processes in DNA. (a) White-light-emitting DNA (WED) with complete energy transfer in the single strand and partial energy transfer in the DNA duplex and (b) electron and energy transfer cascades as competing reactions in triple-modified DNA (blue: Py-≡-dU; red: Nr-≡-dU; green: Pe-≡-dU; black: DNA; for structures see Fig. 1.9) [67, 68].

On the basis of our previous experiments, we decided to investigate Nr-≡-dU and Py-≡-dU as an ET pair in DNA and its potential use in DNA-based artificial light-harvesting systems (Fig. 1.10a) [67]. As mentioned earlier, a major prerequisite for EnT is a spectral overlap between the donor emission and the acceptor absorption, whereby the absorption of donor and acceptor should not overlap so that ground state interactions between the chromophores can be avoided. In our system Py-≡-dU has an absorption maximum of 400 nm and an emission maximum of 440 nm and should be a suitable donor for the Nr-≡-dU acceptor, which has an absorption maximum of 615 nm. We synthesized doubly modified single strands with Py-≡-dU and Nr-≡-dU adjacent to each other. It is noteworthy that the consecutive incorporation of the dyes had only a very minor destabilizing effect or no effect at all on the respective double strands. Excitation of the doubly modified strands at 380 nm allowed selective excitation of the donor. After excitation, quenching of the pyrene fluorescence at 400 nm could be observed. At the same time acceptor fluorescence at 615 nm was visible, evidencing a successful ET. As expected, the FRET rate in the respective duplex was reduced significantly (5.2×10^9 s in the single strand versus 1.7

× 10^9 s in the double strand). Similar to the work of Asanuma et al. mentioned above [25], the EnT efficiency is heavily dependent on the relative orientation of the chromophores to each other. In the duplex, the dyes are twisted helically, resulting in a less unfavorable dipole orientation and a reduced EnT. Because of that, the emission spectra of the double strands showed increased intensity at 440 nm and a decrease in intensity at 615 nm, resulting in an intensity ratio of almost 1 ($I_{440nm/615nm}$ = 0.96) [67], which produces white emission. Therefore, these strands were named "white-light-emitting DNA" (WED). Potential applications of the WED are based on the ability to control the EnT efficiency by association or dissociation of the duplex. For example, WEDs could be used as temperature sensors. If the temperature is increased, emission changes from white to red, due to thermal dehybridization of the duplex. This process can be fully reversed by lowering the temperature again, which reanneals the single strands.

A desirable aim in the design of artificial LHCs is the exploitation of the whole ultraviolet-visible (UV-Vis) absorption range. We incorporated ethynyl perylene (Pe-≡-dU) as a third chromophore additional to Py-≡-dU and Nr-≡-dU to cover the range from 350 nm to 750 nm [68], while still only using one modified DNA strand (Fig. 1.10b). If the three chromophores were located directly adjacent to each other in the modified strand, nearly complete quenching of the fluorescence was observed. Ground state interactions were obtained by absorption shifts and biphasic CD signals for the three building blocks. To avoid these ground state interactions, we increased the distance between the chromophores by intervening A-T pairs. If Py-≡-dU was excited selectively, a strong nile red fluorescence was observed after approximately 1 ns. By time-resolved fluorescence spectroscopy we could show that upon excitation of Py-≡-dU an EnT cascade from Py-≡-dU over Pe-≡-dU to Nr-≡-dU takes place. However, if the chromophores are located directly next to each other, ET processes occur subsequently from Nr-≡-dU to Py-≡-dU and from Nr-≡-dU to Pe-≡-dU, resulting in a charge-separated state, which causes fluorescence quenching. With this three-chromophore system, we managed to create a very promising DNA-based LHA. By carefully controlling the sequence, this system can be switched between a directed EnT process and a directed ET process. The understanding of both processes is crucial for the design and assembly of further artificial light-harvesting systems. Furthermore, every wavelength between 400 nm and 700 nm can generate a charge-separated state, which can be used in chemical photocatalysis or other optochemical applications.

1.3 Supramolecular DNA-Chromophore Architectures: Challenging Sequence Control; Facile Synthesis

In the previous section, the covalent incorporation of chromophores in DNA and their use as potential artificial light-harvesting systems was reported. Despite the easy control of chromophore sequence and position, these covalent LHAs have a major drawback. The synthesis and preparation of multiple modified oligonucleotides is rather inefficient and complex. As mentioned above, oligonucleotides are mostly synthesized by phosphoramidite chemistry [6–9]. One cycle in the synthesis involves coupling, capping, oxidation, and deprotection. Artificial building blocks should therefore be stable under acidic and basic conditions and should be inert against oxidation by iodine. To guarantee this, some building blocks require an elaborate protecting group strategy. Furthermore, solubility problems become an issue when oligonucleotides are modified with more than 5 to 10 chromophores. Even unmodified oligonucleotides are only synthesized efficiently up to a length of 150 bases [58]. Beyond this point, the yields are drastically reduced. Although the scope of application is extended through postsynthetical modifications [69] and chemical ligation [70] the use of the phosphoramidite method is still limited by rather expensive building blocks and time-consuming syntheses.

An alternative approach to organize chromophores on a supramolecular level is the noncovalent self-assembly of modified monomers along a template [71]. Single-stranded DNA is a suitable scaffold for the effective assembly of monomers. The hydrogen bridges between the nucleobases offer excellent binding motives for chromophore monomers, which were modified in accordance to that motive, and the rigid sugar-phosphate backbone of the single strand supports a highly ordered assembly. This assembly is then further stabilized through hydrophobic or π-π interactions between the assembled chromophores and finally a supramolecular chromophore stack is formed in the ideal case. The major advantage of this approach is that chromophore arrays can be prepared in a fast manner without the relatively expensive and time-consuming synthetic incorporation of the dyes into single strands. Furthermore, the overall size and geometry of the chromophore are easily controlled by the size and sequence of the single strand as a chiral template. The groups of Albinsson [72], Kumar [73], Yan [74], and Roelfes [75] reported promising DNA-based LHAs using DNA duplexes and quadruplexes to assemble chromophores along the template by means of intercalation. However, if a specific, sequence-controlled aggregation of chromophores is desired, the use of single-stranded DNA as a template is required.

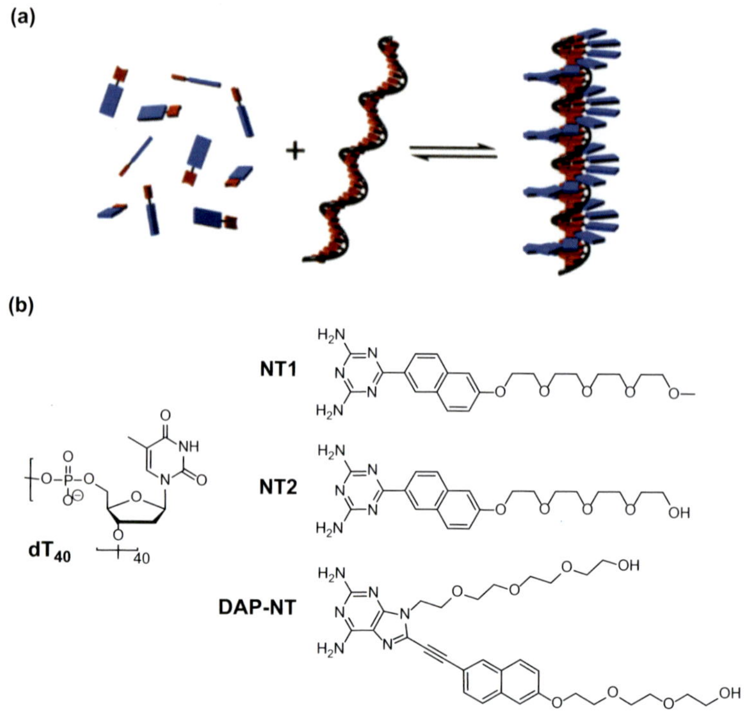

Figure 1.11 Self-assembly of modified naphthalene chromophores by Schenning et al. [76–78]. (a) Schematic representation of single-stranded DNA templated self-assembly (blue: chromophores; red: hydrogen-bonding motives; black: ssDNA) and (b) molecular structures of $\mathbf{dT_{40}}$ and the naphthalene monomers **NT1**, **NT2**, and **DAP-NT**. (a) Reprinted with permission from Ref. [76]. Copyright (2007) American Chemical Society.

One of the first approaches using single-stranded DNA templates was reported by Schenning et al. (Fig. 1.11) [76]. An oligonucleotide containing 40 thymidines (**dT$_{40}$**) was used for the hierarchical aggregation of naphthalene chromophores (**NT1**), which were modified with a diaminotriazine moiety to allow hydrogen bonding to the complementary base in the single strand. CD spectra showed a biphasic signal in the absorptions range of the otherwise achiral monomer **NT1**. This shows a successful transfer of the chirality from the template to the supramolecular assembly. The process of fully reversible self-assembly was investigated by temperature-dependent absorption and CD spectroscopy. The monomers bind to the scaffold via hydrogen bond formation before the helical arrangement is formed. Concentration-dependent studies could show that both monomer-monomer interactions and template-monomer interactions are decisive for the success of the aggregation. Strong monomer-template and weak monomer-monomer interactions hinder the aggregation through missing π-π interactions between the assembled chromophores. Weak monomer-template and strong monomer-monomer interactions, on the other hand, disturb the aggregation as well, since homo-aggregation of the monomers without the template is preferred. To tune these parameters, a new diaminotriazine-modified naphthalene **NT2** was developed, bearing a hydroxyl group to avoid self-aggregation without the template [77]. The new monomer, which formed right-handed assemblies with oligo-2′-deoxythymidines, was used to investigate the influence of the template length. Within this study a strong cooperative effect was found. The naphthalene monomers stabilize themselves and the resulting aggregate through π-π interactions so that the assembly of further monomers is facilitated if a certain number of monomers is already successfully arranged. It could be shown that template lengths under 10 bases required significantly higher monomer concentrations for complete population of the scaffold. A theoretical model suggested that at least eight bases are required in the template strand to provide enough cooperativity and allow successful aggregation. To further increase guest-guest interactions, naphthalenes bearing a 2,4-diaminopurine group with a bigger π system were synthesized (**DAP-NT**) [78]. Surprisingly, the aggregation of the monomers was heavily dependent on the pH of the medium. High pH values (pH > 9) gave the well-known right-handed helix. Lower pH values led to inversion of the chirality through protonation of the monomer, and a left-handed assembly was preferred. The left-handed DNA (Z-DNA) offers larger distances between the bases, so the protonated monomers experience lower electrostatic repulsion. Balaz et al. used 2,4-diaminopurine-modified porphyrins (**Por-DAP**) to further examine the influence of assembly conditions on the resulting helicity (Fig. 1.12) [79, 80]. If a solution of monomers and template strand is prepared at 85°C and then slowly cooled to 20°C, a left-handed M-helix is observed. In the presence of NaCl (c_{NaCl} = 500 mM) the same experiment yields right-handed (P-helix) assemblies. Fast cooling of the sample containing salt again gave a left-handed (M-helix) assembly.

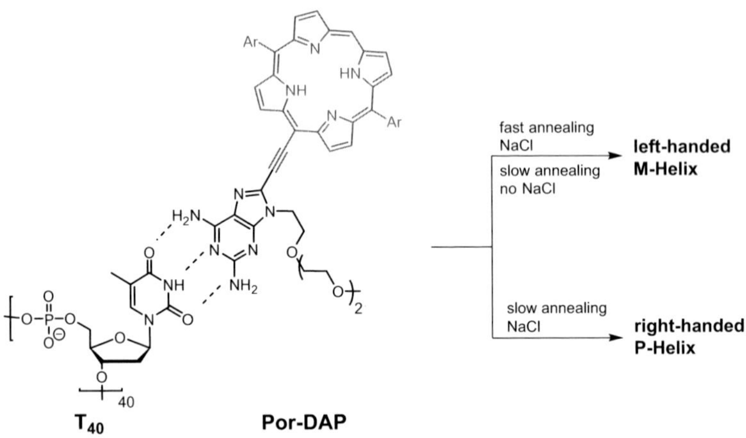

Figure 1.12 Self-assembly of diaminopurine-modified porphyrins along a T_{40} template strand and control of the resulting helicity by ambient conditions (red: chromophore) [79, 80].

We recently studied how if the monomers Py-≡-dU and Nr-≡-dU can be arranged along oligo-2′-dexyadenosines as suitable single strands by self-assembly, the resulting optical properties differ from the respective covalently modified strands with five chromophores adjacent to each other (described above). Firstly, the specific self-assembly of Py-≡-dU was tested with different template strands [81]. While Py-≡-dU stays only soluble in water if the complementary template (dA_{17}) is present, it precipitates in the presence of the noncomplementary template (dT_{17}) or in the absence of DNA.

Therefore, an unspecific binding of the monomer could be excluded. The optical properties of the resulting aggregate were very similar to our covalent system. However, contrary to the covalent approach, which gave a right-handed helix, the CD signal in the pyrene range (350–450 nm) indicated the formation of a left-handed helix in the assembled system. Assembly experiments of Nr-≡-dU along T_{17} gave similar results. While the resulting aggregate had nearly the same properties as the covalently modified DNA, a left-handed helix was observed. Titration experiments revealed that all available binding sites of the template strand were occupied by Nr-≡-dU monomers. If Nr-≡-dU nanoparticles are investigated without the presence of a template, again a left-handed chirality is observed [66]. The intrinsic property of Py-≡-dU and Nr-≡-dU to form left-handed helixes can only be overwritten by the covalent connection of the monomers by phosphodiester bonds [64]. Furthermore, we wanted to know if an EnT is possible within these assemblies (Fig. 1.13). Therefore, mixtures of Py-≡-dU and Nr-≡-dU with different ratios (Py-≡-dU:Nr-≡-dU = 20:0 to 0:20) were assembled along T_{20} [82]. Interestingly, the formation of large Py-≡-dU stacks was already inhibited by only small amounts of Nr-≡-dU, demonstrating the spontaneous assembly of these aggregates. Supramolecular structures with monomer ratios between Py-≡-dU:Nr-≡-dU = 10:10 and 2:18 emitted dual fluorescence, showing an EnT between ethynyl pyrene and ethynyl nile red in the chromophore stack.

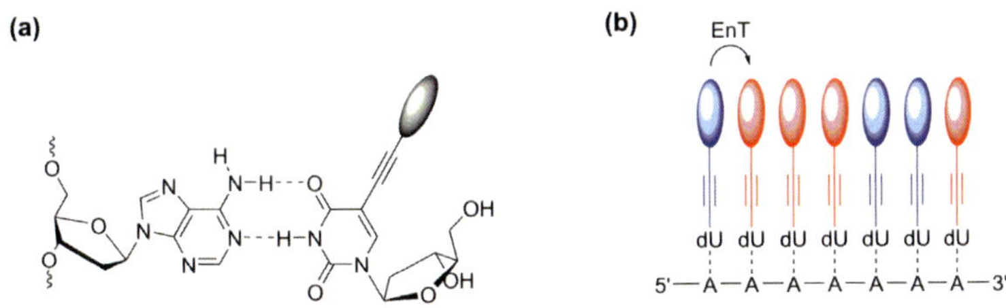

Figure 1.13 Self-assembly of Py-≡-dU and Nr-≡-dU along an oligo-2′deoxyadenosine template strand. (a) Hydrogen-binding motive of the chromophore nucleosides and (b) assembled chromophores with energy transfer from ethynyl pyrene to ethynyl nile red (blue: Py-≡-dU; red: Nr-≡-dU; for structures see Fig. 1.9) [81, 82].

The previous examples impressively show that self-assembly of chromophores along DNA as the template presents an easy approach toward synthesis and preparation of large multichromophore systems. Within these systems, directed ET and EnT processes are possible, making them promising candidates for the design of artificial DNA-based LHAs. To the best of our knowledge there are only very few examples that used specific assembly along a DNA single strand to prepare light-harvesting systems. Schenning et al. refined their previously mentioned system by covalently attaching a cyanine dye (Cy3.5) to the 5′-end of the oligothymidine template strand (T_q) (Fig. 1.14) [83]. 2,4-Diaminopurine-modified naphthalenes **DAP-NT** formed the supramolecular chromophore stack. Since the fluorescence of the naphthalenes overlaps with the absorption of Cy3.5, an ET should be possible. If a solution of the naphthalene monomers is excited at 400 nm, monomer fluorescence is indeed visible. Upon addition of the dT_{40}/Cy3.5 template strand, a time-resolved decrease of the monomer emission and an increase of the Cy3.5 emission at 625 nm are observed. Excitation of the template strand alone only gave a minor emission of the cyanine dye. These two observations combined clearly demonstrate that an EnT from the assembled chromophores to the covalently bound acceptor is occurring. Different template lengths (q = 10, 20, 30, and 40) were used to investigate the EnT efficiency. All strands exhibited an increased acceptor fluorescence, with a maximum at 30 base pairs. At this length, the optimal ratio between the average donor-acceptor distance and binding efficiency of the monomers is obtained. With this work, Schenning and coworkers could successfully use the assembly of chromophores to achieve a directed EnT along a one-dimensional chromophore stack.

For the promising use of covalent and noncovalent DNA assemblies in opto- and nanoelectronic applications, it is necessary that hybrid structures with carbon-based materials (e.g., fullerenes) are constructed. Fullerenes are typically used as electron acceptors and transport domains within organic solar cells [84]. So far only very few DNA-fullerene conjugates are known. Hélène et al. used fullerenes to cleave a DNA strand sequence specifically with light. Further examples were published for similar photobiological applications [85–88]. We used one of the manifold chemical possibilities—a phenylene-bridged maleic acid ester linker—to attach fullerenes to DNA (Fig. 1.15) [89]. With this so-called C6 linker a fullerene was attached to the 5′-end of an oligo-2′deoxyadenosine strand. The modified strands were used to assemble Py-≡-dU Nr-≡-dU mixtures, analogous to our experiments mentioned above. To understand the influence of the fullerene on the DNA conjugates, we compared the optical properties of the fullerene to the nonfullerene assemblies. As aforementioned, an EnT followed by an ET in the opposite direction is taking place in the ethynyl pyrene/ethynyl nile red architectures. This process generates a charge-separated state, a so-called exciton, which causes quenching of the chromophore emission. In the presence of fullerenes, a stronger fluorescence quenching is observed. Not only is the exciton located in the chromophore stack, it also gets dissociated and an electron passes over to the fullerene. Direct charge separation from the excited state of the chromophores can be excluded since the average chromophore-fullerene distance is too large for an effective charge-separation process.

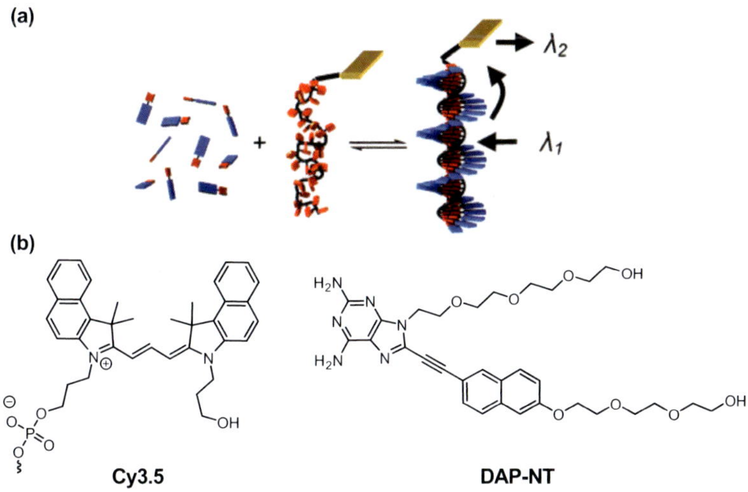

Figure 1.14 Self-assembled light-harvesting complex (LHC) from Schenning et al. [83] (a) Schematic representation of ssDNA template LHC (blue: chromophores; red: hydrogen-bonding motives; black: ssDNA; yellow: cyanine dye Cy3.5) and (b) molecular structures of **Cy3.5** and the naphthalene monomer **DAP-NT**. (a) Reprinted from Ref. [83] with permission of The Royal Society of Chemistry.

In cooperation with the Colsmann group, we tested our new DNA-based hybrid materials as a photoactive layer in solar cells. For the solar cell, we used an inverted architecture (tin oxide/zinc oxide cathode; molybdenum oxide/silver anode). The photoactive layer consisted of a 100 nm thick, representative mixture of Py-≡-dU:Nr-≡-dU = 8:12. Upon one sun irradiation, the device exhibited a photovoltage of 670 mV, which equals a quantum efficiency of 2%. Although, it is clear that the external quantum efficiency is far too low for any real applications, this work shows the potential of synthetic DNA-based nanomaterials.

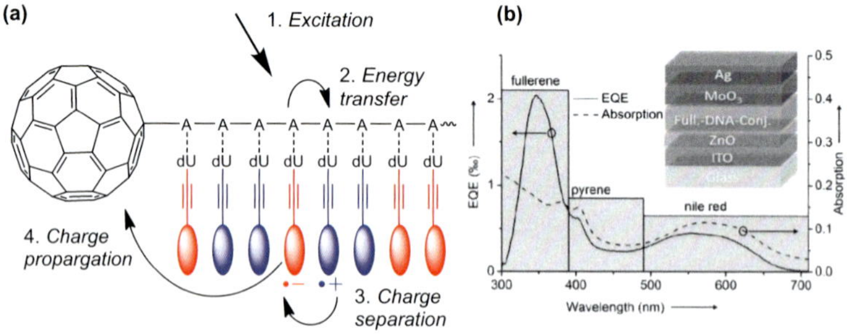

Figure 1.15 Fullerene-DNA conjugates as a photoactive layer in solar cells [89]. (a) Assembly of Py-≡-dU and Nr-≡-dU along a fullerene-dA_{20} strand and illustration of one of the major photophysical pathways (blue: Py-≡-dU; red: Nr-≡-dU; for structures see Fig. 1.9). (b) Absorption spectrum of fullerene-DNA-conjugate layer on glass and the external quantum efficiency (ECE) of a typical solar cell. Inset: Solar cell architecture. (b) Reprinted with permission from Ref. [89]. Copyright (2015) John Wiley and Sons.

1.4 Conclusion

The examples presented herein show that DNA as a template is a powerful tool and scaffold for the hierarchical organization of chromophores at a supramolecular level. The chromophores can be attached either covalently, through synthesis of the corresponding building block for solid phase synthesis, or noncovalently, by self-assembly of appropriately modified chromophores. In both cases, DNA as an architecture is able to transfer the template chirality on the otherwise achiral chromophores, making these assemblies interesting for generation or detection of circularly polarized light. Currently the research mainly focuses on the elucidation and control of ET and EnT processes in the chromophore stacks. For these studies, the covalent approach is the preferred method, since sequence and distance of the chromophores can be controlled in a very precise way. Although the noncovalent assemblies can be prepared more easily, the self-assembly approach lacks the precise control over the sequence for the chromophore arrangement. Different patterns of hydrogen-bonding motives could be used for specific recognition of monomers by choosing the proper nucleobases. In the optimal case, a fourfold specificity, based on the four different nucleobases, is achieved, similar to canonical base pairing in double-stranded DNA, which allows self-controlled assembly of modified monomers to complex supramolecular architectures with new optoelectronic properties. The mentioned examples impressively show, that, in principle, such multichromophore assemblies can be created based on DNA templates. However, few research studies beyond the design and spectroscopic characterization of the DNA-chromophore arrangements have been done so far. Yan et al. coupled a DNA three-way junction fluorophore nanostructure to a bacterial reaction center [90], and the group of Sotzing used dye-modified nanofibers for the creation of a white-light emitting diode [91]. The next important step is the chemical usage or storage of the collected light energy. Further research should therefore include the coupling of DNA-based LHAs to reaction centers for hydrogen generation or chemical photocatalysis.

Notes

The authors declare that there is no conflict of interest.

Acknowledgment

Financial support from the Karlsruhe Institute of Technology (KIT), the Karlsruhe School of Optics & Photonics (KSOP), and the Deutsche Forschungsgemeinschaft (DFG, grant Wa 1386/20-1) is gratefully acknowledged.

References

1. Kühlbrandt, W. and Wang, D. N. (1991). *Nature*, **350**, 130.
2. Horton, P., Ruban, A. V. and Walters, R. G. (1996). *Annu. Rev. Plant Physiol. Plant Mol. Biol.*, **47**, 655–684.

3. Seeman, N. C. (1998). *Angew. Chem. Int. Ed.*, **37**, 3220–3228.
4. Rothemund, P. W. K. (2006). *Nature*, **440**, 297–302.
5. Reese, C. B. (2005). *Org. Biomol. Chem.*, **3**, 3851–3851.
6. Beaucage, S. L. and Caruthers, M. H. (1981). *Tetrahedron Lett.*, **22**, 1859–1862.
7. Caruthers, M. (1985). *Science*, **230**, 281–285.
8. Caruthers, M. H., Barone, A. D., Beaucage, S. L., Dodds, D. R., Fisher, E. F., McBride, L. J., Matteucci, M., Stabinsky, Z. and Tang, J. Y. (1987). In *Methods in Enzymology*, Vol. 154, Academic Press, pp. 287–313.
9. LeProust, E. M., Peck, B. J., Spirin, K., McCuen, H. B., Moore, B., Namsaraev, E. and Caruthers, M. H. (2010). *Nucleic Acids Res.*, **38**, 2522–2540.
10. Schmucker, W. and Wagenknecht, H.-A. (2012). *Synlett*, **23**, 2435–2448.
11. Asanuma, H., Toda, T., Murayama, K., Liang, X. and Kashida, H. (2010). *J. Am. Chem. Soc.*, **132**, 14702–14703.
12. Schärer, O. D. (2003). *Angew. Chem. Int. Ed.*, **42**, 2946–2974.
13. Amann, N. and Wagenknecht, H.-A. (2003). *Tetrahedron Lett.*, **44**, 1685–1690.
14. Kashida, H., Murayama, K., Toda, T. and Asanuma, H. (2011). *Angew. Chem. Int. Ed.*, **50**, 1285–1288.
15. Meggers, E. and Zhang, L. (2010). *Acc. Chem. Res.*, **43**, 1092–1102.
16. Christensen, U. B. and Pedersen, E. B. (2002). *Nucleic Acids Res.*, **30**, 4918–4925.
17. Amann, N., Huber, R. and Wagenknecht, H.-A. (2004). *Angew. Chem. Int. Ed.*, **43**, 1845–1847.
18. Köhler, O., Jarikote, D. V. and Seitz, O. (2005). *ChemBioChem*, **6**, 69–77.
19. Socher, E., Knoll, A. and Seitz, O. (2012). *Org. Biomol. Chem.*, **10**, 7363–7371.
20. Asanuma, H., Shirasuka, K., Takarada, T., Kashida, H. and Komiyama, M. (2003). *J. Am. Chem. Soc.*, **125**, 2217–2223.
21. Asanuma, H., Shirasuka, K. and Komiyama, M. (2002). *Chem. Lett.*, 490–491.
22. Kashida, H., Tanaka, M., Baba, S., Sakamoto, T., Kawai, G., Asanuma, H. and Komiyama, M. (2006). *Chem. Eur. J.*, **12**, 777–784.
23. Asanuma, H., Kashida, H., Liang, X. and Komiyama, M. (2003). *Chem. Commun.*, 1536–1537.
24. Kashida, H., Takatsu, T., Sekiguchi, K. and Asanuma, H. (2010). *Chem. Eur. J.*, **16**, 2479–2486.

25. Kato, T., Kashida, H., Kishida, H., Yada, H., Okamoto, H. and Asanuma, H. (2013). *J. Am. Chem. Soc.*, **135**, 741–750.

26. Zheng, J. (2010). In *Biomedical Applications of Biophysics* (ed. Jue, T.), Humana Press, Totowa, NJ, pp. 119–136.

27. Wilhelmsson, L. M. (2016). In *Fluorescent Analogues of Biomolecular Building Blocks: Design and Applications*, John Wiley & Sons, Inc., pp. 224–241.

28. Wranne, M. S., Füchtbauer, A. F., Dumat, B., Bood, M., El-Sagheer, A. H., Brown, T., Gradén, H., Grøtli, M. and Wilhelmsson, L. M. (2017). *J. Am. Chem. Soc.*, **139**, 9271–9280.

29. Langenegger, S. M. and Häner, R. (2004). *Chem. Commun.*, 2792–2793.

30. Langenegger, S. M. and Häner, R. (2004). *Tetrahedron Lett.*, **45**, 9273–9276.

31. Langenegger, S. M. and Häner, R. (2005). *ChemBioChem*, **6**, 848–851.

32. Malinovskii, V. L., Samain, F. and Häner, R. (2007). *Angew. Chem. Int. Ed.*, **46**, 4464–4467.

33. Garo, F. and Häner, R. (2012). *Angew. Chem. Int. Ed.*, **51**, 916–919.

34. Probst, M., Langenegger, S. M. and Häner, R. (2014). *Chem. Commun.*, **50**, 159–161.

35. Würthner, F. (2004). *Chem. Commun.*, 1564–1579.

36. Wagner, C. and Wagenknecht, H.-A. (2006). *Org. Lett.*, **8**, 4191–4194.

37. Baumstark, D. and Wagenknecht, H.-A. (2008). *Angew. Chem. Int. Ed.*, **47**, 2652–2654.

38. Menacher, F., Stepanenko, V., Würthner, F. and Wagenknecht, H.-A. (2011). *Chem. Eur. J.*, **17**, 6683–6688.

39. Baumstark, D. and Wagenknecht, H.-A. (2008). *Chem. Eur. J.*, **14**, 6640–6645.

40. Ren, R. X.-F., Chaudhuri, N. C., Paris, P. L., Rumney, S. and Kool, E. T. (1996). *J. Am. Chem. Soc.*, **118**, 7671–7678.

41. Gao, J., Strässler, C., Tahmassebi, D. and Kool, E. T. (2002). *J. Am. Chem. Soc.*, **124**, 11590–11591.

42. Gao, J., Watanabe, S. and Kool, E. T. (2004). *J. Am. Chem. Soc.*, **126**, 12748–12749.

43. Teo, Y. N., Wilson, J. N. and Kool, E. T. (2009). *J. Am. Chem. Soc.*, **131**, 3923–3933.

44. Teo, Y. N. and Kool, E. T. (2009). *Bioconj. Chem.*, **20**, 2371–2380.

45. Dai, N., Teo, Y. N. and Kool, E. T. (2010). *Chem. Commun.*, **46**, 1221–1223.
46. Brotschi, C. and Leumann, C. J. (2003). *Angew. Chem. Int. Ed.*, **42**, 1655–1658.
47. Zahn, A. and Leumann, C. J. (2008). *Chem. Eur. J.*, **14**, 1087–1094.
48. Singh, I., Hecker, W., Prasad, A. K., Parmar, V. S. and Seitz, O. (2002). *Chem. Commun.*, 500–501.
49. Beuck, C., Singh, I., Bhattacharya, A., Hecker, W., Parmar, V. S., Seitz, O. and Weinhold, E. (2003). *Angew. Chem. Int. Ed.*, **42**, 3958–3960.
50. Hainke, S. and Seitz, O. (2009). *Angew. Chem. Int. Ed.*, **48**, 8250–8253.
51. Nakamura, M., Ohtoshi, Y. and Yamana, K. (2005). *Chem. Commun.*, 5163–5165.
52. Nakamura, M., Shimomura, Y., Ohtoshi, Y., Sasa, K., Hayashi, H., Nakano, H. and Yamana, K. (2007). *Org. Biomol. Chem.*, **5**, 1945–1951.
53. Nakamura, M., Murakami, Y., Sasa, K., Hayashi, H. and Yamana, K. (2008). *J. Am. Chem. Soc.*, **130**, 6904–6905.
54. Dioubankova, N. N., Malakhov, A. D., Stetsenko, D. A., Gait, M. J., Volynsky, P. E., Efremov, R. G. and Korshun, V. A. (2003). *ChemBioChem*, **4**, 841–847.
55. Wanninger-Weiß, C., Valis, L., Barbaric, J. and Wagenknecht, H.-A. (2008). *Bioorg. Med. Chem.*, **16**, 100–106.
56. Seela, F., Xiong, H., Leonard, P. and Budow, S. (2009). *Org. Biomol. Chem.*, **7**, 1374–1387.
57. Seela, F. and Zulauf, M. (1999). *Helv. Chim. Acta*, **82**, 1878–1898.
58. Teo, Y. N. and Kool, E. T. (2012). *Chem. Rev.*, **112**, 4221–4245.
59. Bouamaied, I., Nguyen, T., Rühl, T. and Stulz, E. (2008). *Org. Biomol. Chem.*, **6**, 3888–3891.
60. Fendt, L. A., Bouamaied, I., Thöni, S., Amiot, N. and Stulz, E. (2007). *J. Am. Chem. Soc.*, **129**, 15319–15329.
61. Nguyen, T., Brewer, A. and Stulz, E. (2009). *Angew. Chem. Int. Ed.*, **48**, 1974–1977.
62. Mayer-Enthart, E., Wagner, C., Barbaric, J. and Wagenknecht, H.-A. (2007). *Tetrahedron*, **63**, 3434–3439.
63. Mayer-Enthart, E. and Wagenknecht, H.-A. (2006). *Angew. Chem. Int. Ed.*, **45**, 3372–3375.
64. Barbaric, J. and Wagenknecht, H.-A. (2006). *Org. Biomol. Chem.*, **4**, 2088–2090.

65. Varghese, R., Gajula, P. K., Chakraborty, T. K. and Wagenknecht, H.-A. (2009). *Synlett*, 3253–3257.

66. Varghese, R. and Wagenknecht, H.-A. (2010). *Chem. Eur. J.*, **16**, 9040–9046.

67. Varghese, R. and Wagenknecht, H. A. (2009). *Chem. Eur. J.*, **15**, 9307–9310.

68. Ensslen, P., Brandl, F., Sezi, S., Varghese, R., Kutta, R.-J., Dick, B. and Wagenknecht, H.-A. (2015). *Chem. Eur. J.*, **21**, 9349–9354.

69. Weisbrod, S. H. and Marx, A. (2008). *Chem. Commun.*, 5675–5675.

70. Qiu, J., El-Sagheer, A. H. and Brown, T. (2013). *Chem. Commun.*, **49**, 6959–6961.

71. Anderson, S., Anderson, H. L. and Sanders, J. K. M. (1993). *Acc. Chem. Res.*, **26**, 469–475.

72. Woller, J. G., Hannestad, J. K. and Albinsson, B. (2013). *J. Am. Chem. Soc.*, **135**, 2759–2768.

73. Kumar, C. V. and Duff, M. R., Jr. (2009). *J. Am. Chem. Soc.*, **131**, 16024–16026.

74. Dutta, P. K., Varghese, R., Nangreave, J., Lin, S., Yan, H. and Liu, Y. (2011). *J. Am. Chem. Soc.*, **133**, 11985–11933.

75. Oltra, N. S., Browne, W. R. and Roelfes, G. (2013). *Chem. Eur. J.*, **19**, 2457–2461.

76. Janssen, P. G. A., Vandenbergh, J., van Dongen, J. L. J., Meijer, E. W. and Schenning, A. P. H. J. (2007). *J. Am. Chem. Soc.*, **129**, 6078–6079.

77. Janssen, P. G. A., Jabbari-Farouji, S., Surin, M., Vila, X., Gielen, J. C., de Greef, T. F. A., Vos, M. R. J., Bomans, P. H. H., Sommerdijk, N. A. J. M., Christianen, P. C. M., Leclère, P., Lazzaroni, R., van der Schoot, P., Meijer, E. W. and Schenning, A. P. H. J. (2009). *J. Am. Chem. Soc.*, **131**, 1222–1231.

78. Janssen, P. G. A., Ruiz-Carretero, A., González-Rodríguez, D., Meijer, E. W. and Schenning, A. P. H. J. (2009). *Angew. Chem. Int. Ed.*, **48**, 8103–8106.

79. Sargsyan, G., Schatz, A., Kubelka, J. and Balaz, M. (2013). *Chem. Commun.*, **49**, 1020–1022.

80. Sargsyan, G., Leonard, B. M., Kubelka, J. and Balaz, M. (2014). *Chem. Eur. J.*, **20**, 1878–1892.

81. Sezi, S. and Wagenknecht, H.-A. (2013). *Chem. Commun.*, **49**, 9257–9257.

82. Ensslen, P., Fritz, Y. and Wagenknecht, H.-A. (2015). *Org. Biomol. Chem.*, **13**, 487–492.

83. Ruiz-Carretero, A., Janssen, P. G. A., Stevens, A. L., Surin, M., Herz, L. M. and Schenning, A. P. H. J. (2011). *Chem. Commun.*, **47**, 884–886.

84. Brabec, C., Scherf, U. and Dyakonov, V. (2011). *Organic Photovoltaics: Materials, Device Physics, and Manufacturing Technologies*, Wiley.

85. Boutorine, A. S., Tokuyama, H., Takasugi, M., Isobe, H., Nakamura, E. and Hélène, C. (1994). *Angew. Chem.*, **106**, 2526–2529.

86. Yang, X., Meng, X., Li, B., Chen, Z., Zhao, D., Tan, X. and Yu, Q. (2008). *Biologicals*, **36**, 223–226.

87. An, Y.-Z., Chen, C.-H. B., Anderson, J. L., Sigman, D. S., Foote, C. S. and Robin, Y. (1996). *Tetrahedron*, **52**, 5179–5189.

88. Ros, T. D., Bergamin, M., Vázquez, E., Spalluto, G., Baiti, B., Moro, S., Boutorine, A. and Prato, M. (2002). *Eur. J. Org. Chem.*, 405–413.

89. Ensslen, P., Gärtner, S., Glaser, K., Colsmann, A. and Wagenknecht, H. A. (2016). *Angew. Chem. Int. Ed.*, **55**, 1904–1908.

90. Dutta, P. K., Levenberg, S., Loskutov, A., Jun, D., Saer, R., Beatty, J. T., Lin, S., Liu, Y., Woodbury, N. W. and Yan, H. (2014). *J. Am. Chem. Soc.*, **136**, 16618–16625.

91. Ner, Y., Grote, J. G., Stuart, J. A. and Sotzing, G. A. (2009). *Angew. Chem. Int. Ed.*, **48**, 5134–5138.

Chapter 2

DNA-Programmed Nanoscale Assemblies of Covalently Linked Functional Monomers

Prolay Das and Seema Singh
Department of Chemistry, Indian Institute of Technology, Patna, India
prolay@iitp.ac.in

2.1 Introduction: DNA-Programmed Assemblies

Ever since the discovery of the structure of DNA by Watson and Crick in 1953, DNA has motivated scientists to study its immense physical and chemical properties by virtue of its fascinating structure. Apart from being the genetic material of any living organism, DNA has been employed for multifaceted applications. Today, DNA is considered as one of the most promising functional nanomaterials. It is easily programmable, and the nanostructures are predictable due to the base pairing fidelity of DNA. Taking advantage of this fact, Seeman introduced the concept of DNA nanotechnology in 1982 [1]. He observed porous nanostructures formed due to interconnected DNA nanorods that are capable of encapsulating a protein. Seeman introduced the idea of using a branched oligomer as a scaffold for

Templated DNA Nanotechnology: Functional DNA Nanoarchitectonics
Edited by Thimmaiah Govindaraju

ISBN 978-981-4800-21-1 (Hardcover), 978-0-429-42866-1 (eBook)
www.panstanford.com

the construction of nanostructures in two and three dimensions [2]. Interweaving of several DNA strands together led to the formation of versatile extended nanostructures that are more rigid compared to normal B-DNA. Rothemund reported a new strategy to form smiley faces using long DNA pieces folded at regular intervals with the help of staple small DNA strands [3]. The term "DNA origami" was coined to signify its similarity with Japanese "paper origami." Paper origami refers to the art of creating 3D shapes and structures without the use of any glue. Similarly, DNA origami can produce different structures without the use of covalent bonding. The DNA strands are held together by traditional hydrogen bonds, yet capable of generating 3D structures by virtue of base pairing through hybridization. Seeman and Chen created a topological cube and a truncated octahedron from DNA oligomers following the concept of DNA self-assembly [4].

With the development of DNA nanotechnology, a plethora of academically interesting and elegant DNA nanostructures has been generated [5]. However, the main challenge for the extension of DNA nanotechnology to real-life applications still exists. DNA nanotechnology is moving beyond the design and structure, to focus on function [6]. Moreover, having gained the expertise to use DNA as a scaffold for the generation of nanostructured materials, scientists got interested in creating novel DNA hybrid materials because of the physicochemical stability, mechanical rigidity, and high precision programmability of DNA [7]. The availability of automated solid phase synthesis and the abundance of synthetic modifications of DNA have inspired new interest. Taking advantages of these facts, researchers have been successful in incorporating or inserting synthetic molecules, vertexes, and linkers directly into the DNA sequences. However, many research groups go for solution phase coupling of organic molecules to the termini of the DNA strands via covalent strategies that involve the use of appropriate derivatives of molecules of choice. It is an important achievement since it opens the door for introducing new construction as well as functional motifs for nanotechnology [8]. Recent advances in DNA nanotechnology have provided tools for biophysics, molecular sensors, enzymatic cascades, drug delivery, tissue engineering, and device fabrication at the nanoscale [9]. Apart from these, structure determination of membrane proteins, DNA robot, DNA-based cancer-targeting ligands, artificial DNA-origami membrane channel, and novel DNA-based drug delivery vehicles are some of the important real-life applications of DNA nanotechnology [10].

2.2 DNA–Synthetic Molecule Hybrid for Supramolecular DNA Nanotechnology

Weaving together of multiple DNA strands into "tiles" or the stapling of a long DNA strand into "origami" assemblies, wireframe structures, 2D and 3D DNA crystal structures, and DNA bricks use DNA as the sole material to guide the assembly process. These nanostructures are DNA-dense with limited rigidity [11]. An interesting and alternative approach to building more versatile DNA nanostructures is the combined use of DNA and synthetic molecules [12]. This strategy merges the diverse structural features and functionalities of organic and inorganic molecules or transition metal complexes with the programmability of DNA. Unmistakably, it creates DNA hybrid nanostructures with novel functions. The availability of a large number of supramolecular motifs provides great potential toward expanding DNA nanotechnology into completely new structures. These supramolecular motifs bring additional interactions, and thus more structural varieties have been achieved [13]. This incites significant research interest as it provides an easy way to modulate their biological properties, such as stability against degradation, cellular uptake, and targeting [14].

Different types of supramolecular interactions play a vital role in the construction of complex DNA nanostructures, which is determined by supramolecular chemistry of the components involved in the construction. In structural DNA nanotechnology, assembly outcome depends upon the sequence encoded in each of the DNA strands whereas in DNA hybrid nanostructures, the assembly process can be fine-tuned by means of DNA sequence, modification, and synthetic molecule insertion and by some additional agents. Several strategies have been introduced for the creation of predictable DNA nanostructures based on DNA-organic hybrid molecular building blocks having discrete shape, size, and geometry. The selection of organic or inorganic molecules and DNA sequence have been found to be the most important consideration to direct the self-assembly of the hybrid structures. The replacement of natural DNA bases with supramolecular building blocks and covalent conjugation of synthetic molecules with DNA strand are the most common strategies for the construction of hybrid DNA nanostructures [15, 16]. Since the repertoire of organic molecules is virtually infinite, the possibilities of developing novel DNA–organic molecular hybrid structures are endless. Judicious selection of organic molecules can give rise to unique DNA-based nanostructures for potential applications. Synthetic molecules having reactive functional groups with rigid and high molecular recognition ability can be excellent synthons for construction of DNA hybrid building blocks. These DNA nanostructures have been used in biosensors, electronic switches, lipid membrane channels, enzyme cascades, metal ion sensors, etc. (Fig. 2.1) [17–19].

2.2.1 Advantages of Covalent Conjugation of Molecules with DNA

DNA hybrid building blocks introduce new assembly features into DNA that complement or possibly even supersede DNA's own base pairing and create completely new structures that have great potential for expanding the field of DNA nanotechnology [20]. The presence of rigid organic or inorganic molecules at junctions of two or more DNA strands has been recognized as a potential alternative to interweaving DNA strands in a "DNA economic" way. In supramolecular DNA assemblies, synthetic molecules bring a number of additional interactions and desirable functions to DNA scaffolds, including base pairing fidelity. DNA alone is thermally susceptible to melting and lacks the required structural rigidity as demanded by 3D functional nanostructures. DNA molecules covalently attached to organic molecules can pave the way for nanostructures that are structurally rigid yet carrying the versatility of DNA self-assembly. Apart from structural rigidity, it has been found that such structures also offer tunable electronic and optical properties for important molecular electronics and optoelectronics applications [21, 22]. The DNA assembly can be dynamically controlled and in certain cases are corrected by using small DNA binder molecules. Covalent conjugation of DNA with synthetic molecules ensures that the precise location of the molecules is known in the nanostructure. One interesting observation is that many aqueous insoluble molecules when conjugated with DNA are rendered soluble along with the DNA in water owing to the hydrophilicity of the DNA.

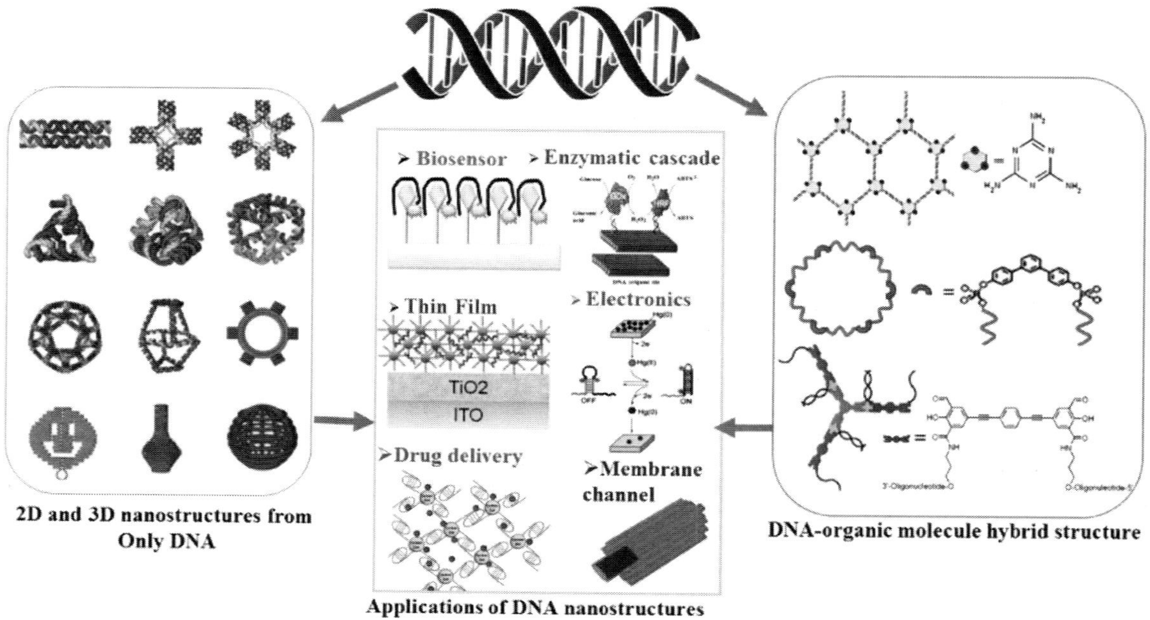

Figure 2.1 DNA-only and DNA hybrid nanostructures and their areas of potential applications [3, 5, 14, 15, 17–19, 21, 22]. Reprinted by permission from Springer Nature Customer Service Centre GmbH: Springer Nature, *Nature*, Ref. [3], Copyright (2006). Reprinted from Ref. [14], Copyright (2016), with permission from Elsevier. Reproduced from Ref. [15] with permission of The Royal Society of Chemistry. Reprinted with permission from Ref. [17]. Copyright (2012) American Chemical Society. Reprinted with permission from Ref. [19]. Copyright (2012) American Chemical Society. Reproduced from Ref. [22] with permission of The Royal Society of Chemistry.

2.3 Conjugation Strategies

2.3.1 DNA Synthesizer-Based Insertion

DNA synthesizer–based modification in the nucleobases has been done by inserting functional groups through either the phosphate backbone or nucleobases modification (Fig. 2.2) [23]. An oligonucleotide can be modified by alkynyl derivatives like amines, biotin, and metal complexes using solid phase nucleic acid synthesis and Sonogashira coupling [24].

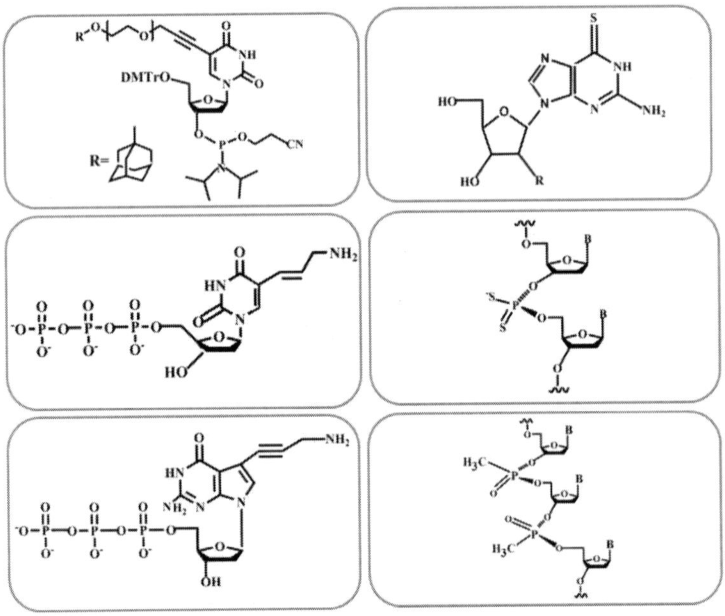

Figure 2.2 Some examples of modified nucleosides used in DNA synthesizer–based insertion [23, 27, 28]. Adapted from Ref. [28] with permission of The Royal Society of Chemistry.

Charge injectors and other specialized organic molecules can be inserted into single-stranded DNA (ssDNA) oligomers by modification of any of the four nucleotides by phosphoramidite chemistry. A particularly interesting yet relatively unexplored chemistry to incorporate organic molecule into DNA is to use triazole derivatives under mild reaction conditions at room temperature.

An oligonucleotide can be modified with O_4-triazoyldeoxyuridine phosphoramidite in a DNA synthesizer. This oligonucleotide can be attached to amine functional groups of organic monomers using postsynthetic modification by replacing the triazole group of convertible nucleobases in DNA strands. On the basis of this concept, some monomer units, like 3-(2,5-di(biphenyl-4-yl)-1H-pyrrol-1-yl) propan-1-amine, have been conjugated to this triazole-modified DNA and can be elongated for the production of conducting polymer [25]. The modification of nucleotides by oxidation to yield 8-oxoG and 8-oxoA has been investigated through a polymerase chain reaction [26]. Fluorophores have been covalently attached to the amino linker of 5'-triphosphates of nucleobases and incorporated in DNA through enzymatic incorporation. The modified DNA chains were further extended by the polymerase chain reaction [27]. C5-modified triphosphates and phosphoramidites with diamondoid moieties were synthesized and linked to the oligonucleotide using a DNA synthesizer (Fig. 2.2) [28].

2.3.1.1 Advantages and disadvantages of synthesizer-based insertion

The modification of nucleobases through synthesizer-based insertion has some advantages, such as less number of synthetic steps, a variety of functional group tolerance, easy purification, and efficient yield. However, the techniques have some limitations. The synthesis may be compromised due to a high error rate and continuation of errors once programmed through synthesizer [29]. Also, the organic molecule should be stable in sufficiently basic condition that is used to detach the DNA oligomer from the beads. Moreover, most of the modified phosphoramidites are susceptible to aerial oxidation and have limited stability, leading to inherent difficulty in obtaining the highly pure version of the compound with chromatographic column purification.

2.3.2 Postsynthetic Modifications

2.3.2.1 Amine–phosphate coupling

Orgel and coworkers in 1983 described a method to conjugate amines to the phosphate group of an oligonucleotide using 1-ethyl-

3-(3-dimethylaminopropyl) carbodiimide (EDC) and imidazole-based phosphoramidation reaction. Gilham and coworkers conjugated nucleic acids to paper via a 5'-phosphate group using the carbodiimide method. This method has been used to conjugate heterocyclic aromatic amines like melamine. This is also an example of zero linker conjugation where an amine can be directly conjoined with a terminal phosphate of DNA. Further, this method has been used to conjugate oligonucleotides with peptides, proteins, and fluorophores. This method has the potential to label nucleic acids for research in structural determination, quantification, localization, and distribution of nucleic acids in vivo and in vitro [30].

Example: The heteroatomic molecule melamine was coupled to ssDNA using phosphoramidate coupling. The conjugation was carried out by adding 2 μL (1.0 nmol) of an aqueous solution of the 40-base-long DNA to a solution of 0.5 mg of EDC and 10 μL of a 0.1 M imidazole buffer solution at pH 6.0 with continuous stirring at 37°C for 90 min. The resulting reaction mixture was purified by ethanol precipitation and dissolved in 10 μL of a 0.1 M 4-(2-hydroxyethyl)-1-piperazineethanesulfonic acid (HEPES-EDTA) buffer. Further, 2 μL of melamine solution (0.2 nmol) was added and incubated at 55°C with stirring for 18 h. The crude product was purified by dialysis using a 1000 Da MW membrane and preparative denaturing polyacrylamide gel electrophoresis (PAGE) (Fig. 2.3) [31].

2.3.2.2 Carboxylic acid–amine coupling

Carboxylic acid–amine coupling through carbodiimide chemistry is also a well-known zero-length cross-linking reaction. The reaction can take place with no addition of chemical structures between conjugating moieties. EDC for aqueous phase and *N,N'*-dicyclohexylcarbodiimide (DCC) for nonaqueous phase are the most common coupling reagents used for the conjugation of biomolecules and others [32]. EDC activates the carboxylic group and forms O-acylisourea as an intermediate that further reacts with the amine groups, leading to the formation of an amide bond. However, instability of the intermediate results in failure of the reaction. *N*-hydroxysuccinimide (NHS) or sulfo-NHS is often used along with EDC to improve the reaction efficiency through stable intermediates [33]. The EDC/NHS-mediated amide coupling has been employed for the conjugation of DNA with an organic molecule for the construction of DNA nanostructures [34]. Further, the method has been used for the conjugation of quantum dots (QDs) and nanoparticles to DNA for various biomedical applications [35, 36]. It is to be noted that amines of the bases are not the ones that react; it is the amines linked to the terminal of DNA often through a linker that participate in the reaction due to a higher reactivity.

Example: Carboxyl-terminated ZnSe/ZnS QDs were conjugated to amine-terminated ssDNA using amide coupling. QDs (20 μL, 1 nM) were activated by adding EDC (10 μL, 10 μM) and NHS (20 μL, 20 μM) in a 1:1 ethanol:sodium phosphate buffer (200 mM) at pH 6. The solution was centrifuged and the supernatant was discarded. The pellets of QDs were washed with water, and ssDNA (10 μL, 5 nM) was added to it with stirring for 3 h at 4°C. The product could be purified by dialysis (MWCO-5 KDa) as well as denaturing PAGE [35].

2.3.2.3 Thiol-based coupling

Thiol-based coupling of oligonucleotide has been explored for conjugation of organic molecules with DNA. The reactivity of the sulfhydryl group of an oligonucleotide with the maleimide group of an enzyme was studied by Ghosh et al. by two different methods [37]. Thiolene Michael addition has been employed for the covalent conjugation of thiol-terminated DNA to poly(*N*-isopropylacrylamide) in an organic solvent [38]. Direct attachment of QDs to DNA was done via thiol conjugation. DNA was modified with thiol and coordinated to QD via dative disulfide linkage [39, 40]. A methodology was developed to conjugate the DNA site specifically to antibodies through thiol coupling and was used in the immune-PCR method to detect Her2+ cells [41]. The conjugation of DNA methyltransferase SsoII to sequence-specific decoys was demonstrated and used as a tool to study the biological system [42].

2.3.2.4 Click chemistry

In 1999 Sharpless was the first to initiate click chemistry where Cu catalyzed [3+2] cycloaddition of azide and alkyne under mild conditions was established [43]. Fluorescein alkynyl 6-carboxyfluorescein (FAM) was conjugated to an azide-modified oligonucleotide using click chemistry (Fig. 2.3) [44]. Azide modification uses an NHS ester functional group to attach an azide moiety at the 5', 3', or any internal position in an oligo. Glyco-oligonucleotides coupling containing mannose and galactose moieties were synthesized by microwave-mediated 1,3-dipolar cycloadditions [45]. The construction of a closed circular nanostructure was realized using ssDNA labeled with 3'-azide and 5'-alkyne [46]. G-quadruplex DNA plays a key role in telomere maintenance and can act as a scaffold for different types of Cu-catalyzed click reactions [47]. One of the examples is conjugation of hemin to G-quadruplex DNA using click chemistry [48].

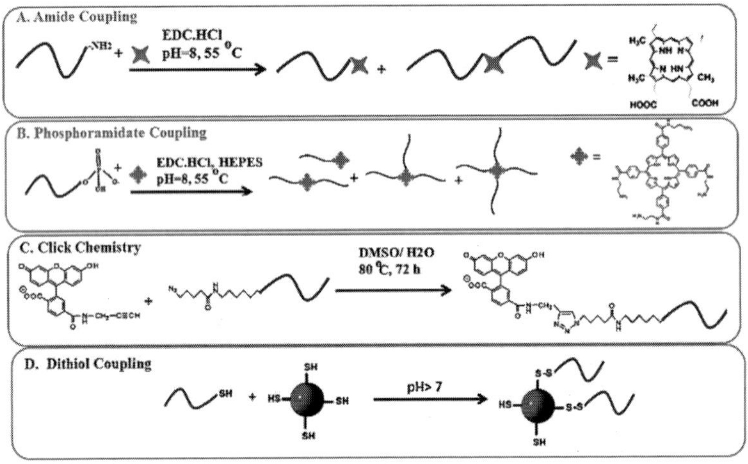

Figure 2.3 Common strategies for postsynthetic modification for DNA–organic molecule conjugation [31, 35, 39, 44].

2.4 Analytical Methods to Purify and Characterize DNA-Organic Hybrid Monomers and Nanostructures

2.4.1 Gel Electrophoresis

2.4.1.1 Agarose gel electrophoresis

Agarose gel electrophoresis is one of the most effective ways to separate DNA sequences ranging from 100 bp to 25 kbp. 1%–1.5% agarose gel electrophoresis can be performed to visualize the DNA samples with 1X tris taurine EDTA (TAE) as a running buffer at 80–120 V. The separation takes place mostly on the basis of mass-by-charge ratio. The DNA samples can be visualized by staining with ethidium bromide (EtBr) or SYBR gold® as a staining agent and captured through a charge-coupled device (CCD) camera. Agarose gel has been employed to separate many DNA-organic hybrids, for example, DNA organic molecule, DNA triblocks, poly(p-phenylene ethynylene) (PPE) ssDNA, and benzenetricarboxylate DNA (Fig. 2.4) [46, 47, 49]. The association complex formed by tris-DNA connected through 1,3,5 trisubstituted benzene was analyzed on agarose gel to separate the monomer, dimer, trimer, and tetramer [50].

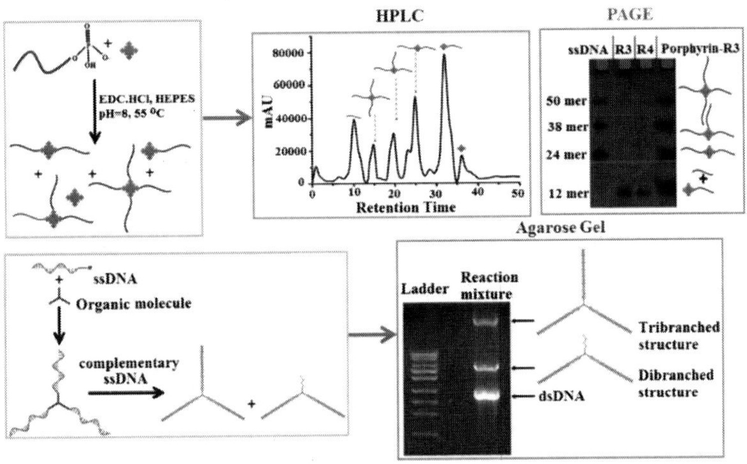

Figure 2.4 Characterization of DNA-organic hybrid monomers by HPLC and gel electrophoresis [49, 55]. Reprinted from Ref. [55], Copyright (2017), with permission from Elsevier.

2.4.1.2 Polyacrylamide gel electrophoresis

Two types of PAGE, native and denaturing, are generally employed for analysis of DNA samples from 5 to 2000 kDa. Denaturing PAGE of 6%–25% (29:1 acrylamide bis-acrylamide cross-linking ratio) in 7 M urea, at pH 8 (tris buffer), can be performed to separate individual moieties in a mixture of the sample. However, the conditions for native PAGE are the same as that for denaturing gel, excluding urea. The samples can be run through a 1X tris taurine EDTA (TTE) buffer at 150–200 V. Sample visualization is usually done after the DNA is stained with either EtBr or SYBR gold®. PAGE has been employed to visualize different organic macrocycle–DNA conjugates from a crude reaction mixture [51]. Sleiman and Aldaye used 10% native PAGE to analyze the triangular DNA nanostructure formed by sequential self-assembly of DNA–organic molecule hybrids [52].

2.4.2 HPLC

High-performance liquid chromatography (HPLC) has been utilized to separate, analyze, and quantify different products in a reaction mixture. The DNA samples can be analyzed using a C-18 column (25 × 21.4 mm) and 5%–20% acetonitrile in a 0.5 M triethylammonium acetate buffer as a solvent at pH 7. The method has been used to separate different moieties in the reaction mixture of DNA–organic molecule conjugate. It relies on separation of the mixture dissolved in solvents and passes it through the column under pressure. The separation occurs on the basis of the affinity of different components with adsorbent, causing different flow rates of each component [53, 54]. It has been used for the characterization of DNA-attached heterocycles, DNA-protoporphyrin hybrids, and DNA-triptycene hybrid nanostructures (Fig. 2.4) [55, 34].

2.4.3 ESI-MS

Electrospray ionization mass spectrometry (ESI-MS) has emerged as a sensitive and robust analytical technique reliable in clinical laboratories. The technique has been used for the characterization of DNA-organic hybrids even at femto-mole concentration in microliter volume. Typically for analysis, samples can be solubilized in aqueous 50% acetonitrile (v/v) with 1% triethylamine to give a final concentration of 20–50 pmol/μL. The sample can be introduced at a flow rate of 5 μL/min., a cone voltage of 50 V, and the source temperature maintained at 70°C. A mass scale spectrum can be deconvoluted by using software like MassLynx to reconstitute a real spectrum from a multiply charged ion mass spectrum obtained by ESI-MS [56]. The method has been employed for characterization of DNA–organic molecule–DNA triblock oligomers, benzenetricarboxylate-DNA hybrids, polyethylene glycol–DNA conjugates, DNA-melamine hybrids, etc. (Fig. 2.5) [31, 49, 57].

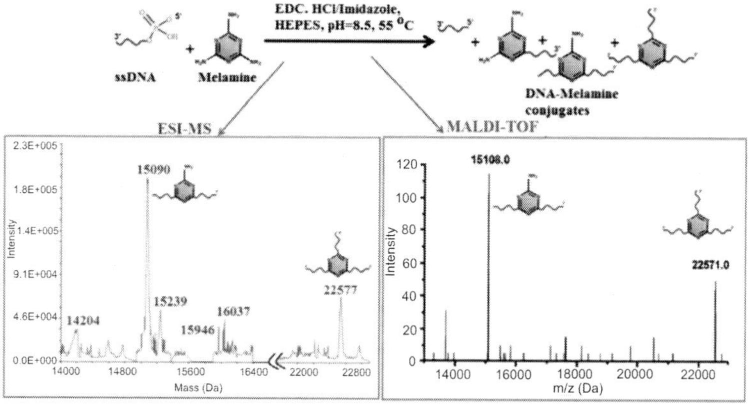

Figure 2.5 Characterization of DNA–organic monomers by mass spectrometric techniques. Reprinted from Ref. [64], Copyright (2016), with permission from Elsevier.

2.4.4 MALDI-TOF

Matrix-assisted laser desorption/ionization–time of flight (MALDI-TOF) is a mass spectrometry technique that utilizes a laser absorbing matrix. Usually picolonic acid and dibasic sodium citrate are used as a matrix for ionization of molecules with minimum fragmentation. The method is convenient for ionization and characterization of fragile DNA-organic hybrids. The DNA sample is mixed with the matrix and applied to a metal plate, followed by irradiation. The ionized sample can be accelerated and analyzed in a mass spectrometer [58, 35]. Detection of high-molecular-weight compounds can be performed by linear mode with a high mass detector (7–9 kV) at a laser voltage of 3500 V. The mixture of the oligonucleotide can be analyzed using positive or negative ion mode. MALDI (+) mode was found to show a better resolution than MALDI (–) at the same laser power. However, MALDI (–) showed better sensitivity for the negatively charged species [59]. MALDI-TOF has been used for the identification of DNA-melamine, DNA-pyrene, DNA-triphenylphosphine, DNA-porphyrin, and DNA-protoporphyrin hybrid molecules [60, 61, 31, 34].

2.4.5 AFM

Atomic force microscopy (AFM) is one of the reliable methods for the imaging of DNA-organic hybrids. AFM images can be obtained by drop-casting the solution of biomolecules on freshly prepared ammonium persulfate (APS)-modified mica and captured in intermittent contact mode atomic force microscopy (ACAFM) [62]. The imaging can be done with a silicon nitride cantilever having a force constant of 20–80 N/m, a typical frequency range of 240–400 KHz, and a nominal scanning rate of 1–3 Hz (Fig. 2.6) [63, 64]. DNA nanotubes with organic molecules as a linker have been prepared, and imaging was done with AFM in noncontact mode [65]. DNA-melamine hybrid linear and network nanostructures have been visualized and differentiated with the help of AFM [31]. The technique has been utilized for the visualization of DNA directed self-assembly of QDs [66].

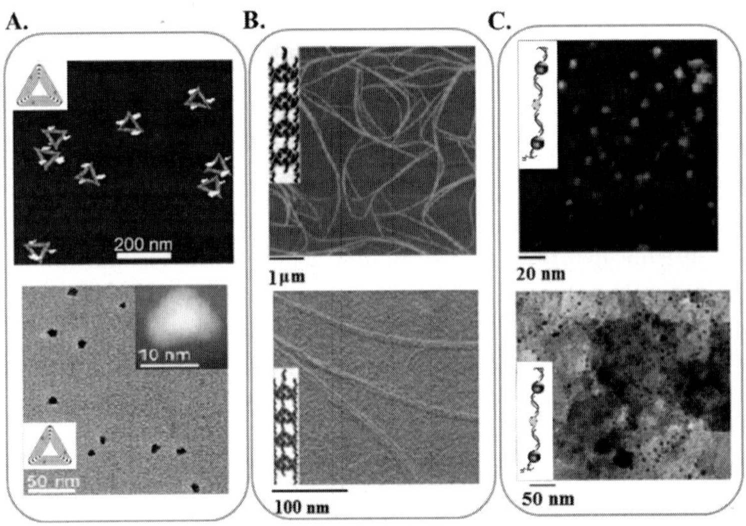

Figure 2.6 AFM (top) and TEM (bottom) characterization of DNA-organic hybrid nanostructures. Reprinted from Ref. [64], Copyright (2015), with permission from Elsevier. Reprinted by permission from Springer Nature Customer Service Centre GmbH: Springer Nature, *Nature Chemistry*, Ref. [65], Copyright (2010). Reprinted with permission from Ref. [66]. Copyright (2012) American Chemical Society.

2.4.6 TEM

Imaging of a biological molecule by transmission electron microscopy (TEM) is carried out by negative staining of the sample. Biological molecules with C, H, or N elements require contrast between background and biomolecule; uranyl acetate or ammonium molybdate is generally used to scatter the electrons. This heavy metal salt covers the background around the biomolecule and outlines the shape of the specimen [67]. TEM images can be prepared by placing ~5 μL of sample solution onto a carbon-coated grid and imaging at an operating voltage of 80 kV. The technique has been used to visualize DNA plasmonic nanostructures, DNA-melamine-mediated QD arrays, and reconfigurable topological structures of DNA [68–70, 31]. A DNA nanostructure containing a cavity for a cargo molecule and developed by covalently linking photolabile o-nitrobenzyl (O-NB) to the oligonucleotide has been characterized by TEM to visualize the dimension of the nanostructure [71]. Ancona et al. developed a facile approach for imaging of the DNA nanostructure without staining, using graphene and a Si layer as a support [72].

2.5 Role of Synthetic Molecules in DNA-Assembled Structures

Synthetic molecules in a DNA-programmed nanostructure were initially used for scaffolding and structural considerations. However, recent research has shown that these molecules could impart important functionality to the DNA nanostructures. Such functional molecules on conjugation with DNA have the potential to be used for light harvesting, molecular electronics, antibacterial activity, photodynamic therapy (PDT), etc.

2.5.1 Organic Molecules as Linkers

It was Shi and Bergstrom who synthesized the first organic-based building blocks for assembling DNA into unique nanostructures, in 1997 [73]. These constructs have branched DNA architectures. Flexible organic tetrahedral branch points were used having flanking p-(2-hydroxyethyl) phenylethynylphenyl moieties that were connected with two oligonucleotide strands using phosphodiester conjugation. Tanaka et al. were the first to optimize azobenzene molecule insertion into the sticky ends of three-point-star motifs [74]. This facilitates their assembly into nanocapsules having great stability. The research groups of von Kiedrowski et al. and Schepinov used DNA substrates to attach with organic molecules [75]. The DNA-organic hybrid molecules showed a promising approach toward self-assembly and formation of nanostructures [76]. Terpyridine-like organic molecules have been conjugated to DNA to form the center for metal coordination, leading to DNA-templated nanostructures [77–79].

Sleiman and group connected DNA duplexes to flexible linkers, namely flexible hexane diol, four thymidine, and rigid linkers 2,9-diphenyl-1,10-phenanthroline and m-triphenylene to position ssDNA at particular angles that ultimately give rise to 3D structures upon self-assembly. The linkers not only provide the required twist and angular disposition from a flat structure to a cyclic structure to the scaffold, they also impart rigidity and stability by increasing the thermal denaturation temperature of duplexes by up to 10°C in the DNA self-assembled structures [80]. The supramolecular cages were prepared by combining complementary small-molecule-DNA hybrids (SMDH) even at high concentrations of DNA. The small organic rigid molecules, such as 1,3,5-triphenylbenzene or 1,3,5-tris(methylphenyl)benzene, were covalently conjugated to three DNA arms of moderate length and hybridized to complementary SMDH to give a cage structure [81–83].

Amine-modified DNA was covalently conjugated to protoporphyrin IX cage-like nanostructures [34]. DNA-porphyrin hybrid units were developed to construct DNA nanostructures that display enhanced antimicrobial activity of the nanonetwork [55]. The self-assembly of DNA has been conjugated to a rigid molecule like triptycene that self-assembles to form a well-defined DNA nanostructure in the form of fibers. This structure was found to host zinc protoporphyrin (Zn Pp IX), which not only corrects the nanostructure but also shows enhanced ROS generation for PDT applications [58]. Heterocyclic compounds like melamine have also been used as an organic linker to create DNA-organic nanostructures in one and two dimensions [31]. A 3D robust nanostructure of DNA capsules has been prepared using azobenzene as a linker [84].

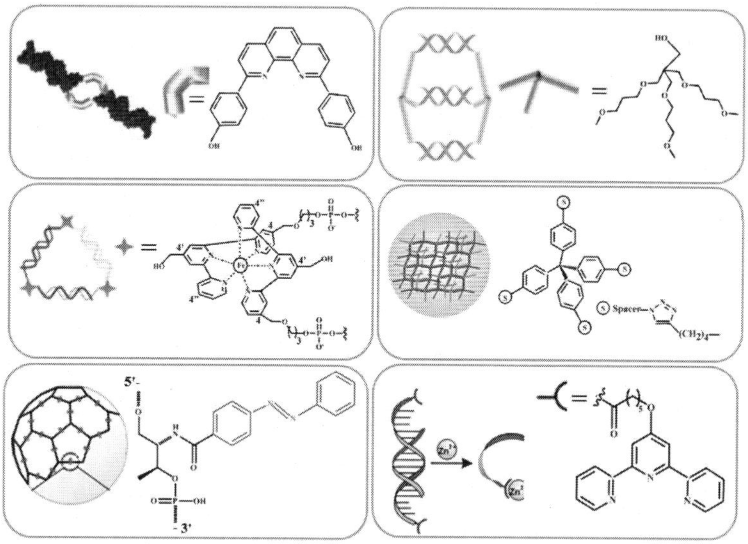

Figure 2.7 Examples of organic molecules used as linkers in conjugation with DNA as monomers of nanostructures [77, 78, 80–82, 84]. Reprinted with permission from Ref. [77]. Copyright (2004) American Chemical Society. Reprinted with permission from Ref. [78]. Copyright (2004) American Chemical Society. Reprinted with permission from Ref. [80]. Copyright (2012) American Chemical Society. Reprinted with permission from Ref. [81]. Copyright (2015) American Chemical Society. Reprinted with permission from Ref. [82]. Copyright (2015) American Chemical Society. Reprinted with permission from Ref. [84]. Copyright (2010) American Chemical Society.

2.5.2 Organic Molecules as Charge Injectors in DNA

Considerable attention has been paid in the last few decades to study the charge transfer in DNA. Often, charge injectors are covalently attached to DNA that trigger formation and subsequent transport of a hole in the DNA. Various techniques have been adopted to analyze the charge transfer through DNA, which includes conductivity measurement, redox product analysis, time-resolved spectroscopy, and autoradiography [85]. The close proximity of organic molecule and DNA via end capping, intercalation, groove binding, or covalent conjugation is a necessary criterion for the transfer of electron or hole through the DNA. Among the organic molecules, various photosensitizers have been used, such as riboflavin, anthraquinones, and naphthalimides, which are excellent charge injectors in DNA [25]. The anthraquinone moiety can be attached to the DNA by covalent conjugation using phosphoramidite chemistry, or it can be tethered to the DNA at the terminal or intermediate position in the form of uridine-anthraquinone (UAQ) [86].

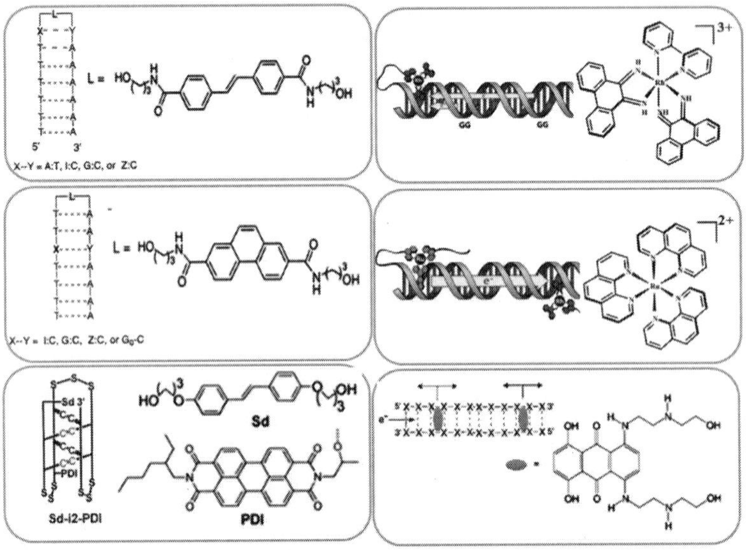

Figure 2.8 Organic molecules covalently conjugated to DNA for charge injection [91–95]. Reprinted from Ref. [91], Copyright (2010), with permission from Elsevier. Reprinted with permission from Ref. [92]. Copyright (2017) American Chemical Society. Reprinted with permission from Ref. [94]. Copyright (2010) American Chemical Society.

Lewis et al. reported the hole injection process in DNA through stilbene dicarboxamide [87–89]. In 2003, Wagenknecht investigated the electron injection in pyrene conjugated to U and C [90]. Barton et al. covalently attached Ru(phen')$_2$dppz^{2+} as an electron donor and Rh(phi)$_2$(phen')$^{3+}$ (phen' = 5-amido-glutaric acid-1,10-phenanthroline) as an acceptor to DNA strands complementary to each other [91]. In 2017, Lewis and coworkers studied the dynamics of electron injection in the i-motif structure of DNA with stilbenediether and perylenediimide chromophores. The two chromophores were separated by four to six hemiprotonated C–C base pairs assembled via a synthetic molecule loop and the dynamics of charge transfer were analyzed by femtosecond transient absorption spectroscopy [92]. Similarly, charge transfer through DNA has been investigated using stilbene dicarboxamide, chromophores, and pyrimidine radical anions as intermediates [93–95].

2.5.3 Organic Molecule–DNA Monomers for Templated Polymerization

DNA-templated synthetic polymers allow the formation of an information-encoded polymer having unique properties of self-assembly, folding, allosteric control, and construction of 3D and 2D supramolecular structures [96]. Covalent attachment of the monomer units with DNA helps to accomplish several purposes.

- The DNA sequence acts as a template for the self-assembly of the monomer units in the polymerization process.

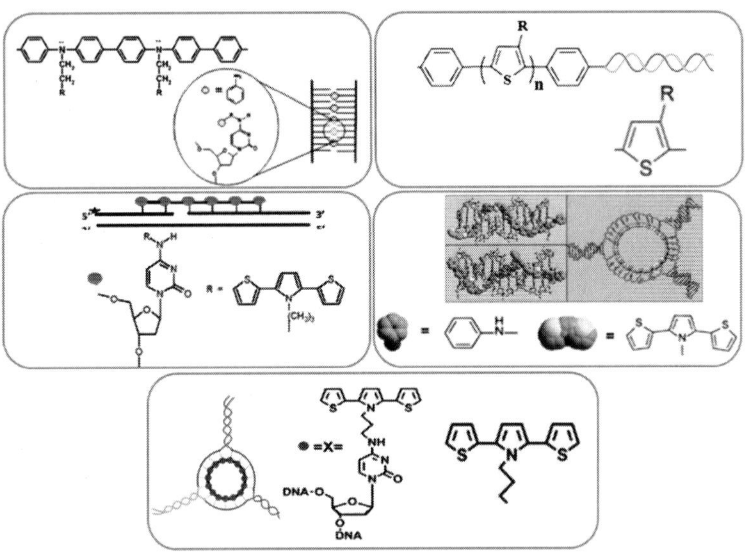

Figure 2.9 DNA-templated polymerization of organic monomers [97–101]. Reprinted with permission from Ref. [97]. Copyright (2008) American Chemical Society. Reprinted with permission from Ref. [98]. Copyright (2010) American Chemical Society. Reprinted with permission from Ref. [99]. Copyright (2012) American Chemical Society. Reprinted with permission from Ref. [100]. Copyright (2013) American Chemical Society.

- The polymerization process can be programmed and controlled in any direction by utilizing the sequence specificity, intrinsic to DNA.
- The parasitic branching of the growing polymer chain can be completely eliminated, resulting in the formation of a well-defined linear polymer.
- The length of the polymer chain can be controlled with very high precision using the exact number of desired DNA-linked monomer units.
- DNA-linked monomer units are water soluble, which makes them easier to process.

Schuster and coworkers synthesized polyaniline or 4-aminobiphenyl-modified oligomers by using O_4-triazolyl-deoxyuridine phosphoramidate coupling by the nucleotide modification approach. These modified DNA-heterocycle polymers show properties of a conducting polymer, with monomers aligned along the major grooves [97]. 2,5-bis(2-thienyl)pyrrole was covalently conjugated to nucleobases of ssDNA and annealed to a complementary DNA strand. The reaction of this duplex modified with monomers with horseradish peroxidase and H_2O_2 yields polymers having fascinating optical and conducting properties, like conjugated polymers [98]. Cyclic co-oligomers of aniline and 2,5-bis(2-thienyl)pyrrole with double-stranded DNA (dsDNA) were developed to create DNA-copolymers of a well-defined structure [99, 100]. Bifunctionalized polythiophene having thiol and azide end functional groups was synthesized by Suzuki–Miyaura polymerization, and the azide group of polymers was conjugated to DNA via click chemistry. This method can be used for the synthesis of a multiblock copolymer and biomolecule-based polymer [101]. Wilks and O'Reilly explored the covalent coupling of DNA-poly(*N*-isopropylacrylamide) in organic solvents. Modification of DNA and polymer was done with norbornene (Nb) or tetrazine (Tz) and an inverse electron-demand Diels–Alder reaction was used to couple Nb and Tz to conjugate DNA to polymer, with a good yield of product [102]. O_4-triazoyl-deoxyuridine phosphoramidite presents an excellent possibility as a modified base for the attachment of monomer units of a polymer. This convertible nucleotide approach is suitable for covalent attachment of amine containing monomers at room temperature and under mild conditions in aqueous phase. The monomer units could align themselves on the DNA template and be polymerized using enzymatic reactions [29]. This is an example where presynthetic as well as postsynthetic approaches are combined to tether monomers to DNA-templated polymerization.

2.5.4 Organic Molecules and Nanoparticles for Light Harvesting

DNA-mediated nanostructures as a light-harvesting tool have been used for incorporation of chromophores for light harvesting and energy conversion. The sequence design of DNA allows one to control the definite distance and orientation between the chromophores [103]. DNA-templated light-harvesting systems have been made with three distinct chromophores: ethylenepyrene (Py), cyanine-derived dye (Cy3), and Alexa Fluor 647 (AF) arranged in different configurations with varying ratios of donor and acceptor. The triad showed efficient energy transfer in a stepwise manner from the primary donor (Py) to the acceptor (AF) through the intermediate chromophore (Cy3) on excitation of Py [104]. Three chromophores, namely ethynyl pyrene, ethynyl perylene, and ethynyl red, were covalently conjugated to the 5' position of 2'-deoxyuridine via acetylene to dsDNA using phosphoramidate chemistry. The versatility of the system lies in the fact that an emission range from 350 nm to 700 nm was covered, which indicates a potential use of DNA for a light-harvesting system [105]. A compact, covalently coupled QD-DNA probe was designed for specific detection of target DNA via FRET between QD and dye-conjugated DNA [106]. Controlled generation of ROS was obtained by assembling the FRET probe protoporphyrin IX (PpIX) and ZnSe/ZnS QDs that are covalently conjugated to DNA [35]. Such DNA-conjugated FRET pairs were found to target cancer cells through cancer biomarker–specific aptamers, making them attractive candidates for PDT application [37]. Olejko et al. have prepared the DNA origami nanostructure for the characterization of a switchable multicolor FRET cascade [107]. DNA has been used as an antenna to study the long-range FRET through chromophores, dyes, and QDs [108, 109].

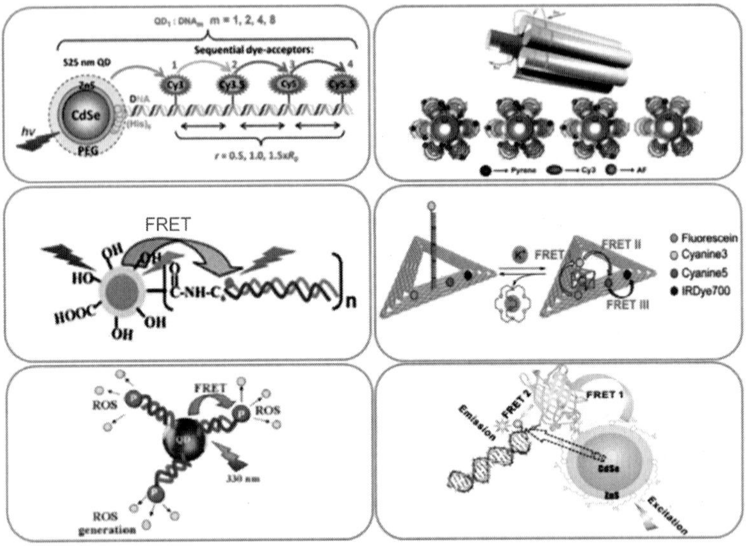

Figure 2.10 Organic-DNA monomers for potential applications in light harvesting [35, 104, 106–109]. Reprinted with permission from Ref. [104]. Copyright (2011) American Chemical Society. Reprinted with permission from Ref. [106]. Copyright (2008) American Chemical Society. Reprinted with permission from Ref. [108]. Copyright (2008) American Chemical Society. Reprinted with permission from Ref. [109]. Copyright (2013) American Chemical Society.

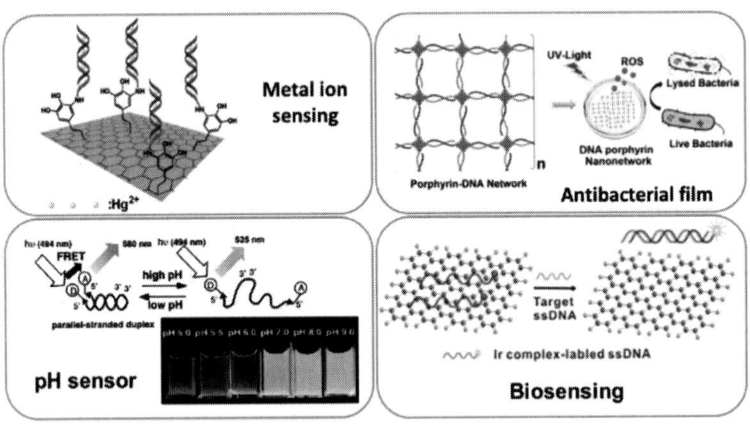

Figure 2.11 Various potential applications of nanostructures made from DNA–organic monomer units [55, 111, 113, 116]. Reprinted from Ref. [111], Copyright (2013), with permission from Elsevier. Reprinted with permission from Ref. [113]. Copyright (2016) American Chemical Society. Reprinted with permission from Ref. [116]. Copyright (2005) American Chemical Society.

2.5.5 Antibacterial, Biosensing, and Miscellaneous Functions

A porphyrin derivative with four pendant amine groups has been covalently conjugated to ssDNA and self-assembled to create a DNA-porphyrin hybrid nanonetwork. The nanonetwork was found to show coherent ROS generation and enhanced antibacterial activity against staphylococcus aureus upon irradiation [55]. Graphene oxide (GO)-chitosan (CHI)-based nanocomposites have been developed and successfully used for fabricating a DNA-based biosensor to detect typhoid. The biosensor was prepared by covalent conjugation of Salmonella typhi–specific 5′-NH_2-modified ssDNA probe on GO-CHI via glutaraldehyde [110]. A DNA-grafted graphene was prepared by reducing graphene with dopamine and then conjugated to amine-terminated ssDNA by Michael addition reaction. The DNA-graphene composite was used as a biosensor for the detection of mercury (II) ions (Hg^{2+}) [111]. A biosensor for the detection of arsenic trioxide based on surface plasmon resonance has been prepared by immobilizing the calf thymus DNA onto a mercaptoethanol/gold electrode using EDC/NHS chemistry [112]. An iridium (III) complex with a carboxyl group was synthesized and conjugated to amine-modified ssDNA and further combined to a GO nanosheet to detect target ssDNA. A DNA microarray was developed to monitor and screen target-specific stress responsive genes [113]. Hu et al. developed an efficient method using fluorophore-labeled DNA for sensitive and selective detection of serious diseases. They have used a perylenediimide derivative as a quencher of fluorophore-labeled DNA (FAM, tetramethylrhodamine [TAMRA], or Cy5) for multiplexed fluorescence detection of DNA [114]. ZnO modified with reduced diazonium salt has been covalently conjugated to DNA using sulfosuccinimidyl-4-(*N*-maleimidomethyl) cyclohexane-1-carboxylate (SMCC) as a cross-linker. The DNA-functionalized ZnO showed good biosensing to distinguish one- to four-base mismatch to detect target DNA [115]. Two fluorophores—fluorescein isothiocyanate (FITC) as donor and TAMRA as acceptor—were conjugated to 5′ terminal of two complementary DNA sequences using phosphoramidite chemistry. Depending on the parallel or antiparallel conformation of dsDNA, the energy transfer between the fluorophores changes, which enables monitoring of intracellular pH [116].

2.5.6 Synthetic Molecules Used in Molecular Electronics

In 1999, the passage of electric current and charge transfer through DNA was first measured by Fink and Schonenberger by using a gold-coated carbon grid [117]. An electroactive oligonucleotide functionalized-polypyrrole electrode has been used for the sensitive detection of oligonucleotide recognition. The polypyrrole-DNA hybrid was used as a biosensor, where specific hybridization of the target DNA induced significant modification in the electrochemical response of the polypyrrole [118]. A tetrahedral nanostructure of DNA was utilized to investigate the redox molecules methylene blue (MB) and ferrocene (Fc) bound to the surface of the gold electrode [119]. Precise localization of single-walled carbon nanotubes (SWCNT) in electronic devices has been achieved using an oligonucleotide as a scaffold. The SWCNT-DNA was prepared by binding streptavidin-modified SWCNT and biotin-modified DNA [120]. PPE, a widely used species for molecular electronics, was modified to NHS-PPE and conjugated to DNA via amide coupling. The strategy has been proposed for the construction of electronic devices based on two-terminal or three-terminal oligonucleotides [50]. An array of QD-melamine-DNA was developed by covalently conjugating QD and melamine to complementary DNA sequences, followed by hybridization-mediated self-assembly for potential application in nanoelectronics [64].

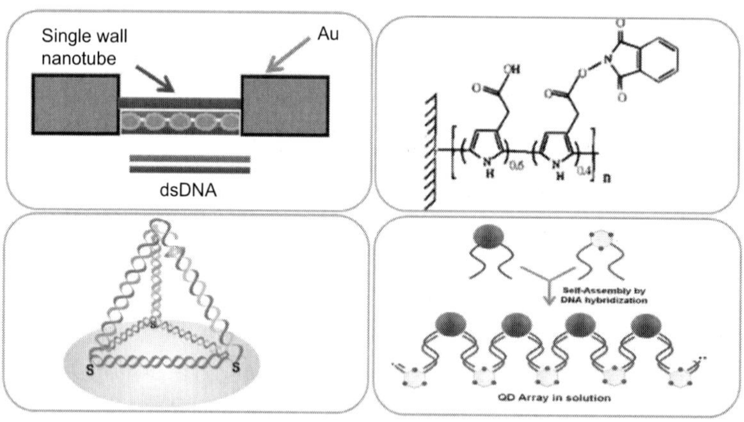

Figure 2.12 Applications of DNA-organic hybrid nanostructures for molecular electronics [64, 118–120]. Reprinted with permission from Ref. [118]. Copyright (1997) American Chemical Society. Reprinted with permission from Ref. [119]. Copyright (2012) American Chemical Society.

2.6 Self-Assembly as a Tool to Engineer DNA-Programmed Nanostructures from a DNA-Monomer Hybrid

Self-assembly is the process by which individual components form an organized structure spontaneously due to local interaction between the components. Self-assembly of molecules involves noncovalent interactions, which include hydrophobic forces, van der Waals forces, hydrogen bonding, electrostatic interaction, and π-π interaction [15, 50]. Self-assembly of DNA is key to developing branched DNA molecules for the construction of DNA nanostructures. DNA nanotechnology utilizes self-assembly as a bottom-up approach that does not suffer from size limitation of top-down approaches, like lithography miniaturization [121]. This bottom-up approach of DNA nanostructure formation can be used in the construction of billions of molecular devices [122, 123]. Lee and coworkers covalently conjugated three organic molecules, benzenetricarboxylate, polyethylene glycol (PEG), and PPE, to ssDNA. These DNA-organic hybrids were elongated through complementary DNA hybridization through self-assembly or ligation to create micrometer-sized hybrid structures [50]. A nanostructure has been prepared by using branched tris-DNA in which three oligonucleotides were conjugated covalently to an organic molecule as a rigid linker. Among these three oligonucleotides, two were identical sequences and one was different and the nanostructure was formed by self-assembly of complementary tris-DNA unit [59]. The organic linker 1,3-bis (4-hydroxyphenyl) benzene has been used as a vertex and covalently conjugated to three different nucleic acids. The organic molecule–DNA units were self-assembled to complementary DNA sequences to generate a triangular DNA nanostructure [124]. A porphyrin derivative with four amine groups was covalently conjugated to ssDNA and self-assembled to a complementary DNA-porphyrin unit to give a DNA-porphyrin nanonetwork, as mentioned before [55]. Similarly, DNA nanocages have been developed by self-assembly of DNA-protoporphyrin IX conjugates [34]. Glycans have been covalently conjugated to DNA templates with different chemistries to construct DNA glycoconjugates [125].

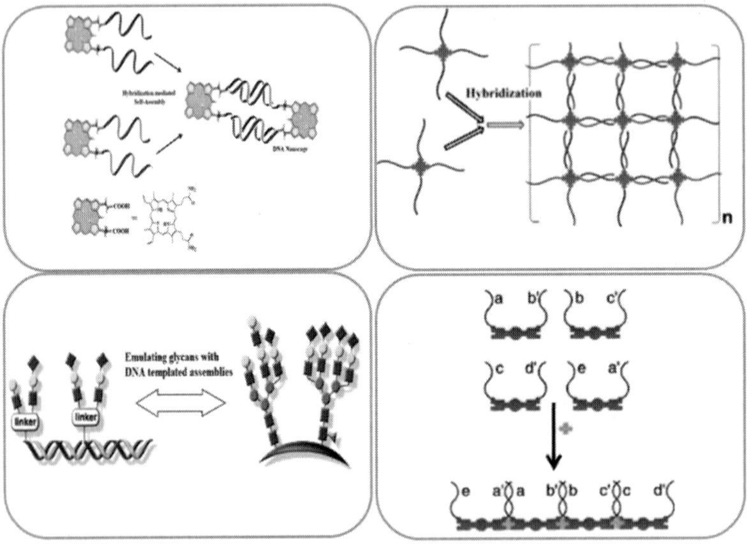

Figure 2.13 Self-assembly as a tool to create nanostructures from DNA–organic monomer units [15, 34, 55, 125]. Reproduced from Ref. [15] with permission of The Royal Society of Chemistry.

2.7 Conclusions

DNA-programmed nanostructures that contain covalently linked organic monomers to DNA are described here. DNA-only nanostructures have limited applications due to issues with rigidity, stability, and others. Introduction of DNA–synthetic molecular hybrids could help to overcome these issues by offering extra structural stability, base pairing fidelity, and other interesting properties. Covalent attachment of molecules to DNA has paved the way to create nanodevices with integrated properties of both the DNA and the attached organic and inorganic molecules, nanoparticles, proteins, etc. Various strategies have been adopted for the conjugation of synthetic molecules to DNA, which include presynthetic and postsynthetic modifications of DNA. Presynthetic modification involves modification of a nucleotide by phosphoramidite, followed by its insertion into the DNA oligomer via a DNA synthesizer. Postsynthetic modification of DNA is done by the covalent conjugation of the molecule of interest to DNA via phosphoramidate linkage, amide bond, thiol coupling, or click chemistry. The DNA–synthetic molecule hybrids and the resulting nanostructures are characterized by various analytical techniques, such as mass spectrometry, gel electrophoresis, HPLC, and electron microscopy. Various advantages of these synthetic molecules in the DNA nanostructure have been observed, which include improved electronic and optical properties, energy harvesting, biological applications, and polymer formation.

References

1. Seeman, N. C. (1982). Nucleic acid junctions and lattices, *J. Theor. Biol.*, **99**, 237–247.
2. Winfree, E., Liu, F., Wenzler, L. A. and Seeman, N. C. (1998). Design and self-assembly of two-dimensional DNA crystals, *Nature*, **394**, 539–544.
3. Rothemund, P. W. (2006). Folding DNA to create nanoscale shapes and patterns, *Nature*, **440**, 297–302.
4. Chen, J. H. and Seeman, N. C. (1991). Synthesis from DNA of a molecule with the connectivity of a cube, *Nature*, **350**, 631–633.
5. Zhang, F., Nangreave, J., Liu, Y. and Yan, H. (2014). Structural DNA nanotechnology: state of the art and future perspective, *J. Am. Chem. Soc.*, **136**, 11198–11211.
6. Service, R. F. (2011). DNA nanotechnology. DNA nanotechnology grows up, *Science*, **332**, 1140–1143.
7. Pinheiro, A. V., Han, D., Shih, W. M. and Yan, H. (2011). Challenges and opportunities for structural DNA nanotechnology, *Nat. Nanotechnol.*, **6**, 763–772.
8. Pei, H., Zuo, X., Zhu, D., Huang, Q. and Fan, C. (2014). Functional DNA nanostructures for theranostic applications, *Acc. Chem. Res.*, **47**, 550–559.
9. Fu, J., Liu, M., Liu, Y. and Yan, H. (2012). Spatially-interactive biomolecular networks organized by nucleic acid nanostructures, *Acc. Chem. Res.*, **45**, 1215–1226.
10. Torring, T. and Gothelf, K. V. (2013). HIV and innate immunity–a genomics perspective, *F1000 Prime Rep.*, **5**, 1–4.
11. Ke, Y. (2014). Designer three dimensional DNA architectures, *Curr. Opin. Struct. Biol.*, **27**, 122–128.

12. Aldaye, F. A., Palmer, A. L. and Sleiman, H. F. (2008). Assembling materials with DNA as the guide, *Science*, **321**, 1795–1799.

13. Bhatia, D., Sharma, S. and Krishnan, Y. (2011). Synthetic, biofunctional nucleic acid-based molecular devices, *Curr. Opin. Chem. Biol.*, **22**, 475–484.

14. Singh, S., Mishra, A., Kumari, R., Sinha, K. K., Singh, M. K. and Das, P. (2016). Carbon dot assisted formation of DNA hydrogel for sustained release of drug, *Carbon*, **114**, 169–176.

15. Gothelf, K. V. and LaBean, T. H. (2005). DNA programmed assembly of nanostructures, *Org. Biomol. Chem.*, **3**, 4023–4037.

16. Li, H., LaBean, T. H. and Leong, K. W. (2011). Nucleic acid-based nanoengineering: novel structures for biomedical applications, *Interface Focus*, **1**, 702–724.

17. Yildirim, N., Long, F., Gao, C., He, M., Shi, H.-C. and Gu, A. Z. (2012). Aptamer-based optical biosensor for rapid and sensitive detection of 17β-estradiol in water sample, *Environ. Sci. Technol.*, **46**, 3288–3294.

18. Langecker, M., Arnaut, V., Martin, T. G., List, J., Renner, S., Mayer, M., Dietz, H. and Simmel, F. C. (2012). Synthetic lipid membrane channels formed by designed DNA nanostructures, *Science*, **338**, 932–936.

19. Fu, J., Liu, M., Liu, Y., Woodbury, N. W. and Yan, H. (2012). Interenzyme substrate diffusion for an enzyme cascade organized on spatially addressable DNA nanostructures, *J. Am. Chem. Soc.*, **134**, 5516–5519.

20. Lo, P. K., Metera, K. L. and Sleiman, H. F. (2010). Self-assembly of three-dimensional DNA nanostructures and potential biological applications, *Curr. Opin. Struct. Biol.*, **14**, 597–607.

21. Ranallo, S., Amodio, A., Idili, A., Porchetta, A. and Ricci, F. (2016). Electronic control of DNA-based nanoswitches and nanodevices, *Chem. Sci.*, **7**, 66–71.

22. Noh, H., Goodman, S. M., Mohan, P., Goodwin, A. P., Nagpal, P. and Cha, J. N. (2014). Direct conjugation of DNA to quantum dots for scalable assembly of photoactive thin films, *RSC Adv.*, **4**, 8064–8071.

23. Verma, S. and Eckstein, F. (1998). Modified oligonucleotides: synthesis and strategy for users, *Annu. Rev. Biochem.*, **67**, 99–134.

24. Khan, S. I. and Grinstaff, M. W. (1999). Palladium(0)-catalyzed modification of oligonucleotides during automated solid-phase synthesis, *J. Am. Chem. Soc.*, **121**, 4704–4705.

25. Das, P. (2007). Long-Range Charge Transfer In Plasmid DNA Condensates and DNA-Directed Assembly of Conducting Polymers

(Doctoral Thesis), School of Chemistry and Biochemistry, Georgia Institute of Technology, Atlanta.

26. Duarte, V., Muller, J. G. and Burrows, C. J. (1999). Insertion of dGMP and dAMP during in vitro DNA synthesis opposite an oxidized form of 7,8-dihydro-8-oxoguanine, *Nucleic Acids Res.*, **27**, 496–502.
27. Giller, G., Tasara, T., Angerer, B., Muhlegger, K., Amacker, M. and Winter, H. (2003). Incorporation of reporter molecule-labeled nucleotides by DNA polymerases. I. Chemical synthesis of various reporter group-labeled 2′-deoxyribonucleoside-5′-triphosphates, *Nucleic Acids Res.*, **31**, 2630–2635.
28. Wang, Y., Tkachenko, B. A., Schreiner, P. R. and Marx, A. (2011). Diamondoid-modified DNA, *Org. Biomol. Chem.*, **9**, 7482–7490.
29. Ma, S., Tang, N. and Tian, J. (2012). DNA synthesis, assembly and applications in synthetic biology, *Curr. Opin. Chem. Biol.*, **16**, 260–267.
30. Su, Y.-C., Chen, H.-Y., Ko, N. C., Hwang, C. C., Wu, M. H., Wang, L. F., Wang, Y.-M., Chang, S.-N., Wang, E.-C. and Wang, T.-P. (2014). Effective and site-specific phosphoramidation reaction for universally labeling nucleic acids, *Anal. Biochem.*, **449**, 118–128.
31. Kumari, R., Banerjee, S. S., Bhowmick, A. K. and Das, P. (2015). DNA-melamine hybrid molecules: from self-assembly to nanostructures, *Beilstein J. Nanotechnol.*, **6**, 1432–1438.
32. Sonawane, M. D. and Nimse, S. B. (2016). Surface modification chemistries of materials used in diagnostic platforms with biomolecules, *J. Chem.*, **2016**, 9241378.
33. Wang, C., Yan, Q., Liu, H.-B., Zhou, X.-H. and Xiao, S.-J. (2011). Different EDC/NHS activation mechanisms between PAA and PMAA brushes and the following amidation reactions, *Langmuir*, **27**, 12058–12068.
34. Singh, V., Mohan, M., Anindya, R. and Das, P. (2015). Self-assembled nanocages from DNA–protoporphyrin hybrid molecules, *RSC Adv.*, **5**, 89025–89029.
35. Singh, S., Chakraborty, A., Singh, V., Molla, A., Hussain, S., Singh, M. K. and Das, P. (2015). DNA mediated assembly of quantum dot–protoporphyrin IX FRET probes and the effect of FRET efficiency on ROS generation, *Phys. Chem. Chem. Phys.*, **17**, 5973–5981.
36. Singh, S., Jha, P., Singh, V., Hussain, S., Singh, M. K. and Das, P. (2016). A quantum dot–MUC1 aptamer conjugate for targeted delivery of protoporphyrin IX and specific photokilling of cancer cells through ROS generation, *Integr. Biol.*, **8**, 1040–1048.

37. Ghosh, S. S., Kao, P. M., Mccue, A. W. and Chappelle, H. L. (1990). Use of maleimide-thiol coupling chemistry for efficient syntheses of oligonucleotide-enzyme conjugate hybridization probes, Bioconjugate Chem., **1**, 71–76.

38. Wilks, T. R. and O'reilly, R. K. (2016). Efficient DNA–polymer coupling in organic solvents: a survey of amide coupling, thiol-ene and tetrazine–norbornene chemistries applied to conjugation of poly(*N*-isopropylacrylamide), *Sci. Rep.*, **6**, 39192.

39. Medintz, I. L., Uyeda, H. T., Goldman, E. R. and Mattoussi, H. (2005). Quantum dot bioconjugates for imaging, labelling and sensing, *Nat. Mater.*, **4**, 435–446.

40. Klostranec, J. M. and Chan, W. C. W. (2006). Quantum dots in biological and biomedical research: recent progress and present challenges, *Adv. Mater.* **18**, 1953–1964.

41. Kazane, S. A., Sok, D., Cho, E. H., Uson, M. L., Kuhn, P., Schultz, P. G. and Smider, V. V. (2012). Site-specific DNA-antibody conjugates for specific and sensitive immuno-PCR, *Proc. Natl. Acad. Sci. U.S.A.*, **109**, 3731–3736.

42. Metelev, V. G., Kubareva, E. A., Vorob'eva, O. V., Romanenkov, A. S. and Oretskava, T. S. (2003). Specific conjugation of DNA binding proteins to DNA templates through thiol–disulfide exchange, *FEBS Lett.*, **538**, 48–52.

43. Rostovtsev, V. V., Green, L. G., Fokin, V. V. and Sharpless K. B. (2002). A stepwise huisgen cycloaddition process: copper(I)-catalyzed regioselective "ligation" of azides and terminal alkynes, *Angew. Chem. Int. Ed.*, **41**, 2596–2599.

44. Seo, T. S., Li, Z., Ruparel, H. and Jingyue J. (2003). Click chemistry to construct fluorescent oligonucleotides for DNA sequencing, *J. Org. Chem.*, **68**, 609–612.

45. Pourceau, G., Meyer, A., Vasseur, J. J. and Morvan, F. (2009). Synthesis of mannose and galactose oligonucleotide conjugates by bi-click chemistry, *J. Org. Chem.*, **74**, 1218–1222.

46. Kumar, R., El-Sagheer, A., Tumpane, J., Lincoln, P., Wilhelmsson, M. and Brown, T. (2007). Template-directed oligonucleotide strand ligation, covalent intramolecular DNA circularization and catenation using click chemistry, *J. Am. Chem. Soc.*, **129**, 6859–6864.

47. Xu, Y., Suzuki, Y. and Komiyama, M. (2009). Click chemistry for the identification of G-quadruplex structures: discovery of a DNA–RNA G-quadruplex, *Angew. Chem. Int. Ed.*, **48**, 3281–3284.

48. Kosman, J., Stanislawska, A., Gluszynska, A. and Juskowiak, B. (2017). Dataset on characterization of hemin-azide derivative and DNA oligonucleotide-hemin conjugate, *Int. J. Biol. Macromol.*, **13**, 53–59.

49. Lee, J. K., Jung, Y. H., Tok, J. B.-H. and Bao, Z. (2011). Syntheses of organic molecule-DNA hybrid structures, *ACS Nano*, **5**, 2067–2074.

50. Kuroda, T., Sakurai, Y., Suzuki, Y., Nakamura, A. O., Kuwahara, M., Ozaki, H. and Sawai, H. (2006). Assembly of DNA nanostructures with branched tris-DNA, *Chem. Asian J.*, **1**, 575–580.

51. Okochi, K. D., Monfregola, L., Dickerson, S. M., Mccaffrey, R., Domaille, D. W., Yu, C., Hafenstine, G. R., Jin, Y., Cha, J. N., Kuchta, R. D., Caruthers, M. and Zhang, W. (2017). Synthesis of small-molecule/DNA hybrids through on-bead amide-coupling approach, *J. Org. Chem.*, **82**(20), 10803–10811.

52. Aldaye, F. A. and Sleiman, H. F. (2007). Dynamic DNA templates for discrete gold nanoparticle assemblies: control of geometry, modularity, write/erase and structural switching, *J. Am. Chem. Soc.*, **129**, 4130–4131.

53. Gerber, F., Krummen, M., Potfeter, H., Roth A., Siffrin, C. and Spoendlin, C. (2004). Practical aspects of fast reversed-phase high-performance liquid chromatography using 3 μm particle packed columns and monolithic columns in pharmaceutical development and production working under current good manufacturing practice, *J. Chromatogr. A*, **1036**, 127–133.

54. Lou, C., Samuelsent, S. V., Christensen, N. J., Vester, B. and Wenge, J. (2017). Oligonucleotides containing aminated 2′-amino-LNA nucleotides: synthesis and strong binding to complementary DNA and RNA, *Bioconjugate Chem.*, **28**, 1076–1083.

55. Kumari, R., Singh, S., Monisha, M., Bhowmick, S., Roy, A., Das, N. and Das, P. (2016). Hierarchical coassembly of DNA–triptycene hybrid molecular building blocks and zinc protoporphyrin IX, *Beilstein J. Nanotechnol.*, **7**, 697–707.

56. Potier, N., Dorsselaer, A. V., Cordier, Y., Roch, O. and Bischoff, R. (1994). Negative electrospray ionization mass spectrometry of synthetic and chemically modified oligonucleotides, *Nucleic Acids Res.*, **22**, 3895–3903.

57. Lee, J. K., Jung, Y. H., Stoltenberg, R. M., Tok, J. B.-H. and Bao, Z. (2008). Synthesis of DNA–organic molecule–DNA triblock oligomers using the amide coupling reaction and their enzymatic amplification, *J. Am. Chem. Soc.*, **130**, 12854–12855.

58. Egelhofer, V., Bussow, K., Luebbert, C., Lehrach, H. and Nordhoff, E. (2000). Improvements in protein identification by MALDI-TOF-MS peptide mapping, *Anal. Chem.*, **72**, 2741–2750.

59. Zagorevskii, D. V., Aldersley, M. F. and Ferris, J. P. (2006). MALDI analysis of oligonucleotides directly from montmorillonite, *J. Am. Soc. Mass Spectrom.*, **17**, 1265–1270.

60. Liu, K., Zheng, L., Liu, Q., De Vries, J. W., Gerasimov, J. Y. and Herrmann, A. (2014). Nucleic acid chemistry in the organic phase: from functionalized oligonucleotides to DNA side chain polymers, *J. Am. Chem. Soc.*, **136**, 14255–14262.

61. Kumari, R., Khan, M. I., Bhowmick, S., Sinha, K. K., Das, N. and Das, P. (2017). Self-assembly of DNA-porphyrin hybrid molecules for the creation of antimicrobial nanonetwork, *J. Photochem. Photobiol. B*, **172**, 28–35.

62. Willemsen, O. H., Snel, M. M. E., Cambi, A., Greve, J., Grooth, B. G. D. and Figdor, C. G. (2000). Biomolecular interactions measured by atomic force microscopy, *Biophys. J.*, **79**, 3267–3281.

63. Shlyakhtenko, L. S., Gall, A. A., Weimer, J. J., Hawn, D. D. and Lyubchenko, Y. L. (1999). Atomic force microscopy imaging of DNA covalently immobilized on a functionalized mica substrate, *Biophys. J.*, **77**, 568–576.

64. Singh, S., Kumari, R., Chakraborty, A., Hussain, S., Singh, M. K. and Das, P. (2016). Melamine–DNA encoded periodicity of quantum dot arrays, *J. Colloid Interface Sci.*, **461**, 45–49.

65. Lo, P. K., Karam, P., Aldaye, F. A., Mclaughlin, C. K., Hamblin, G. D., Cosa, G. and Sleiman, H. F. (2010). Loading and selective release of cargo in DNA nanotubes with longitudinal variation, *Nat. Chem.*, **2**, 319–328.

66. Deng, Z., Samanta, A., Nangreave, J., Yan, H. and Liu, Y. (2012). Robust DNA-functionalized core/shell quantum dots with fluorescent emission spanning from UV–vis to near-IR and compatible with DNA-directed self-assembly, *J. Am. Chem. Soc.*, **134**, 17424–17427.

67. Padalkar, S., Zhao, J., Stuart, K., Panitch, A., Rickus, J. and Stanciu, L. (2008). Preparation of biomolecule gel matrices for electron microscopy, *Ultramicroscopy*, **108**, 309–313.

68. Knudsen, J. B., Liu, L., Kodal, A. L. B., Madsen, M., Li, Q., Song, J., Woehrstein, J. B., Wickham, S. F. J., Strauss, M. T., Schueder, F., Vinther, J., Krissanaprasit, A., Gudnason, D., Smith, A. A. A., Ogaki, R., Zelikin, A. N., Besenbacher, F., Birkedal, V., Yin, P., Shih, W. M., Jungmann, R., Dong, M. and Gothelf, K. (2015). Routing of individual polymers in designed patterns, *Nat. Nanotechnol.*, **10**, 892–898.

69. Bellot, G., Mcclintock, M. A., Lin, C. and Shih, W. M. (2011). Recovery of intact DNA nanostructures after agarose gel–based separation, *Nat. Methods*, **8**, 192–194.

70. Han, D., Pal, S., Liu, Y. and Yan, H. (2010). Folding and cutting DNA into reconfigurable topological nanostructures, *Nat. Nanotechnol.*, **5**, 712–717.

71. Kohman, R. E., Cha, S. S., Man, H.-Y. and Han, X. (2016). Light-triggered release of bioactive molecules from DNA nanostructures, *Nano Lett.*, **16**, 2781–2785.

72. Buckhout-White, S., Robinson, J. T., Bassim, N. D., Goldman, E. R., Medintz, I. L. and Ancona, M. G. (2013). TEM imaging of unstained DNA nanostructures using suspended graphene, *Soft Matter*, **9**, 1414–1417.

73. Shi, J. and Bergstrom, D. E. (1997). Assembly of novel DNA cycles with rigid tetrahedral linkers, *Angew. Chem. Int. Ed.*, **36**, 111–113.

74. Tanaka, F., Mochizuki, T., Liang, X., Asanuma, H., Tanaka, S., Suzuki, K., Kitamura, S., Nishikawa, A., Ui-Tei, K. and Hagiya, M. (2010). Robust and photocontrollable DNA capsules using azobenzenes, *Nano Lett.*, **10**, 3560–3565.

75. Scheffler, M., Dorenbeck, A., Jordan, S., Wustefeld, M. and Von Kiedrowski, G. (1999). Self-assembly of trisoligonucleotidyls: the case for nano-acetylene and nano-cyclobutadiene, *Angew. Chem. Int. Ed.*, **38**, 3311–3315.

76. Shchepinov, M. S., Mir, K. U., Elder, J. K., Frank-Kamenetskii, M. D. and Southern, E. M. (1999). Oligonucleotide dendrimers: stable nano-structures, *Nucleic Acids Res.*, **27**, 3035–3041.

77. Choi, J. S., Kang, C. W., Jung, K., Yang, J. W., Kim, Y.-G. and Han, H. (2004). Synthesis of DNA triangles with vertexes of bis (terpyridine) iron(II) complexes, *J. Am. Chem. Soc.*, **126**, 8606–8607.

78. Goritz, M. and Kramer, R. (2005). Allosteric control of oligonucleotide hybridization by metal-induced cyclization, *J. Am. Chem. Soc.*, **127**, 18016–18017.

79. Kalek, M., Madsen, A. S. and Wengel, J. (2007). Effective modulation of DNA duplex stability by reversible transition metal complex formation in the minor groove, *J. Am. Chem. Soc.*, **129**, 9392–9400.

80. Greschner, A. A., Toader, V. and Sleiman, H. F. (2012). The role of organic linkers in directing DNA self-assembly and significantly stabilizing DNA duplexes, *J. Am. Chem. Soc.*, **134**, 14382–14389.

81. Hong B. J., Eryazici, I., Bleher, R., Thaner, R. V., Mirkin, C. A. and Nguyen, S. T. (2015). Directed assembly of nucleic acid-based polymeric

nanoparticles from molecular tetravalent cores, *J. Am. Chem. Soc.*, **137**, 8184–8191.

82. Hong, B. J., Cho, V. Y., Bleher, R., Schatz, G. C. and Nguyen, S. T. (2015). Enhancing DNA-mediated assemblies of supramolecular cage dimers through tuning core flexibility and DNA length: a combined experimental-modeling study, *J. Am. Chem. Soc.*, **137**, 13381–13388.

83. Eryazici, I., Prytkova, T. R., Schatz, G. C. and Nguyen, S. (2010). Cooperative melting in caged dimers with only two DNA duplexes, *J. Am. Chem. Soc.*, **132**, 17068–17070.

84. Tanaka, F., Mochizuki, T., Liang, X., Asanuma, H., Tanaka, S., Suzuki, K., Kitamura, S.-I., Nishikawa, A., Ui-Tei, K. and Hagiya, M. (2010). Robust and photocontrollable DNA capsules using azobenzenes, *Nano Lett.*, **10**, 3560–3565.

85. Fujitsuka, M. and Majima, T. (2017). Charge transfer dynamics in DNA revealed by time resolved spectroscopy, *Chem. Sci.*, **8**, 1752–1762.

86. Deshmukh, H. M., Joglekar, S. P. and Broom, A. D. (1995). Self-complementary oligodeoxyribonucleotides containing 2'-O-(anthraquinone-2-methyl)adenosine, *Bioconjugate Chem.*, **6**, 578–586.

87. Lewis, F. D., Wu, T., Zhang, Y., Letsinger, R. L., Greenfield, S. R. and Wasielewski, M. (1997). Distance-dependent electron transfer in DNA hairpins, *Science*, **277**, 673–676.

88. Lewis, F. D., Wu, T., Liu, X., Letsinger, R. L., Greenfield, S. R., Miller, S. E. and Wasielewski, M. (2000). Dynamics of photoinduced charge separation and charge recombination in synthetic DNA hairpins with stilbenedicarboxamide linkers, *J. Am. Chem. Soc.*, **122**, 2889–2902.

89. Lewis, F. D., Kalgutkar, R. S., Wu, Y., Liu, X., Liu, J., Hayes, R. T., Miller, S. E. and Wasielewski, M. (2000). Driving force dependence of electron transfer dynamics in synthetic DNA hairpins, *J. Am. Chem. Soc.*, **122**, 12346–12351.

90. Wagenknecht, H.-A. (2003). Reductive electron transfer and transport of excess electrons in DNA, *Angew. Chem. Int. Ed.*, **42**, 2454–2460.

91. Barton, J. K., Olmon, E. D. and Sontz, P. A. (2011). Metal complexes for DNA-mediated charge transport, *Coord. Chem. Rev.*, **255**, 619–634.

92. Fujii, T., Thazhathveetil, A. K., Yildirim, I., Young, R. M., Wasielewski, M. R., Schatz, G. C. and Lewis, F. D. (2017). Structure and dynamics of electron injection and charge recombination in i-motif DNA conjugates, *J. Phys. Chem. B*, **121**, 8058–8068.

93. Lewis, F. D., Wu, T., Zhang, Y., Letsinger, R. L., Greenfield, S. R. and Wasielewski, M. R. (1997). Distance-dependent electron transfer in DNA hairpins, *Science*, **277**, 673–676.

94. Lewis, F. D., Kalgutkar, R. S., Wu, Y., Liu, X., Liu, J., Hayes, R. T., Miller, S. E. and Wasielewski, M. R. (2000). Driving force dependence of electron transfer dynamics in synthetic DNA hairpins, *J. Am. Chem. Soc.*, **122**, 12346–12351.

95. Wagenknecht, H.-A. (2003). Reductive electron transfer and transport of excess electrons in DNA, *Angew. Chem. Int. Ed.*, **42**, 2454–2460.

96. Surin, M. (2016). From nucleobase to DNA templates for precision supramolecular assemblies and synthetic polymers, *Polym. Chem.*, **7**, 4137–4150.

97. Datta, B. and Schuster, G. B. (2008). DNA-directed synthesis of aniline and 4-aminobiphenyl oligomers: programmed transfer of sequence information to a conjoined polymer nanowire, *J. Am. Chem. Soc.*, **130**, 2965–2973.

98. Chen, W., Guler, G., Kuruvilla, E. and Schuster, G. B. (2010). Development of self-organizing, self-directing molecular nanowires: synthesis and characterization of conjoined DNA–2,5-bis(2-thienyl)pyrrole oligomers, *Macromolecules*, **43**, 4032–4040.

99. Chen, W. and Schuster, G. B. (2012). DNA-programmed modular assembly of cyclic and linear nanoarrays for the synthesis of two-dimensional conducting polymers, *J. Am. Chem. Soc.*, **134**, 840–843.

100. Chen, W. and Schuster, G. B. (2013). Precise sequence control in linear and cyclic copolymers of 2,5-bis(2-thienyl)pyrrole and aniline by DNA-programmed assembly, *J. Am. Chem. Soc.*, **135**, 4438–4449.

101. Lee, J. K., Ko, S. and Bao, Z. (2012). In situ hetero end-functionalized polythiophene and subsequent "click" chemistry with DNA, *Macromol. Rapid Commun.*, **33**, 938–942.

102. Wilks, T. R. and O'reilly, R. K. (2016). Efficient DNA–polymer coupling in organic solvents: a survey of amide coupling, thiol-ene and tetrazine–norbornene chemistries applied to conjugation of poly(N-isopropylacrylamide), *Sci. Rep.*, **6**, 39192.

103. Ramamurthy, V. (2015). Photochemistry within a water-soluble organic capsule, *Acc. Chem. Res.*, **48**, 2724–2733.

104. Dutta, P. K., Varghese, R., Nangreave, J., Lin, S., Yan, H. and Liu, Y. (2011). DNA-directed artificial light-harvesting antenna, *J. Am. Chem. Soc.*, **133**, 11985–11993.

105. Ensslen, P., Brandl, F., Sezi, S., Varghese, R., Kutta, R.-J., Dick, B. and Wagenknecht, H.-A. (2015). DNA-based oligochromophores as light-harvesting systems, *Chem. Eur. J.*, **21**, 9349–9354.

106. Zhou, D., Ying, L., Hong, X., Hall, E. A., Abell, C. and Klenerman, D. (2008). A compact functional quantum dot–DNA conjugate: preparation, hybridization, and specific label-free DNA detection, *Langmuir*, **24**, 1659–1664.

107. Olejko, L., Cywinski, P. J. and Bald, I. (2016). An ion-controlled four-color fluorescent telomeric switch on DNA origami structures, *Nanoscale*, **8**, 10339–10347.

108. Lu, H., Schöps, O., Waggon, U. and Neimeyer, C. M. (2008). Self-assembled donor comprising quantum dots and fluorescent proteins for long-range fluorescence resonance energy transfer, *J. Am. Chem. Soc.*, **130**, 4815–4827.

109. Spillmann, C. M., Ancona, M. G., Buckhout-White, S., Algar, W. R., Stewart, M. H., Susumu, K., Huston, A. L., Goldman, E. R. and Medintz, I. L. (2013). Achieving effective terminal exciton delivery in quantum dot antenna-sensitized multistep DNA photonic wires, *ACS Nano*, **7**, 7101–7118.

110. Singh, A., Sinsinbar, G., Choudhary, M., Kumar, V., Pasricha, R., Verma, H. N., Singh, S. and Arora, K. (2013). Graphene oxide-chitosan nanocomposites based electrochemical DNA biosensor for detection of typhoid, *Sens. Actuators, B*, **185**, 675–684.

111. Zhang, Y., Zhao, H., Wu, W., Xue, Y., Zhang, X., He, Y., Li, X. and Yuan, Z. (2013). A novel graphene-DNA biosensor for selective detection of mercury ions, *Biosens. Bioelectron.*, **48**, 180–187.

112. Solanki, P. R., Prabhakar, N., Pandey, M. K. and Malhotra, B. D. (2009). Surface plasmon resonance-based DNA biosensor for arsenic trioxide detection, *Intern. J. Environ. Anal. Chem.*, **89**, 49–57.

113. Zhao, Q., Zhou, Y., Li, Y., Gu, W., Zhang, Q. and Liu, J. (2016). Luminescent iridium(III) complex labeled DNA for graphene oxide based biosensors, *Anal. Chem.*, **88**, 1892–1899.

114. Hu, R., Liu, T., Zhang, X.-B., Huan, S.-Y., Wu, C., Fu, T. and Tan, W. (2014). Multicolor fluorescent biosensor for multiplexed detection of DNA, *Anal. Chem.*, **86**, 5009–5016.

115. Wang, C., Huang, N., Zhuang, H. and Jiang, X. (2015). Enhanced performance of nanocrystalline ZnO DNA biosensor via introducing electrochemical covalent biolinkers, *ACS Appl. Mater. Interfaces*, **7**, 7605–7612.

116. Ohmichi, T., Kawamoto, Y., Wu, P., Miyoshi, D., Karimata, H. and Sugimoto, N. (2005). DNA-based biosensor for monitoring pH *in vitro* and in living cells, *Biochemistry*, **44**, 7125–7130.

117. Fink, H. W. and Schonenberger, C. (1999). Electrical conduction through DNA molecules, *Nature*, **398**, 407–410.

118. Korri-Youssoufi, H., Garnier, F., Srivastava, P., Godillot, P. and Yassar, A. (1997). Toward bioelectronics: specific DNA recognition based on an oligonucleotide-functionalized polypyrrole, *J. Am. Chem. Soc.*, **119**, 7388–7389.

119. Lu, N., Pei, H., Ge, Z., Simmons, C. R., Yan, H. and Fan, C. (2012). Charge transport within a three-dimensional DNA nanostructure framework, *J. Am. Chem. Soc.*, **134**, 13148–13151.

120. Keren, K., Berman, R. S., Buchstab, E., Sivan, U. and Braun, E. (2003). DNA-templated carbon nanotube field-effect transistor, *Science*, **302**, 1380–1382.

121. Lehn, J. M. (1990). Perspectives in supramolecular chemistry-from molecular recognition towards molecular information processing and self-organization, *Angew. Chem. Int. Ed. Engl.*, **29**, 1304–1319.

122. Seeman, N. C. and Lukeman, P. S. (2005). Nucleic acid nanostructures: bottom-up control of geometry on the nanoscale, *Rep. Prog. Phys.*, **68**, 237–270.

123. Ke, Y., Ong, L. L., Shih, W. M. and Yin, P. (2012). Three-dimensional structures self-assembled from DNA bricks, *Science*, **338**, 1177–1183.

124. Wilner, O. I. and Willner, I. (2012). Functionalised DNA nanostructures, *Chem. Rev.*, **112**, 2528–2556.

125. Novoa, A. and Winssinger, N. (2015). DNA display of glycoconjugates to emulate oligomeric interactions of glycans, *Beilstein J. Org. Chem.*, **11**, 709–719.

Chapter 3

Functional Molecule–Templated DNA Nanoarchitectures

Bappaditya Roy, Debasis Ghosh, and Thimmaiah Govindaraju
Bioorganic Chemistry Laboratory, New Chemistry Unit, Jawaharlal Nehru Centre for Advanced Scientific Research, Jakkur P.O., Bengaluru 560064, Karnataka, India
tgraju@jncasr.ac.in

3.1 Introduction

Biomolecular structures are fascinating at all length scales as their molecular organization and structure-property correlations engender numerous exciting phenomena [1, 2]. The simplest, most elegant, and information-rich biomolecular organization is the double-stranded DNA (dsDNA). Two complementary single-stranded DNA (ssDNA) molecules spontaneously hybridize through canonical Watson–Crick (WC) base pairing to form a right-handed DNA double helix that is ~2 nm thick and exhibits a persistence length of ~50 nm, making it a perfect nanomaterial (Fig. 3.1) [3–8]. Noncovalent interactions, such as hydrogen bonding, electrostatic interaction, hydrophobic forces, and nucleobase stacking (π-π

Templated DNA Nanotechnology: Functional DNA Nanoarchitectonics
Edited by Thimmaiah Govindaraju

ISBN 978-981-4800-21-1 (Hardcover), 978-0-429-42866-1 (eBook)
www.panstanford.com

interaction), are primarily responsible for maintaining the giant DNA helical structure. In addition to conventional WC base pairing, unconventional hydrogen bonding drives the assembly of ssDNA into double, triple, or quadruple strands, hairpins, and branched structures [4, 5]. DNA has structurally evolved over billions of years to effectively store and communicate the genetic information in a majority of the living organisms [6–8]. The remarkable specificity of molecular recognition between complementary nucleobases has made DNA an attractive biomolecule for micro- and nanofabrications. In recent times, numerous efforts have been directed toward utilizing DNA as a potential biomolecule in the celebrated nanotechnological advances; the concept was originally introduced by Seeman [9–13]. A number of unusual DNA-based structural arrangements and shapes are designed and illustrated as DNA nanotechnology [14–16]. Currently, the complex computer-based design, programming, use of very long DNA sequences, reproducibility, and cost involved in the processes limit the utility of DNA nanotechnology in its original form.

The concept of molecular architectonics is considered a state of the art in engineering molecular organization for functional applications [17, 18]. In fact, the elegant molecular design, through the integration of biomolecules and synthetic functional molecules, provides control over their organization with nanometer precision, which in turn leads to the realm of nanoarchitectonics [17–23]. The field of nanoarchitectonics deals with controlled organization of atoms and molecules to construct defined nanoarchitectures with functional properties and applications [19–23]. Notably, in the preparation of nanomaterials with functional properties, "nano" is not a direct extension of either the bottom-up or the top-down approach and, hence, the concept of molecular architectonics is required to integrate the realm of molecules to (nano)materials to applications [17, 19–23]. In this context, functional molecule–DNA nanoarchitectonics is aimed at developing fascinating (nano) materials through unambiguous design strategies to extract precise structural and functional characteristics. The advances in the field are expected to make rapid growth, far beyond the expectations in areas as diverse as chemistry, biology, functional advanced materials, and analytical and environmental sciences. Several approaches have been adapted to design and facilitate artificial DNA self-assembly. Among them, molecular self-assembly is a powerful tool to generate well-defined functional nanoarchitectures [24–26]. A bottom-up approach through self-assembly to arrange components at the molecular level to develop material at the nanoscale, that is molecular architectonics to nanoarchitectonics, is a challenging task. For this, short ssDNA sequences or oligonucleotides (dB_n, where B = nucleobase and $n < 40–50$, which corresponds to the persistence length of DNA) can form ensembles with functional small molecules (SMs). A SM with a polycyclic aromatic core and its assembly structures through strong π-π interaction can be used as templates to assemble dB_n. In addition, dB_n with multiple interaction sites and an intrinsic nature to form a helical structure can act as a mutual template to form well-ordered architectures. The main advantages of SM-DNA nanoarchitectonics are the use of functional molecules and short ssDNAs, which ensures built-in functional features and cost-effective technological platforms for practical applications.

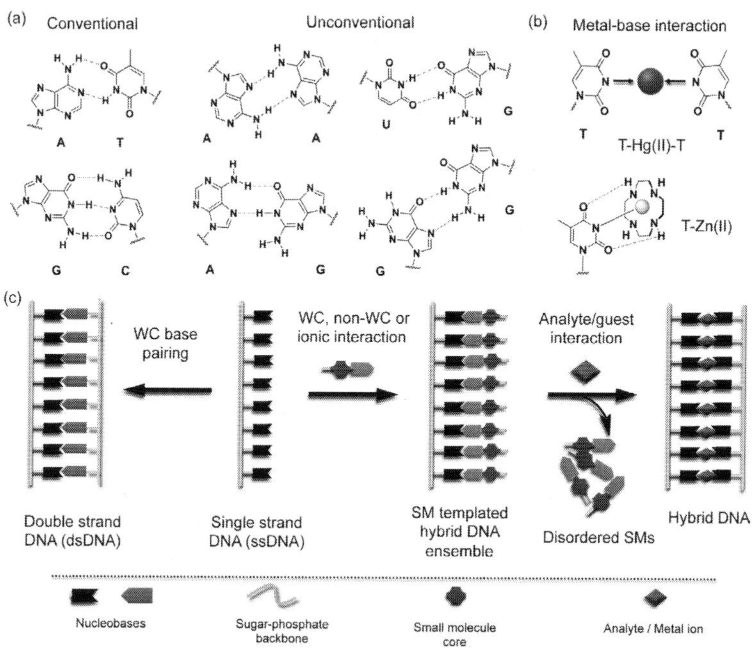

Figure 3.1 (a) Different types of conventional and unconventional hydrogen bonding, (b) different nucleobase-metal ion interactions, and (c) schematic representation of dsDNA and small molecule (SM) or metal-templated hybrid DNA ensembles supported by canonical and noncanonical hydrogen bonding and ionic and metal ion interactions.

In this chapter, we present mutually templated SM-DNA nanoarchitectures supported by canonical (WC type) and noncanonical (non-WC type) hydrogen bonding and ionic and metal interactions of nucleobases and their analogs (Fig. 3.1). The concept of SM-DNA nanoarchitectonics is introduced and in particular, applications in the fields ranging from biology to materials science are discussed. Current progress and future prospects are essential parts of this chapter.

3.2 Perspective: Small Molecule and DNA Hybrid Ensemble

The process of templated and bottom-up chromophoric self-assembly has vast implications in the field of nanoarchitectonics for generating versatile structural and functional properties [24, 26]. The flat polycyclic aromatic chromophores as SMs are capable of undergoing self-assembly through aromatic π-π interactions and serve as host or template for guests via host-guest binding interactions. Strong noncovalent interactions, especially hydrogen bonding and ionic, metal, and hydrophobic interaction, play a major role in the formation of functional molecule–templated nanoarchitectures. The short oligonucleotides (ssDNA) have the capacity to induce programmed molecular organization within the templated assembly through well-defined molecular recognition. In a process of "all or nothing" the complete template of either type, SM or DNA, is filled one by one by the right host-host or guest-guest and host-guest interactions [27]. If the host-guest interaction is stronger enough than the host-host or guest-guest interaction, an uneven dispersion of the guest molecules over the template will form an inhomogeneous distribution. However, separately, the strong host-host or guest-guest interaction will potentially lead to nontemplated self-assembly of the guest molecules that will be larger than the average template size, that is, uncontrolled self-assembly. Therefore, control over the molecular interactions among the host SM (or the host SM assembly) and guest (ssDNA) is necessary to develop a perfect templated host-guest assembly. The π-conjugated systems with appended recognition units (to provide hydrogen bonding, ionic interaction, or metal binding) can play an important role in designing templated nanoarchitectures. There have been several important findings reported along this line to create functional nanoarchitectures. Our research group used π-conjugated arylene molecules and their assembly strucutres, such as naphthalenediimide (NDI) and perylenediimide (PDI), tethered with nucleobases that act as templates for ssDNA through canonical or noncanonical hydrogen-bonding interactions [24, 28–30]. The canonical hydrogen-bonding mode of templating follows WC hydrogen bonding, viz A-T and G-C base pair interactions [8, 31]. Conversely, the non-WC or noncanonical bonding mode follows the mismatch interaction between base pairs such as A-A, A-G, G-G, and G-U [32]. Further, noncanonical-type templating includes non-WC-type hydrogen bonding and ionic interactions between (a) cationic organic motifs, such as a guanidium cation with an anionic phosphodiester (5′ and 3′) of a ssDNA chain, and (b) metal ion (M^{n+}) with a nucleobase such as strong and specific thymine-Zn(II)/Hg(II)-thymine (T- Hg/Zn-T) interactions.

3.3 Templated DNA Nanoarchitectures Via Canonical Hydrogen-Bonding Interaction

Our research group aims to construct SM-templated DNA nanoarchitectures and tools for numerous applications, including developing stimuli-responsive drug delivery systems. We envisioned the assembly of π-conjugated organic dyes tethered with nucleobases and their use as templates for the organization of ssDNA through canonical and noncanonical hydrogen-bonding interactions. To create templated DNA nanoarchitectures, a PDI derivative **(APA)** symmetrically functionalized with adenine (A) was designed and synthesized [29]. The intriguing property of A to form canonical hydrogen bonding with thymine (T), that is an A-T pair, and noncanonical hydrogen bonding with itself and guanine (G) (A-A and A-G) inspired the construction of hybrid DNA ensembles [29, 32–35]. PDI is a π-conjugated aromatic dye that attracted the attention of researchers owing to its interesting photophysical properties [24, 36–38]. The self-assembled **APA** template oligothymidine used canonical (A-T) hydrogen bonding to form a [dT_n-**(APA)**$_n$-dT_n, n = 10 or 20]-type hybrid DNA ensemble (Fig. 3.2a). Spectroscopic tools, such as absorbance (ultraviolet-visible [UV-Vis]), fluorescence (photoluminescence [PL]), and circular dichroism (CD), were used to monitor the templated assembly formation. The UV-Vis spectrum of **APA** at 50 mM in a PBS buffer solution (10 mM, water/DMSO = 90:10, v/v %) exhibited a hypsochromic shift in the presence of dT_{10} (1–10 μM). This revealed the existence of mutual interactions between A of **APA** and the complementary T of dT_n via WC interaction and the hydrophobic force–induced aggregation via π-π of the perylene core in **APA** to form a templated hybrid structure. The mutual interaction of **APA** and T of dT_n also produced a bisignated CD signal for the PDI chromophore at 400–600 nm and a negative band at 250–300 nm corresponding to nucleobase absorption. These CD data confirmed mutual imprinting of chirality (helical) on the achiral PDI stacks and random coil dT_n in the coassembly structure. The plot of CD intensity at 561 nm for the PDI chromophore against the A-T ratio showed saturation at 1:1, suggesting a stoichiometry of 10:2 (**APA**:dT_{10}) for the formation of [dT_{10}:(**APA**)$_{10}$:dT_{10}]-type ensemble. The spectroscopic studies revealed the formation of mutually templated and chiral-imprinted ordered ensembles of the bicomponent system (**APA** and dT_n) through a "double zipper" helical assembly, viz [dT_n:(**APA**)$_n$:dT_n] (Fig. 3.2a). The hybrid ensemble structure was visualized using atomic force microscopy (AFM) (Fig. 3.2d). The nanoarchitecture of the [dT_{20}:**APA**$_{20}$:dT_{20}] ensemble showed the formation of an ordered left-handed helical (*M*-helix) assembly with a lateral thickness (~5 nm) corresponding to the overall thickness of the molecular ensemble (Fig. 3.2e). The left-handed helical architecture is well corroborated with the observed negative Cotton effect in the CD spectrum of the [dT_{20}:(**APA**)$_{20}$:dT_{20}] ensemble. Further study revealed that the dA_{20} and dG_{20} formed stable *P*-helix and *M*-helix, respectively (see Section 3.4). This indicated mutual templating between **APA** and dA_{10}/dG_{10} through unconventional or noncanonical hydrogen bonding to form *P*- and *M*-helical hybrid DNA ensembles. Furthermore, the pH-dependent study revealed the collapse of the "double zipper" assembly at pH < 6. This stimuli-responsive (pH responsive) behavior owing to the hydrogen-bonded SM-ssDNA ensemble can be used as a tool for the development of pH-triggered drug delivery systems for therapeutic SMs and oligonucleotides (Fig. 3.2b).

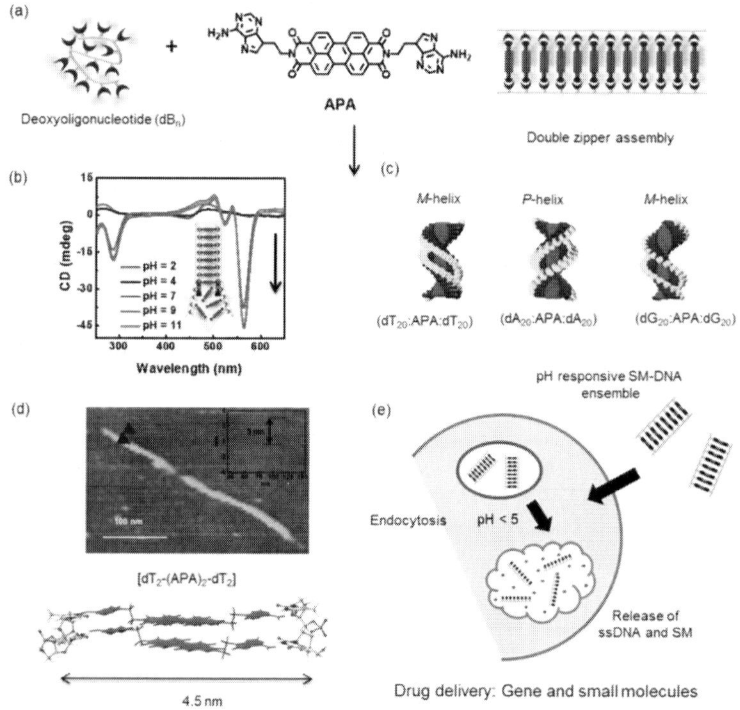

Figure 3.2 (a) Schematic presentation of the mutually templated "double zipper" assembly of **APA** and dT_{20}, [dT_{20}-(**APA**)$_{20}$-dT_{20}], through canonical and noncanonical hydrogen-bonding interactions. (b) pH-dependent dissociation of the ensemble structure was confirmed by CD measurement. (c) P- and M-type double-zipper helical assembly. (d) AFM image of a single-ensembled helical nanofiber and its height profile data (5 nm, inset); typical thickness (4.5 nm) of [dT_2-(**APA**)$_2$-dT_2] calculated from the theoretical model. (e) A hypothetical model for a proposed drug delivery system design using SM-ssDNA ensembles. (a–d) Adapted from Ref [29] with permission of The Royal Society of Chemistry.

In another interesting work from our group, mutually templated SM-DNA nanoarchitectures were constructed by employing NDI functionalized with a butyl chain and A (**BNA**) and oligothymidines through canonical hydrogen-bonding interactions of the A-T base pair [30]. NDIs are well known for their semiconductor properties, and we envisioned developing nanoarchitectures of ordered NDI and ssDNA assembly that would help to achieve conductivity through the ensemble network [24, 39–47]. The mutual templating of **BNA** and dT_n (n = 6, 10, 20) was designed to form highly ordered **BNA**$_n$-dT_n

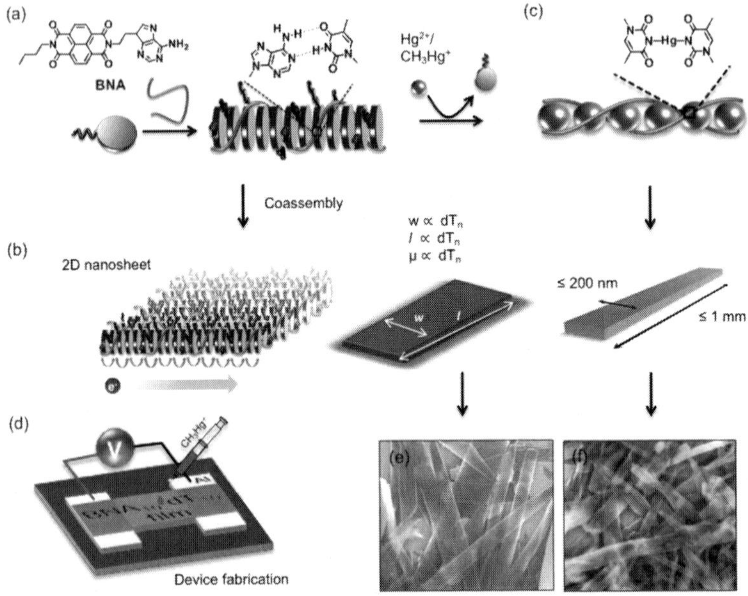

Figure 3.3 (a) 2D sheet-like nanoarchitectures were formed in a mutually templated assembly of **BNA** and dT_n, (b) the lateral dimensions (width *w*, length *l*, and conductivity *μ*) of the $\mathbf{BNA}_n$-dT_n of 2D sheets were directly proportional to the chain length of the dT_n used, (c) mercury-mediated displacement of **BNA** from the $\mathbf{BNA}_n$-dT_n ensemble and subsequent formation of the metallo-DNA duplex [dT-Hg-dT]$_n$, (d) schematic representation of FET device fabricated using $\mathbf{BNA}_{10}$-dT_{10} for the detection of Hg^{2+} or CH_3Hg^+, and (e, f) FESEM images of the $\mathbf{BNA}_n$-dT_n 2D sheet and **BNA** 1D tapes. (a–f) Adapted with permission from Ref [30]. Copyright (2016) American Chemical Society.

ensembles owing to increased cooperative interactions between the constituent partners supported by noncovalent A-T base pairing and π-π interactions (Fig. 3.3a). Interestingly, the mutually templated BNA_n-dT_n ensembles form 2D sheets with variable lateral dimensions (width *w* and length *l*) depending on the dT_n sequence length, which demonstrated the tunability of the 2D nanoarchitectures. This templated 2D assembly allowed us to modulate the chiroptical, morphological, and conductivity properties of BNA_n-dT_n. The insights into conductivity properties of the ensembles helped us to construct a field-effect transistor (FET) device. The charge-transport property of **BNA** was examined in the presence and absence of dT_n. For instance, the $\mathbf{BNA}_{10}$-dT_{10} coassembly resulted in the formation of 2D sheet-like nanoarchitecture (Fig. 3.3b). However, a 1D tape was obtained from the self-assembly of **BNA** alone. It is well known that the 2D sheet structure is more tolerant to molecular disorder compared to 1D tapes [18, 21, 39, 41]. The role of various nanostructure morphologies in the transport property at a dielectric-semiconductor interface was elucidated by comparing the doping level of the 1D tapes and 2D sheets. The property measurement showed a capacitance value to be 4 nF/cm^2, and the doping levels were determined as 1×10^9 for 1D tapes consisting of **BNA** and 1×10^{11} for 2D sheets obtained from **BNA**$_{20}$-dT$_{20}$. The 2D nanoarchitectures and proper aromatic stacking contributed to the better transport property of the **BNA**$_n$-dT$_n$ 2D sheets compared to those of the 1D tapes consisting of only **BNA**. Such development strategy of templated nanoarchitectonics is useful in designing potential technological platforms for bio-optoelectronics and sensor applications. The **BNA**$_{20}$-dT$_{20}$ system was further used for a chiro-optical and conductivity-based rapid and subnanomolar ultrasensitive (0.1 nM, 0.02 ppb) detection of mercury in water (see Section 3.6).

Iizawa and coworkers used A-T base pair interaction for the development of a helical nanofiber [48]. They employed an anthracene moiety as the guest building block, which is capable of undergoing self-assembly and has excellent optoelectronic property. The anthracene building block was tethered with two thymidylic acids as molecular recognition sites. Two anthracene-based guest molecules, **1** and **2**, were actually employed, where the recognition sites (thymidylic acids) are attached to naphthalene unit at 1,5- and 2,6- positions, respectively (Fig. 3.4). These guest molecules were anticipated to form complementary WC interaction with oligoadenylic acid (dA$_n$). Together, UV-Vis and CD spectroscopy data and AFM observations suggested the formation of helical J-aggregates from the bicomponent self-assembly of **2**-dA$_{20}$, (Fig. 3.5). Generally, 1D J-aggregates could exist in a brickwork-, ladder-, or staircase-type arrangement [49]. The UV-Vis absorption spectra of the stable bicomponent self-assembly of **2** (2,6) with dA$_{20}$ showed a small, thermally unstable J-band around 420 nm at 25°C, which disappeared at 50°C. Thus, the A-T base pair interaction induces a head-to-tail arrangement of the transition moments along the anthracene short axis to form a staircase-type arrangement in the bicomponent self-assembly (Fig. 3.5). However, no such J-aggregation of the anthracene was observed for the bicomponent self-assembly of **1** and dA_{20}. CD spectra of the binary self-assembly of **2**-dA_{20} strongly indicated a helical arrangement of the anthracene moieties. It showed negative and positive cotton bands at 426 and 415 nm, respectively, with a zero crossing at 422 nm, thus directly indicating the helical arrangement of the transition dipoles along the anthracene short axis. The proposed structure was supported by the fact that the two thymidylic acid moieties at the 2- and 6- positions disturb the head-to-tail arrangement along the short axis of the anthracene. This study revealed the importance of small structural differences that play a key role in the construction of DNA-templated nanostructures. AFM data revealed the width of the nanofiber to be ~3.3 nm, which is consistent with the total length of **1** (l = 3.3 nm) and the thickness of two adenylic acid dA_{20} molecules (~1.8 nm). Thus, it was assumed that the A-T base pairs were aligned helically at both ends of **1** in an arrangement similar to the DNA-like helical nanofibers.

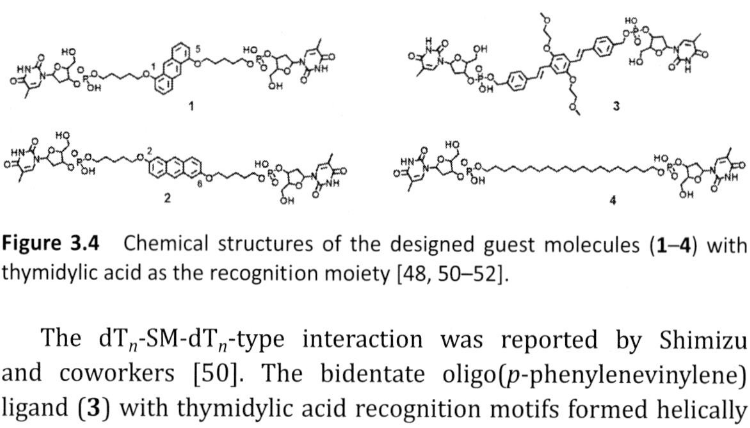

Figure 3.4 Chemical structures of the designed guest molecules (**1–4**) with thymidylic acid as the recognition moiety [48, 50–52].

The dT_n-SM-dT_n-type interaction was reported by Shimizu and coworkers [50]. The bidentate oligo(*p*-phenylenevinylene) ligand (**3**) with thymidylic acid recognition motifs formed helically stacked supramolecular ensemble with dA_{20}. The UV-Vis and CD spectroscopic measurements and AFM observations revealed the formation of helical stack structures (Fig. 3.6a–c). The helical pitch length of the stack along the growth axis was estimated to be 60 nm for the **3**:dA_{20} (T:A = 1:1) ensemble. The fiber widths corresponded to the sum of the SM (**3**), two adenylic acid moieties (or the sum of **3**), and a single adenylic acid. Homo- (T:A = 1:1) and bicomponent (T:A = 1:2) self-assembly properties were studied at the molecular scale to obtain detailed insight into molecular self-assembly. This bottom-up approach to fabricate nanoarchitectures with guest molecules was further explored by varying the oligomer chain length of the dA_n (n = 2, 4, 6, 8, 10, 20, and 40) [51]. For the relatively short templates of dA_n (n = 2, 4, 6, and 8) the designed thymidine-appended bola-amphiphile, **4**, formed nanofibers similar to those formed from the single component alone (only **4**, in the absence of dA_n). Interestingly, increase in the chain length of the dA_n template (n = 10, 20, and 40) improved the formation of better helical nanofibers. The observed discrete helical nanofibers are the result of complementary templated self-assembly of the guest molecule and ssDNA. For templates from dA_2 to dA_{40}, an increase in the length of the nanofiber was observed. For templates with $n \geq 10$, manifestation of the helical structure was observed.

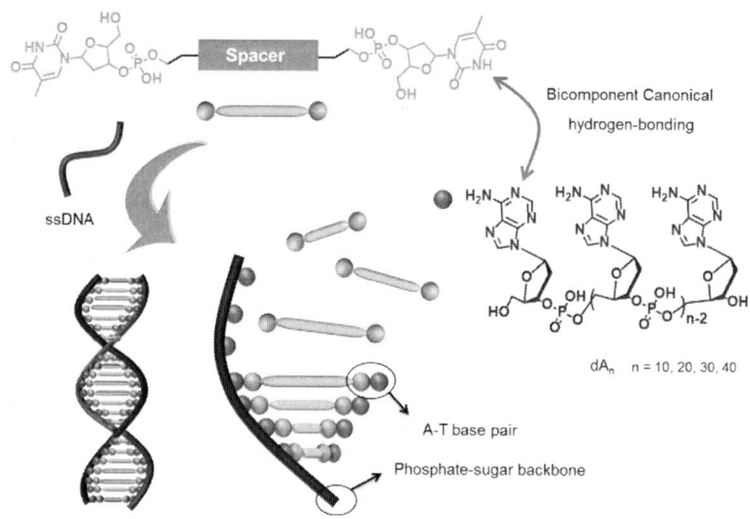

Figure 3.5 Schematic representation of the bicomponent helical structure formation through complementary WC interaction of the A-tethered bidented guest molecules (**1** and **2**) with oligoadenylic acid (dA_n) as the template-binding site. Adapted with permission from Ref. [48]. Copyright 2009, John Wiley and Sons.

It is established that the length of the ssDNA enhances the formation of helical nanofibers. Shimizu and coworkers extended their idea with the nucleotide-based bola-amphiphile **4**, to form an ensemble structure with dA_n of various chain lengths, n = 10, 20,

30, and 40 [52]. All the dA_n formed right-handed helical nanofibers in assembly with **4** in a tris/EDTA buffer solution. The CD spectra of these self-assemblies in solution showed a positive Cotton effect that is consistent with the right-handed helicity observed in the corresponding AFM images (Fig. 3.6d).

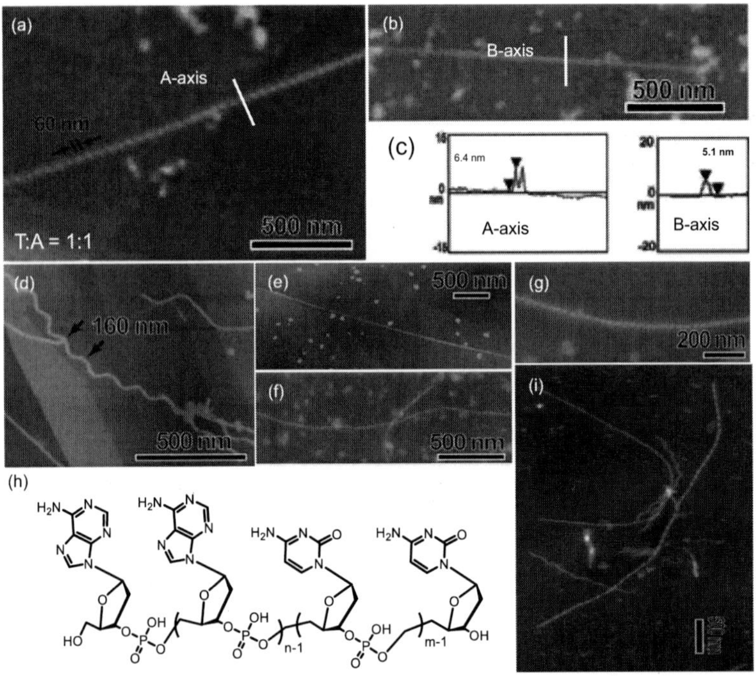

Figure 3.6 AFM images of the binary self-assembly of (a) **3**/dA_{20} (T:A = 1:1, 1.8×10^{-2} M/1.8×10^{-3} M); (b) **3**/dA_{20} (T:A = 1:2, 1.8×10^{-2} M/9×10^{-4} M); (c) section profile along the A and B axes; (d–g) **4**/dA_{10}, (b) **4**/dA_{20}, (c) **4**/dA_{30}, and (d) **4**/dA_{40} formed in $0.1 \times$ tris/EDTA buffer solutions; (h) chemical structure of alternating base sequences, dA_nC_m; and (i) AFM image of the stranded rope structures of $dA_{20}C_{10}$. (a–i) Adapted from Ref. [52] with permission of The Royal Society of Chemistry.

Interestingly, the helical pitches of the bicomponent self-assembly increased with the length of the dA_n template. The continuous CD spectral change was attributed to the large propeller twist of the angle between the planes of the hydrogen-bonded base pairs [53]. Further, the ssDNA sequence with alternating A (complementary) and C (noncomplementary) bases at the 5′- and

Table 3.1 The structural properties of the bicomponent ensembles of **4** and dA_n

	Template		Self-assembled morphology		
Composition	**Total length**[a]	**Complementary bases**[b]	**Pitch/nm**	**Diameter/nm**[d]	**Higher-order structure**
1/2	10	10	Helical fiber/11 nm	6.4	Wavy fiber[e] pitch = 160 nm
1/3	20	20	Helical fiber/18 nm	6.0	—
1/4	30	30	Helical fiber/24 nm	6.4	Double-stranded Rope[f] pitch = 100 nm
1/5	40	40	Helical fiber/40 nm	6.6	—
1/6	20	10	Helical fiber/12 nm	6.1	—
1/7	30	10	Helical fiber/11 nm	6.0	Double stranded Rope[g] pitch = 100 nm
1/8	30	20	Helical fiber/ —[c]	6.7	double-stranded Rope[h] pitch = 100 nm

Source: Adapted from Ref [52] with permission of The Royal Society of Chemistry.
[a]Number of bases in the oligonucleotide template
[b]Number of complementary bases in the oligonucleotide template
[c]Not determined
[d]Diameters of nanofibers estimated by section profiles
Yields of higher-order structures: [e]~20%, [f]100%, [g]~100%, and [h]~20%

3′- ends, respectively, was employed in the bicomponent assembly (Fig. 3.6h). This was an attempt to test the bicomponent self-assembly of the guest with the host ssDNA though both WC- and non-WC-type hydrogen-bonding interactions. Interestingly, the ssDNA composed of alternating base sequences also acted as a template (partially WC interaction) to induce the formation of helical nanofiber structures. The diameter of the nanofibers remained constant (6.0–6.6 nm), irrespective of the template used, **3** or **4**. The results showed that the helical pitches of the resultant nanofibers depend on the length of the complementary region (A-T) but not on the length of the noncomplementary (C-T) region. This was an important finding by the authors in understanding the DNA-templated nanoarchitectures. The templates with 10 complementary A units formed nanofibers. In contrast, the ensemble of **4** with oligonucleotide containing 20 A units and 10 C units produced a wavy morphology coexisting with a stranded rope (Fig. 3.6i). A phenomenon called "end fraying" has been reported, in which hydrogen bonds break and eventually a base pair becomes free at the terminus of the helix [54]. For **4**-dB_n ensembles, the authors proposed a similar "end fraying" phenomenon. The relatively short templates, such as dA_{10}, are more prone to an end fraying effect than assemblies with the designed guest molecule, **4**. This end fraying effect possibly allows the guests to be more mobile and move away from each other in the ensemble structure. The resulting binary assembly state increases the deviation angle θ between free guest molecules, which leads to the formation of short-pitch helical nanofibers. The systematic morphology change with the ssDNA length and the order of sequence is summarized in Table 3.1.

3.4 Templated DNA Nanoarchitectures Via Noncanonical Hydrogen-Bonding Interaction

The noncanonical hydrogen-bonding interactions, such as A-A and A-G interactions, are to be explored for generating novel DNA nanoarchitectures [28, 29]. To accomplish this challenging task, design of a suitable nucleobase-conjugated guest molecule that facilitates noncanonical hydrogen bonding with noncomplementary nucleobases is necessary. Such noncanonical base pairing is reported from our research group employing the PDI derivative **APA** (Fig. 3.2) [29]. In addition to using **APA** for constructing hybrid DNA ensembles through canonical hydrogen bonding (A-T), as discussed in Section 3.3, its potential for forming noncanonical hydrogen bonding with dA_n and dG_n was explored. The CD spectroscopic study revealed the interaction between **APA** and noncomplementary dB_n ($dB_n = dA_n$ and dG_n) to form hybrid ensembles of the type [dB_n:(**APA**)$_n$:dB_n]. Thus, mutual templating between **APA** and dA_n/dG_n through unconventional hydrogen bonding was successfully achieved. Further, CD and computational studies showed the formation of *P*- and *M*-helical imprinting in [dA_n:(**APA**)$_n$:dA_n] and [dG_n:(**APA**)$_n$:dG_n], respectively (Fig. 3.2). Notably, selecting a suitable SM template is paramount in constructing functional molecule–DNA nanoarchitectures, as revealed by our previous study wherein a NDI analog of **APA** failed to form hybrid DNA ensembles with complementary and noncomplementary dB_n [28].

Schenning and coworkers used ssDNA as the template on which a supramolecular arrangement of chromophores (diaminotriazine units) was formed via complementary hydrogen-bonding interactions [27, 55–57, 59]. Two monodentate guest molecules of naphthalene (**5**) and oligo(p-phenylene)vinylene (**6**) with a diaminotriazine moiety were designed (Fig. 3.7). Guest molecules, **5** and **6,** with a diaminotriazine recognition unit were studied for the complexation with ssDNA. A detailed study established the 1:1 binding of **6** and the host T base of dT_n. When **6** was added to dT_{40}, the chromophoric absorption maximum band shifted to the red region from 311 to 318 nm and the corresponding CD spectrum showed an induced positive bisignate Cotton effect at the region corresponding to achiral chromophoric absorption. The spectroscopic data further revealed that the chirality was imprinted from the dT_{40} template to the guest molecule. This study established the DNA-templated nanoarchitectures of SMs. In addition, the molecular modeling and simulations supported the organization of the SMs (**5** or **6**) and ssDNA through hydrogen-bonding interactions [50]. The detailed structural calculation provided key information on the major interactions that drive the formation of SM-DNA ensemble structures. The hydrogen-bonding interaction between the guest and the template ssDNA was found to diminish for long dT_n (n = 40) sequences. The stacking of the guest molecules (**5** or **6**) was found to have a stronger influence on the ssDNA-templated structure. The strong interactions between guest molecules in the stacks led to the extension of the coassembly structure along the strand axis.

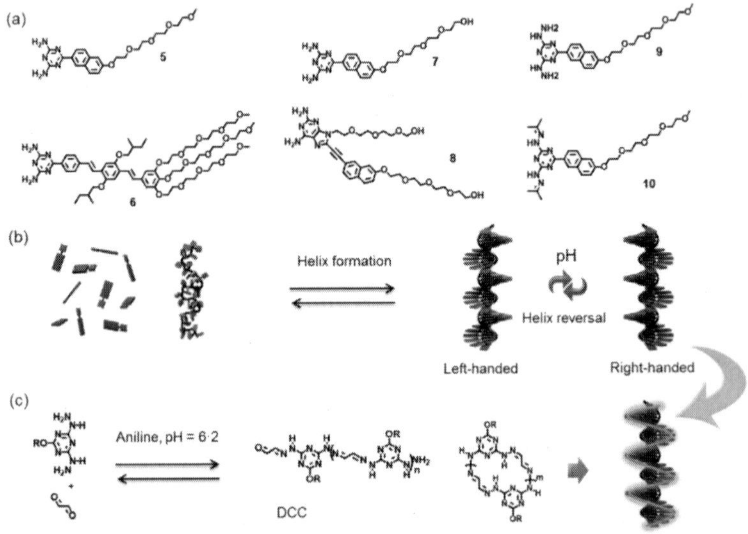

Figure 3.7 (a) Chemical structures of 1,3,5-triazine (diaminotriazine)- and diaminopurine-appended guest derivatives (**5-10**) as recognition chromophore units studied for the complexation with ssDNA templates. (b) Achiral guest derivatives form a helical ensemble in the presence of a chiral ssDNA template. The switchable helicity of the guest-ssDNA assembly in response to the change in pH of the solution. (c) The dynamic polymerization reaction was carried out using **9** and glyoxal at pH 6.2 in the presence of aniline as a nucleophilic catalyst. The linear polymerization product was selectively controlled by using a long ssDNA template. DCC = dynamic covalent chemistry. (a, b) Adapted with permission from Refs. [27, 55]. Copyright (2009 and 2007) American Chemical Society. Adapted with permission from Refs. [56, 57]. Copyright 2009, John Wiley and Sons. (c) Adapted from Ref. [59] with permission of The Royal Society of Chemistry.

For the exploitation of the templated nanofabrication, it is essential to develop a deeper understanding of the binding mechanism between the host and the guest molecules and also among the guest molecules. The guest molecules with large hydrophobic motifs or surfaces undergo self-aggregation in an aqueous solution irrespective of the presence of a template. The naphthalene derivative **5** showed induced chirality in the presence of dT_{40} due to the templated assembly and can also form nontemplated self-assembly at higher concentrations [27]. The authors modified naphthalene derivative **5** to suppress the nontemplated self-assembly of the guest molecule by replacing the methyl-terminated ethylene oxide with a hydroxyl-terminated ethylene oxide, which yielded **7**. The self-assembly behavior was studied by estimating the binding energies for both host-guest and guest-guest interactions. The elongation temperature T_e, at which the self-assembly starts, for nontemplated cooperative self-assembly of **7** was found to be less than **5**. Comparing **5** and **7**, it was revealed that the removal of the methyl group decreased the self-assembly of the guest and slightly increased that of guest-dT_{40}. The concentration range where the nontemplated self-assembly of **7** interferes with the templated self-assembly, therefore, lies at higher concentrations than that for **5**. This study pointed to the guest self-aggregation and showed the way to circumvent the problem in designing suitable guest molecules targeting the ssDNA template nanoarchitectures.

The chiral ssDNA template is a supramolecular stack of achiral chromophores, producing a helical organization of the guest molecules (Fig. 3.7b). The efficiency of this templated self-assembly depends on the host-guest and guest-guest interactions as discussed (*vide supra*). Further, the achiral guest naphthalene derivative **8** functionalized with diaminopurine is designed to bind to the dT_n template [57]. Surprisingly, the helicity of the **8**:dT_n DNA-templated assembly was found to be switchable in response to a change in the pH of the solution, due to protonation of the guest (Fig. 3.7b). At lower pH, the right-handed helical assembly transformed into a left-handed form. A positive Cotton effect in the chromophore absorption region with the zero crossing was detected at pH 7, revealing that the guest (**8**) molecules are arranged in a right-handed fashion along the dT_{40} template. The other homo ssDNA strands, such as dA_{40} and dC_{20}, under similar conditions were unable to induce such Cotton effect. It was concluded that guest binds to dT_{40} through WC-type hydrogen bonding similar to the previously studied derivatives (*vide supra*). Further, the protonated state of **8** controls the helicity of the ensemble while preserving the hydrogen bond–driven pairing interactions. The "helix reversal" occurs in the pH range at which protonation of the guest occurs. The change in pH from 9 to 2 showed a complete reversal of CD spectra (positive to negative signal). Further, it is known that the T units in dT_n, at pH 3, are not protonated and the phosphates are protonated [58]. Overall, it seems that the protonated state controls the helicity of the ensemble while preserving the hydrogen bond pairing of guest and host. The outcome defined a pH-switchable stability, and supramolecular helicity was established with detailed pH-dependent spectroscopic studies. This study recognized unique possibilities to arrange functional molecules equipped with similar hydrogen-bonding recognition units into the ssDNA-templated switchable functional nanoarchitectures.

In another report, ssDNA of an appropriate length was employed to serve as the template for supramolecular polymerization of guests, in which the ssDNA template can be recycled by a simple heating-cooling process [59]. The dynamic combinatorial chemistry (DCC) is a promising approach to generate stimuli-responsive polymer systems [60]. The combination of ssDNA as a template with DCC is an appealing nonenzymatic approach for fabricating well-defined nanostructured materials on a large scale using amplification methods. For this, three dihydrazine-triazine naphthalene guests, **5**, **9**, and **10**, were designed. These hydrazine derivatives reacted in a reversible and dynamic manner with aldehydes, forming hydrazones (Fig. 3.7c). The dynamic polymerization reaction was carried out using **9** and glyoxal at pH 6.2 in the presence of aniline as a nucleophilic catalyst. The hydrazine derivative **9** formed a well-characterized right-handed and 1D templated structure with dT_{40}. Upon addition of glyoxal into the solution of **9**-dT_{40} ensemble at pH 6.2 in the presence of aniline at 25°C, the CD signal at 330 nm gradually changed from a positive to a negative sign within 30 min. of component mixing. In the case of dT_{15} (a shorter chain length) as a template, even after 24 h, adding glyoxal did not show a CD signal at λ = 330 nm. This indicates that DNA-templated polymerization for a smaller template size yields predominately free monomers in solution that form kinetically trapped cyclic dimers (Fig. 3.7c). In this study, authors demonstrated that the chain length of the template is sufficient to control the dynamic polymerization of the functional molecule and to overcome the hydrogen-bonded dimerization/cyclization reaction of the monomeric building blocks.

This approach may be useful in the design of a functional molecule equipped with an appropriate hydrogen-bonding moiety to create uniform and well-organized nanoscale objects [61, 62].

We employed WC- and Hoogsteen-type hydrogen-bonding to exploit the controllable molecular organization of nucleobase-appended NDI derivatives **11** and **12** with peptide nucleic acid (PNA) dimer clamps to create well-defined supramolecular architectures [28]. The PNAs are the DNA mimics with nucleobases attached to the pseudopeptide backbone [63]. In this study, we demonstrated unique structural transformations by exploiting both WC- and Hoogsteen-type hydrogen-bonding potential of the nucleobase A (Fig. 3.1a and Fig. 3.8b). In this study, we controlled the ensemble organization by shuffling the mixing choices of the components in the bicomponent assembly formation into well-defined 2D microstructures, porous spheres, tapes, and petal-like 2D sheets. The obtained nano- and microarchitectures were found to be important and served the basic purpose of designing guest molecules (Fig. 3.8b). This bioinspired strategy comprising molecular recognition of nucleobases can be used to design and develop various templated functional ensembles and nano- and microarchitectures.

A non-WC-type interaction can create distinct structural and functional properties, in contrast to WC interactions. The use of synthetic and unnatural base analogs that present more than one hydrogen-bonding "face" to access higher-order DNA structures has generated great interest [64]. The nucleobase A possesses WC and Hoogsteen hydrogen-bonding faces, each with a two-point donor-acceptor motifs (Fig. 3.1a). In an interesting work, Sleiman and coworkers established that SMs such as cyanuric acid (CA) can be used to induce the assembly of nucleic acids (ssDNA) in water [65]. This concept can lead to the generation of new functional structures from inexpensive molecules and readily allows prospective applications. For the formation of a guest-ssDNA ensemble, the assembly can mediate through cooperative growth, which generates supramolecular polymers with a very high aspect ratio. The non-WC-type orthogonal assembly mode expanded the four-letter DNA alphabets (A, T, G, and C) and increases the range of possible architectures in the field of DNA nanotechnology. Investigations revealed that the CA-dA_{15} ensemble forms fiber structures that involve the proposed "hexameric rosette"-type hydrogen-bonded array (Fig. 3.8c) as the underlying assembly motif. Further, AFM height measurements revealed that the assembly of A analog with CA brought together dA_{15} strands into a triplex mode of organization. Overall, examples discussed in this section show that noncanonical hydrogen-bonding interactions have great potential for utilization in the development of novel templated DNA nanoarchitectures. Strategic utilization of the recognition sites and synthetic clever designs therefore lead to novel molecular systems to develop functional molecular materials.

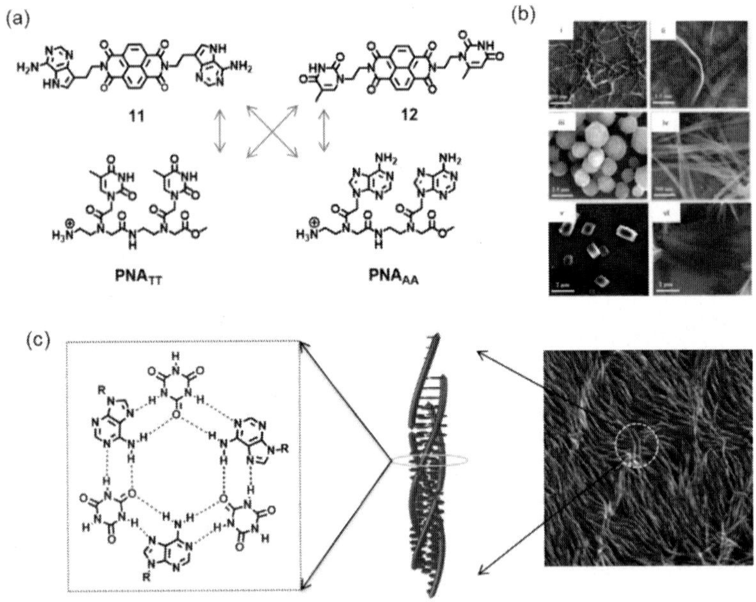

Figure 3.8 (a) Chemical structures of nucleobase-appended NDI derivatives (**11** and **12**) studied for complexation with the peptide nucleic acids (PNAs). (b) FESEM micrographs: (i) fibers obtained from **11** in 100% DMSO solution, (ii) nanoribbons obtained from **11** in 90% CH3CN-DMSO solution, (iii) porous spheres obtained from **12** in 100% DMSO solution, (iv) bundles of belts obtained from **12** in 90% CH3CN-DMSO solution, (v) rectangular 2D microstructures obtained from **11** + $\mathbf{PNA_{TT}}$ (1:1) in DMSO-water solution, and (vi) petal-like 2D sheets obtained from **12** + $\mathbf{PNA_{AA}}$ (1:1) in DMSO-water solution. (c) Hydrogen-bonded "hexameric rosette" structure formation of a CA-dA15 ensemble that leads to the formation of triple helix fiber structures imaged in AFM measurement. (a, b) Adapted from Ref. [28] with permission of The Royal Society of Chemistry. (c) Reprinted by permission from Springer Customer Service Centre GmbH: Springer Nature, *Nature Chemistry*, Ref. [65], Copyright (2016).

3.5 Templated DNA Nanoarchitectures Via Ionic Interaction

Electrostatic interaction is an effective noncovalent interaction (after hydrogen bonding) responsible for maintaining the secondary structure of biomacromolecules, for example, protein, peptide, DNA, and RNA. Among the non-WC interactions used to create templated DNA nanostructures, ionic interaction can play an important role in forming and maintaining the templated coassembly. Initial work from Armitage and coworkers played an important role in explaining the assembly of ionic dye molecules in the presence of ssDNA [66, 67]. They have proposed a model for translation from a monomer to a dimer stack (H-aggregate) along the floor of the DNA minor groove. A tricationic cyanine dye spontaneously assembles into a helical J-aggregate in the presence of a dsDNA template. The cyanine dye aggregate likely consists of dimeric units assembled in an offset, face-to-face (H-type) orientation within the minor groove of the dsDNA template. These results suggest the possible use of charged guests to control the overall structure of the coassembly with DNA. The J-aggregated helical structure was formed with a high degree of cooperativity. Slight changes in environmental conditions, such as temperature and ionic strength, result in the interconversion between J- and H-aggregate states.

Intercalation of planar molecules into DNA [68, 69] and binding of arc-shaped molecules into the minor groove [70, 71] are the two major association mechanisms followed by SM and DNA ensemble formation. There is another binding mechanism, "phosphate clamp," reported by Williams and coworkers (Fig. 3.9a) [72]. This binding mechanism deals with the interactions that exclusively utilize the backbone functional groups. This backbone-binding mode falls outside the previously discussed intercalator and groove-binding class of interactions. This approach does not enable access to sequence-programmed self-assemblies; however, it may be useful for gene delivery applications. Apart from these, adaptation of ionic interaction of cationic compounds with the phosphodiester backbone of ssDNA can lead to the stabilization of DNA nanoarchitectures (Fig. 3.5). Further, incorporation of π-stacking units with cationic-binding groups will be useful for the study of their templated interaction with ssDNA in aqueous solution. Ulrich and coworkers reported the design and synthesis of anion binders featuring aromatic side groups of varying sizes [73]. The guanidinium group, with a well-known ability to interact with oxoanions through salt bridge interactions [74], was symmetrically functionalized with aromatic moieties (**13–15**) that can provide π-stacking interactions during self-assembly with dT_n (Fig. 3.9b) [63]. The fluorescence titration study revealed that the excimer emission (maximum at λ_{em} = 470 nm) of the pyrene molecule in **15** was favored at the expense of monomer emission (λ_{em} = 396 nm) upon addition of dT_n (Fig. 3.9c). This effect can be clearly appreciated when plotting the ratio of excimer and monomer emission (E/M). In a competitive experiment, all three molecules (**13–15**) were added into the dT_{40} solution at a time. The competition study between stoichiometric amounts of guanidiniums was carried out to assess the relative binding affinity between the three different ligands, and the trend was in the order **13** < **14** < **15**. This competition experiment revealed the importance of π-stacking interactions in the stabilization of the DNA-templated assemblies.

Design of functional synthetic molecules for the development of SM- and DNA-templated nanoarchitectures exclusively driven by coassembly through specific interactions has become an area of growing interest. In particular, cation-phosphate binding interaction has inspired many researchers to design non-WC-type DNA-templated ensembles. To explored this concept Barboiu and coworkers designed triphenylamine (TPA)-based molecular scaffolds with guanidinium and trimethylamine cationic-binding sites (**16** and **17**, respectively) (Fig. 3.9c) [75]. TPA molecules containing hydrogen-bonding moieties self-assemble into fiber architectures. Tris(4-formylphenyl)amine and 1,3,5-benzenetrialdehyde aromatic core centers (**16** and **17**, respectively) were armed with amino-guanidine or Girard's reagent cationic sites for specific binding with the anionic phosphodiester linker of DNA. These propeller-shaped molecules were designed to bind with ssDNA. The inherent CD signal of the ssDNA showed drastic amplification specific to helical structures in the presence of the derivatives. Interestingly, the appearance of a new set of CD signals in the absorption region of guest chromophores (300–450 nm) revealed the propeller-shaped tris(4-formylphenyl)amine core, thus inducing an axial molecular chirality. The templated assembly created a columnar superstructure of **16** or **17** with the 25b ssDNA, 5′-CAA-GCC-CTT-AAC-GAA-CTT-CAA-CGT-A-3′, which in turn resulted in the formation of stacks in the hybrid fibers (Fig. 3.9c). It was observed that the stacking of guest chromophores took place only in the presence of the ssDNA, which is assisted by the mutual electrostatic binding between the SMs and ssDNA templates. The stacking interaction brought the tris(4-formylphenyl)amine units in close proximity to favor intermolecular hydrophobic interactions in aqueous solution. Therefore, the authors proposed that the systematic changes in the CD spectra may be attributed to the mutual compensation of the total charge of ssDNA by the finite ratio of tris(4-formylphenyl)amine molecules. Through this concept, the authors perhaps developed a simple method to determine the number of bases or the length of the ssDNA. Furthermore, the strategic use of SMs or building blocks and the interactive/recognition properties for a mutual binding purpose could lead to easy and selective detection of fragments of ssDNA containing different homo- or heteronucleobase sequences.

Recently, Lee and coworkers demonstrated that the achiral pyridine-based molecule **18** and the pyridinium-based cationic molecule **19** can form an ensemble and encapsulate a dsDNA through electrostatic interactions between pyridinium cations and the negatively charged phosphate anions of DNA (Fig. 3.9d) [76]. On adding the mixed solution of **18** and **19** (in a 1/2 mole ratio of 6:4) into native dsDNA at pH 7.4, a virus-like hierarchical assembly was formed. The structure holds the native DNA and a coat of the synthetic molecules. The ensemble showed helicity switching triggered by pH change under physiological conditions (Fig. 3.9d), switching from the left-handed Z form to the right-handed B form of dsDNA when the pH is lowered from 7.4 to 4.5. The CD spectra showed the induction of a negative Cotton effect at higher wavelengths in the spectral range of the associated achiral aromatic segments. The CD signals become more apparent on increasing the content of the coat molecules up to a pyridine/phosphate (positive/negative) ratio of 1.0. This result demonstrates that the SMs and DNA ensemble induce the formation of a responsive helical coat assembly around the DNA chain with a preferred handedness. They extended this idea and demonstrated the helicity switching driven by the low pH inside endosomes during translocation into a cell. This observation also adds some insights into the puzzling biological role of the higher-energy conformer Z DNA.

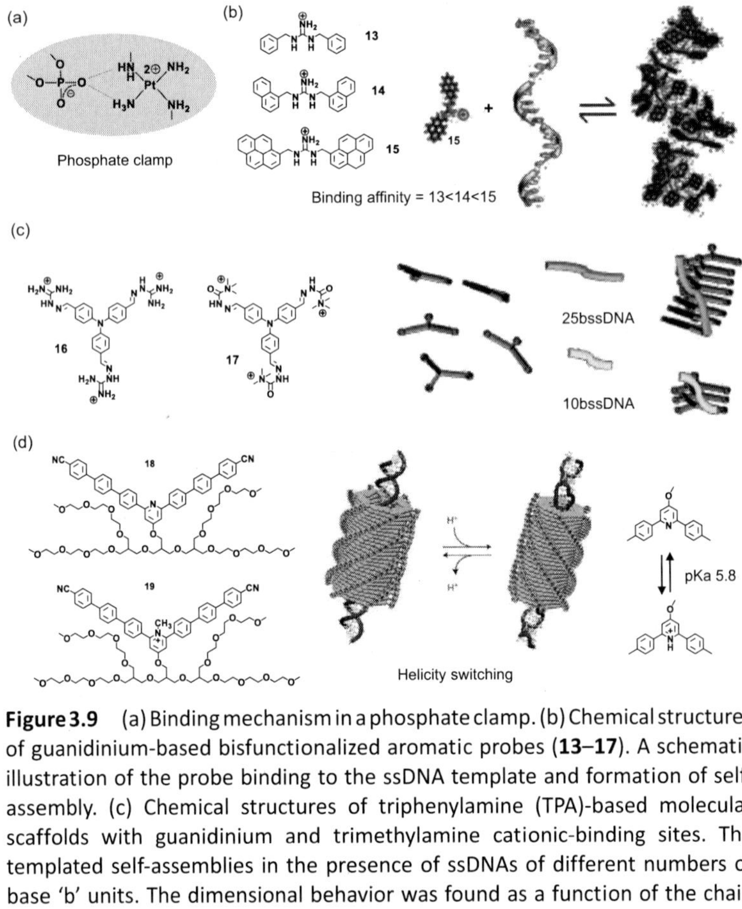

Figure 3.9 (a) Binding mechanism in a phosphate clamp. (b) Chemical structures of guanidinium-based bisfunctionalized aromatic probes (**13–17**). A schematic illustration of the probe binding to the ssDNA template and formation of self-assembly. (c) Chemical structures of triphenylamine (TPA)-based molecular scaffolds with guanidinium and trimethylamine cationic-binding sites. The templated self-assemblies in the presence of ssDNAs of different numbers of base 'b' units. The dimensional behavior was found as a function of the chain length. (d) Molecular structures of synthetic coat molecules **18** and **19** with pyridine and methylpyridinium, respectively. Collective motion in the helicity inversion of DNA and the synthetic coat assembly triggered by a pH change. (a) Adapted with permission from Ref. [72]. Copyright (2006) American Chemical Society. (b) Adapted from Ref. [73] with permission of The Royal Society of Chemistry. (c) Adapted from Ref. [75] with permission of The Royal Society of Chemistry. (d) Reprinted by permission from Springer Customer Service Centre GmbH: Springer Nature, *Nature Nanotechnology*, Ref. [76], Copyright (2017).

3.6 Templated DNA Nanoarchitecture Via Metal-Base Pair Interaction

Recent developments have shown the use of DNA as a construction material for nano-ordered objects. However, DNA-based molecular nanoarchitectures are constructed with mainly unmodified or, at best, end-modified oligonucleotides [77, 78]. Thus, the focus is now shifting toward the development of functionalized DNA structures [79]. One of notable developments in this direction is the substitution of the canonical WC base pair interaction with metal complexation [80]. Specifically, "metal-base pair" interaction is developed, which could potentially impart novel chemical, physical, and biological properties to derived DNA-templated nanoarchitectures (Fig. 3.10b) [81–86].

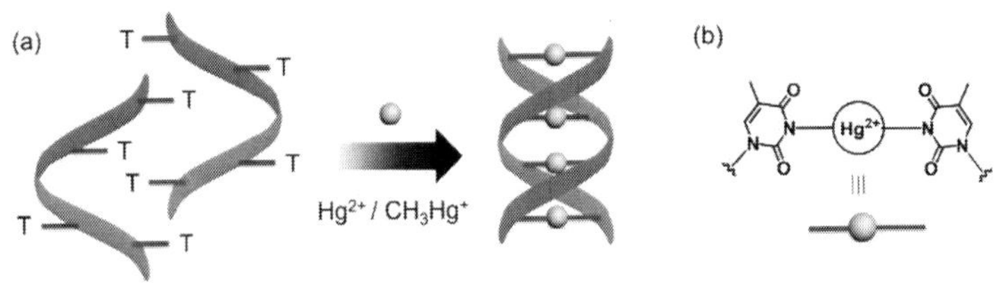

Figure 3.10 T-Hg(II)-T interaction is used in the construction of a DNA duplex, maintaining the 2:1 molar ratio of T and Hg(II). Adapted with permission from Ref. [87]. Copyright (2006) American Chemical Society.

Tanaka and coworkers reported an artificial ligand that is potentially suitable for the coordination of Zn(II) inside a DNA double helix [87]. They have established novel complementary nucleoside recognition by the Zn(II) macrocyclic tetra-amine (cyclen) complex with high selectivity, in aqueous solution (Fig. 3.11b). The Zn(II) center coordinates to the imide nitrogen atom of the nucleobase T, and hydrogen-bonding occurs between the carbonyl groups of the T and the NH groups of the cyclen (Fig. 3.11b). Similarly, binding studies of ssDNA revealed that the Hg(II)):nucleobase ratio is 1:2, that is one equivalent of Hg(II) binds to two equivalent of T units [88]. The imino proton of T dissociates and forms the T-Hg(II)-T metal-nucleobase pairs in the DNA duplex architecture [30, 89, 90]. The templated DNA nanotechnology or functional molecule–DNA nanoarchitectonics has been employed to develop a sensory platform for the detection of organic and inorganic mercury with high selectivity and ultrasensitivity. In a recent study, we have reported a selective and ultrasensitive detection of Hg(II) or $CH_3Hg(I)$ in water using an elegantly designed SM-ssDNA ensemble (see Section 3.3) [30]. This design strategy describes the SM-templated assembly of oligothymidine (dT_n) sequences that acts as a sensor platform for mercury, a highly toxic and global environmental pollutant (Fig. 3.3). Notably, this design strategy and methodology overcome the drawbacks of previous reports of Hg(II) detection using DNA (aptamer) containing few T units [91–97]. The previous reports are limited by the selectivity and sensitivity aspects owing to the use of DNA with few T units. However, our design strategy successfully addressed these drawbacks by constructing a SM-templated ssDNA assembly [SM:dT_n] that is energetically favored toward complete displacement SM units by Hg(II) and formation of a metallo-DNA duplex $[dT\text{-}Hg\text{-}dT]_n$. The use of an oligothymine (dT_n) sequence in the design instead of aptamers with few T units (as with literature reports) and SM with chiroptical and semiconducting properties serves as superior signaling output (sensor mechanism) for Hg(II) detection and needs special mention. In particular, **BNA**, the SM employed in the design, contains a butyl group to support noncovalent hydrophobic interactions and a NDI core for aromatic π-π interaction and to serve as a chiroptical and electrically active moiety for chirality- and conductivity-based measurements. The Hg(II)-assisted selective transformation of **BNA**$_n$-dT_n into a metallo-DNA duplex $[dT\text{-}Hg(II)\text{-}dT]_n$ was used as a chiroptical- and conductivity-based sensory platform for Hg(II) detection (Fig. 3.3). The conductivity (σ) measurements in a fabricated FET device with BNA_{10}-dT_{10} nanoarchitectures was performed upon addition of Hg(II) at different concentrations. A typical σ of ~1×10^{-6} S/cm and current density (J at 5 V) as high as 5×10^3 A/m^2 were obtained for the **BNA**$_{10}$-dT_{10} active layer. However, addition of Hg(II) onto the lateral device showed a decrease in the J value. The preferential binding of Hg(II) to dT_{10} breaks the ensemble structure, which results in the formation of $[dT\text{-}Hg(II)\text{-}dT]_{10}$ and the release of pristine **BNA** molecules (Fig. 3.3). Furthermore, it was observed that the extent of dissociation of the **BNA**$_{10}$-dT_{10} ensemble structure is proportional to the added Hg(II). We found an appreciable variation in the conductivity for Hg(II) concentrations as low as 0.1 nM. The detection limit in conductometric-based mercury sensing was found to be relatively better than the spectroscopic detection of Hg^{2+} [20% and 40% change in CD signal, corresponding to 5 nM and 10 nM of Hg(II), respectively]. Therefore, such design strategy involving a mutually templated SM and ssDNA ensemble structure is expected to aid in the development of novel DNA nanoarchitectonics and be potentially useful in bio-optoelectronics and sensor applications.

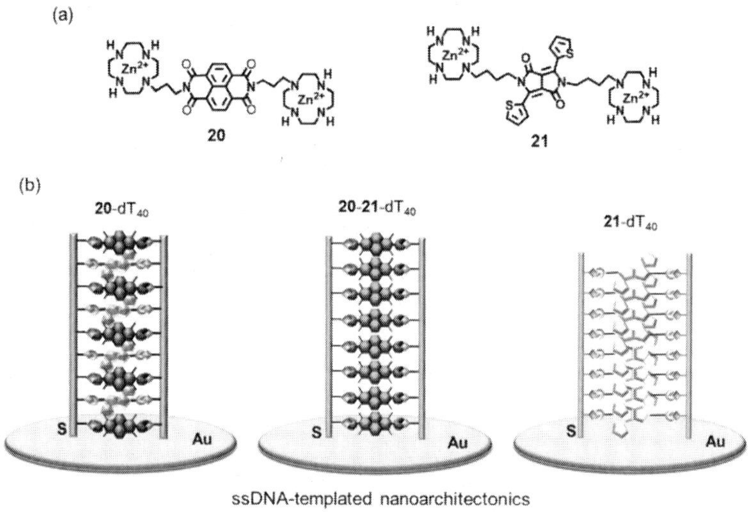

Figure 3.11 (a) Zn(II)-cyclen recognition motif in NDI-tethered derivatives **20** and **21** is used to bind with the T base in a 1:1 molar ratio with ssDNA. (b) Schematic presentation of the ssDNA-templated discrete and co-immobilized stack of aromatic donor and acceptor systems used for the design of nanoarchitectures. Adapted with permission from Refs. [98–100]. Copyright 2012, 2014, and 2015, John Wiley and Sons.

Nakamura and coworkers [98–100] developed a novel approached to assemble metal-ssDNA ensemble using an aromatic chromophore tethered with bis[Zn(II)-cyclen] (**20** and **21**) as the recognition center (Fig. 3.11). The binding interaction of SM and ssDNA increased the possibility to construct ensembles of various lengths of the stacked NDI arrays just by changing the dB_n length. The formation of the stacked array of NDI chromophores was found to be controllable by temperature [98]. The UV-Vis, PL, and CD spectroscopy results suggested selective binding of Zn(II)-cyclens of the guest molecule (**20** and **21**) to the T bases of dT_n to form the

1:1 complex. The NDI chromophoric units of **20** are stacked in an array fashion, and the number of stacked NDIs is consistent with the number of T nucleobases or the length of dT_n. In a similar approach, the authors have used diketopyrrolopyrrole (DPP) **21** with the Zn(II)-centered recognition motif as the guest molecule [92]. The UV-vis spectroscopy study suggested the formation of J-type aggregates in the **21**-dT_n ensemble. Thus, the authors have established two different guest-ssDNA ensembles using two different aromatic moieties (NDI and DPP respectively). The guest molecules with NDI and DPP chromophores were proposed to form a coassembly with ssDNA (Fig. 3.11b) [100]. It was anticipated that the LUMO level of dialkyl-substituted DPP derivative **21** (~2.51 eV) is higher than that of dialkyl-substituted NDI derivative **20** (~3.40 eV), which indicates possibility of photoexcited electron transfer from **21** to **20** [101, 102]. The authors have demonstrated photocurrent generation in a donor-acceptor heterojunction of ssDNA-multichromophore array [100]. The discrete arrays consisting of **21** (donor) and **20** (acceptor) were co-immobilized on a gold (Au) electrode. The co-immobilized electrode exhibited a good photocurrent response because of the efficient charge separation between the **20** and **21** arrays. In contrast, an immobilized electrode consisting of randomly assembled **20** and **21** arrays generated no photocurrent response, as DPP formed ground-state charge-transfer complexes with NDI in the randomly assembled array templated by ssDNA. Overall, functional molecule–templated DNA nanoarchitectures were designed with an accurate understanding of charge separation and used for photocurrent generation that proposed to be useful in constructing photovoltaic cells and biosensors (see more details in Chapter 7).

3.7 Conclusions and Future Perspectives

DNA has become a fast-evolving tool in the realm of molecular architectonics and structural nanoarchitectonics, not only because of its structural characteristics, such as thickness (2 nm) and persistence length (~50 nm), which are in the nanoregime and its chirality trait, but also because it is one of the natural polymers with well-defined multiple and predictive binding sites or sequence in the form of four nucleobases. In modern research, programmable molecular architectonics through functional assembly of designer templates or modules and ssDNA is expected to grow rapidly and provide limitless opportunities to evolve with complex molecular structures, emergent properties, and functions. The molecular architectonics through complementary binding interaction between designer SMs and short DNA sequences is certain to generate well-defined functional nanostructures, ascribed as DNA nanoarchitectonics. The DNA nanoarchitectonics is highly dependent on the conventional or unconventional complementary base pairing and finding suitable SMs as functional templates, which is quite challenging but progressive in nature. The small functional molecules can be used as self-assembled templates, which in turn can be used as templates for ssDNA to develop novel nanoarchitectures with built-in functional properties and applications, an important attribute compared to classical DNA nanotechnology. However, there are many challenges, open questions, and opportunities in this growing field of molecular architectonics to nanoarchitectonics of functional molecules and DNA. The detailed experimental and theoretical or computational investigations into the kinetics and thermodynamics of mutually templated functional molecule–DNA architectures and their annealing mechanics may reveal some interesting solutions to increase the stability of these ensembles under different experimental conditions of functional relevance. The yields of the functional nanoarchitectures and their responsive nature to various stimuli enable better prospects for in vitro and in vivo drug delivery and nanobiotechnological applications, ultrasensitive and selective sensory systems and platforms, chiroptical and conductivity-based applications in neuroscience, and materials and biomaterials science.

Special emphasis must be given to the development of DNA nanoarchitectonics through more complex but cost effective and functional designs by combining the SM-DNA assemblies with other biomolecules and materials. For instance, generating 3D objects using templated assemblies of SMs and DNA with precise control on the nano- meso-, and macroscopic molecular organization without any dispersion is the next challenging task. Structural and functional switching at both molecular and material levels, with reasonable control on the time and response to the chosen stimulus, is the key to success. The future directions for the field would also include use of templated assemblies of SMs and DNA to program the nonbiological systems for controlling biological events. The fields of molecular architectonics and DNA nanoarchitectonics are at a nascent stage and are expected to grow quickly and impact chemistry, biology, photonics, optoelectronics, bioelectronics, (bio) sensors, nanofabrication, and allied areas. We will soon witness the importance and applications of functional molecule–DNA nanoarchitectonics in biology, biomolecular architectonics, and materials and biomaterials sciences.

Acknowledgments

We thank Prof. C. N. R. Rao FRS for constant support and encouragement, Jawaharlal Nehru Centre for Advanced Scientific Research (JNCASR), the Department of Biotechnology (DBT) (Nanotechnology Special Task Force Grant BT/PR10263/NNT/28/711/2013), Government of India, and Sheikh Saqr Laboratory (SSL), ICMS-JNCASR, for financial support.

References

1. Jeffrey G. A. and Saenger, W. (1991). *Hydrogen Bonding in Biological Structures*, Springer, Berlin.
2. Whitsides, G. M. and Grzvbowski, B. (2002). Self-assembly at all scales, *Science*, **295**, 2418–2421.
3. Watson, J. D. and Crick, F. H. (1953). A structure for deoxyribose nucleic acid, *Nature*, **171**, 737–738.
4. Seeman, N. C. (2001). DNA nicks and nodes and nanotechnology, *Nano Lett.*, **1**, 22–26.
5. Seeman, N. C. (2010). Nanomaterials based on DNA, *Annu. Rev. Biochem.*, **79**, 65–87.
6. Hershey, A. and Chase, M. (1952). Independent functions of viral protein and nucleic acid in growth of bacteriophage, *J. Gen. Physiol.*, **36**, 39–56.
7. Clayton, J. and Dennis, C. (2003). *50 Years of DNA*, Nature Publishing Group, 20 Palgrave Macmillan, New York.
8. Blackburn, G. M., Gait, M. J., Loakes, D. and Williams, D. M. (2006). *Nucleic Acids in Chemistry and Biology*, RSC Publishing, Cambridge, UK.
9. Seeman, N. C. (2003). DNA in a material world, *Nature*, **421**, 427–431.

10. Seeman, N. C. (2005). Structural DNA nanotechnology: an overview, *Methods Mol. Biol.*, **303**, 143–166.

11. Rothemund, P. W. K. (2006). Folding DNA to create nanoscale shapes and patterns, *Nature*, **440**, 297–302.

12. Shao, Y., Jia, H., Cao, T. and Liu, D. (2017). Supramolecular hydrogels based on DNA self-assembly, *Acc. Chem. Res.*, **50**, 659–668.

13. Surin, M. (2016). From nucleobase to DNA templates for precision supramolecular assemblies and synthetic polymers, *Polym. Chem.*, **7**, 4137–4150.

14. Tintor, M., Eritja, R. and Fabrega, C. (2014). DNA nanoarchitectures: steps towards biological applications, *ChemBioChem*, **15**, 1374–1390.

15. Seeman, N. C. (1982). Nucleic acid junctions and lattices, *J. Theor. Biol.*, **99**, 237–247.

16. Pinheiro, A. V., Han, D., Shih, W. M. and Yan H. (2011). Challenges and opportunities for structural DNA nanotechnology, *Nat. Nanotechnol.*, **6**, 763–772.

17. Avinash, M. B. and Govindaraju, T. (2018). Architectonics: design of molecular architecture for functional applications, *Acc. Chem. Res.*, **51**(2), 414–426.

18. Avinash, M. B., Narayan, K. S., Swathi, K. and Govindaraju, T. (2016). Molecular architectonics of naphthalenediimides for efficient structure-property correlation, *ACS Appl. Mater. Interfaces*, **8**, 8678–8685.

19. Aono, M., Bando, Y. and Ariga, K. (2012). Nanoarchitectonics: pioneering a new paradigm for nanotechnology in materials development, *Adv. Mater.*, **24**, 150–151.

20. Ariga, K., Ji, Q., Mori, T., Naito, M., Yamauchi, Y., Abe, H. and Hill, J. P. (2013). Enzyme nanoarchitectonics: organization and device application, *Chem. Soc. Rev.*, **42**, 6322–6345.

21. Govindaraju, T. and Avinash, M. B. (2012). Two-dimensional nanoarchitectonics: organic and hybrid materials, *Nanoscale*, **4**, 6102–6117.

22. Govindaraju, T. and Avinash, M. B. (2014). Nanoarchitectonics of biomolecular assemblies for functional applications, *Nanoscale*, **6**, 13348–13369.

23. Komiyama, M., Yoshimoto, K., Sisido M. and Ariga, K. (2017). Chemistry can make strict and fuzzy controls for bio-systems: DNA nanoarchitectonics and cell-macromolecular nanoarchitectonics, *Bull Chem. Soc. Jpn.*, **90**, 967–1004.

24. Avinash, M. B. and Govindaraju, T. (2012). Amino acid derivatized arylenediimides: a versatile modular approach for functional molecular materials, *Adv. Mater.*, **24**, 3905–3922.

25. Aida, T., Meijer, E. W. and Stupp, S. I. (2012). Functional supramolecular polymers, *Science*, **335**, 813–817.

26. Brunsveld, L., Folmer, B. J. B., Meijer, E. W. and Sijbesma, R. P. (2001). Comparisons of indefinite self-association models, *Chem. Rev.*, **101**, 4071–4097.

27. Janssen, P. G. A., Jabbari-Farouji, S., Surin, M., Vila, X., Gielen, J. C., de Greef, T. F. A., Vos, M. R. J., Bomans, P. H. H., Sommerdijk, N. A. J. M., Christianen, P. C. M., Leclère, P., Lazzaroni, R., van der Schoot, P., Meijer, E. W. and Schenning, A. P. H. J. (2009). Insights into templated supramolecular polymerization: binding of naphthalene derivatives to ssDNA templates of different lengths, *J. Am. Chem. Soc.*, **131**, 1222–1231.

28. Narayanaswamy, N., Avinash, M. B. and Govindaraju, T. (2013). Exploring hydrogen bonding and weak aromatic interactions induced assembly of adenine and thymine functionalised naphthalenediimides, *New J. Chem.*, **37**, 1302–1306.

29. Narayanaswamy, N., Suresh, G., Priyakumar, U. D. and Govindaraju, T. (2015). Double zipper helical assembly of deoxyoligonucleotides: mutual templating chiral imprinting to form hybrid DNA ensembles, *Chem. Commun.*, **51**, 5493–5496.

30. Pandeeswar, M., Senanayak, S. P. and Govindaraju, T. (2016). Nanoarchitectonics of small molecule and DNA for ultrasensitive detection of mercury, *ACS Appl. Mater. Interfaces*, **8**, 30362–30371.

31. Watson, J. D. (1968). *The Double Helix: A Personal Account of the Discovery of the Structure of DNA*, 1st ed., Atheneum, New York.

32. Rich, A., Davies, D. R., F. Crick, H. C. and Watson, J. D. (1961). The molecular structure of polyadenylic acid, *J. Mol. Biol.*, **3**, 71–86.

33. Chakraborty, S., Sharma, S., Maiti, P. K. and Krishnan, Y. (2009). The poly dA helix: a new structural motif for high performance DNA-based molecular switches, *Nucleic Acids Res.*, **37**, 2810–2817.

34. Verma, S., Mishra, A. K. and Kumar, J. (2010). The many facets of adenine: coordination, crystal patterns, and catalysis, *Acc. Chem. Res.*, **43**, 79–91.

35. Jissy, A. K. and Datta, A. (2014). Design and applications of noncanonical DNA base pairs, *J. Phys. Chem. Lett.*, **5**, 154–166.

36. Pandeeswar, M., Shilpa, R., Kandjani, A. E., Periasamy, S. R., Sabri, Y. M., Madhu, C., Bhargava, S. K. and Govindaraju, T. (2018). SERS and fluorescence-based ultrasensitive detection of mercury in water, *Biosens. Bioelectron.*, **100**, 556–564.

37. Pandeeswar, M. and Govindaraju, T. (2016). Engineering molecular self-assembly of perylene diimide through pH-responsive chiroptical switching, *Mol. Syst. Des. Eng.*, **1**, 202–207.

38. Avinash, M. B. and Govindaraju, T. (2013). Extremely slow dynamics of an abiotic helical assembly: unusual relevance to the secondary structure of proteins, *J. Phys. Chem. Lett.*, **4**, 583–588.

39. Avinash, M. B. and Govindaraju, T. (2011). Engineering molecular organization of naphthalenediimides: large nanosheets with metallic-conductivity and attoliter containers, *Adv. Funct. Mater.*, **20**, 3875–3882.

40. Pandeeswar, M., Khare, H., Ramakumar, S. and Govindaraju, T. (2014). Biomimetic molecular organization of naphthalene diimide in the solid state: tunable (chiro-) optical, viscoelastic and nanoscale properties, *RSC Adv.*, **4**, 20154–20163.

41. Pandeeswar, M. and Govindaraju, T. (2015). Bioinspired nanoarchitectonics of naphthalene diimide to access 2D sheets of tunable size, shape, and optoelectronic properties, *J. Inorg. Organomet. Polym.*, **25**, 293–300.

42. Pandeeswar, M., Khare, H., Ramakumar, S. and Govindaraju, T. (2015). Crystallographic insight-guided nanoarchitectonics and conductivity modulation of an n-type organic semiconductor through peptide conjugation, *Chem. Commun.*, **51**, 8315–8318.

43. Avinash, M. B., Swathi, K., Narayan, K. S. and Govindaraju, T. (2016). Molecular architectonics of naphthalenediimides for efficient structure-property correlation, *ACS Appl. Mater. Interfaces*, **8**, 8678–8685.

44. Pandeeswar, M., Senanayak, S. P., Narayan K. S. and Govindaraju, T. (2016). Multi-stimuli responsive charge-transfer hydrogel for room temperature organic ferroelectric thin-film devices, *J. Am. Chem. Soc.*, **138**, 8259–8268.

45. Suseela, Y. V., Das, S., Pati S. K. and Govindaraju, T. (2016). Imidazolyl-naphthalenediimide-based threading intercalators of DNA, *ChemBioChem*, **7**, 2162–2171.

46. Bhosale, S. V., Jani, C. H. and Langford, S. J. (2008). Chemistry of naphthalene diimides, *Chem. Soc. Rev.*, **37**, 331–342.

47. Sirringhaus, H., Brown, P. J., Friend, R. H., Nielsen, M., Bechgaard, M., Langeveld-Voss, K. B. M. W., Spiering, A. J. H., Janssen, R. A. J., Meijer, E. W., Herwig, P. and de Leeuw, D. M. (2009). Two-dimensional charge transport in self-organized, high-mobility conjugated polymers, *Nature*, **401**, 685–688.

48. Iwaura, R., Ohnishi-Kameyama, M. and Iizawa, T. (2009). Construction of helical J-aggregates self-assembled from a thymidylic acid appended anthracene dye and DNA as a template, *Chem. Eur. J.*, **15**, 3729–3735.

49. Harrison, W. J., Mateer, D. L. and Tiddy, G. J. T. (1996). Liquid-crystalline J-aggregates formed by aqueous ionic cyanine dyes, *J. Phys. Chem.*, **100**, 2310–2321.

50. Iwaura, R., Hoeben, F. J. M., Masuda, M., Schenning, A. P. H. J., Meijer, E. W. and Shimizu, T. (2006). Molecular-level helical stack of a nucleotide-appended oligo(*p*-phenylenevinylene) directed by supramolecular self-assembly with a complementary oligonucleotide as a template, *J. Am. Chem. Soc.*, **128**, 13298–13304.

51. Iwaura, R., Yoshida, K., Masuda, M., Ohnishi-Kameyama, M., Yoshida, M. and Shimizu, T. (2003). Oligonucleotide-templated self-assembly of nucleotide bolaamphiphiles: DNA-like nanofibers edged by a double-helical arrangement of A–T base pairs, *Angew. Chem. Int. Ed.*, **42**, 1009–1012.

52. Iwaura, R., Kikkawa, Y., Ohnishi-Kameyama M. and Shimizu, T. (2007). Effects of oligoDNA template length and sequence on binary self-assembly of a nucleotide bolaamphiphile, *Org. Biomol. Chem.*, **5**, 3450–3455.

53. Tinoco, I. (1964). Circular dichroism and rotatory dispersion curves for helices, *J. Am. Chem. Soc.*, **86**, 297–298.

54. Andreatta, D., Sen, S., PerezLustres, J. L., Kovalenko, S. A., Ernsting, N. P., Murphy, C. J., Coleman, R. S. and Berg, M. A. (2006). Ultrafast dynamics in DNA: "fraying" at the end of the helix, *J. Am. Chem. Soc.*, **128**, 6885–6892.

55. Janssen, P. G. A., Vandenbergh, J., van Dongen, J. L. J., Meijer, E. W. and Schenning, A. P. H. J. (2007). ssDNA templated self-assembly of chromophores, *J. Am. Chem. Soc.*, **129**, 6078–6079.

56. Surin, M., Janssen, P. G. A., Lazzaroni, R., Leclere, P., Meijer, E. W. and Schenning, A. P. H. J. (2009). Supramolecular organization of ssDNA-templated π-conjugated oligomers via hydrogen bonding, *Adv. Mater.*, **21**, 1126–1130.

57. Janssen, P. G. A., Ruiz-Carretero, A., González-Rodríguez, D., Meijer, E. W. and Schenning, A. P. H. J. (2009). pH-switchable helicity of DNA-templated assemblies, *Angew. Chem. Int. Ed.*, **48**, 8103–8106.

58. Saenger, W. (1984). *Principles of Nucleic Acid Structure*, Springer, New York.

59. Lin, J., Surin, M., Beljonne, D., Lou, X., van Dongen, J. L. J. and Schenning, A. P. H. J. (2012). On the mechanism of dynamic polymerization *via* recycled ss-DNA templated assembly of non-natural bases, *Chem. Sci.*, **3**, 2732–2736.

60. Corbett, P. T., Leclaire, J., Vial, L., K. West, R., Wietor, J.-L., Sanders, J. K. M. and Otto, S. (2006). Dynamic combinatorial chemistry, *Chem. Rev.*, **106**, 3652–3711.

61. Xu, L. and Liu, D. (2010). Functional evolution on the assembled DNA template, *Chem. Soc. Rev.*, **39**, 150–155.

62. Wang, Z. G. and Ding, B. (2014). Engineering DNA self-assemblies as templates for functional nanostructures, *Acc. Chem. Res.*, **47**, 1654–1662.

63. Nielsen, P. E., Egholm, M., Berg, R. H. and Buchardt, O. (1991). Sequence-selective recognition of DNA by strand displacement with a thymine-substituted polyamide, *Science*, **254**, 1497–1501.

64. Largy, E., Liu, W., Hasan, A. and Perrin, D. M. (2013). Base-pairing behavior of a carbocyclic Janus-AT nucleoside analogue capable of recognizing A and T within a DNA duplex, *ChemBioChem*, **14**, 2199–2208.

65. Avakyan, N., Greschner, A. A., Aldaye, Serpell, F., Jhn, C., Toader, V., Petitjean, A. and Sleiman, H. F. (2016). Reprogramming the assembly of unmodified DNA with a small molecule, *Nat. Chem.*, **8**, 368–367.

66. Seifert, J. L., Connor, R. E., Kushon, S. A., Wang, M. and Armitage, B. A. (1999). Spontaneous assembly of helical cyanine dye aggregates on DNA nanotemplates, *J. Am. Chem. Soc.*, **121**, 2987–2995.

67. Wang, M., Silva, G. L. and Armitage, B. A. (2000). DNA-templated formation of a helical cyanine dye J-aggregate, *J. Am. Chem. Soc.*, **122**, 9977–9986.

68. Lerman, L. S. (1961). Structural considerations in the interaction of DNA and acridines, *J. Mol. Biol.*, **3**, 18–30.

69. Narayanaswamy, N., Kumar, M., Das, S., Sharma, R., Samanta, P. K., Pati, S. K., Dhar, S. K., Kundu T. K. and Govindaraju, T. (2014). A thiazole coumarin (TC) turn-on fluorescence probe for AT-base pair detection

and multipurpose applications in different biological systems, *Sci. Rep.*, **4**, 6476.

70. Wartell, R. M., Larson, J. E. and Wells, R. D. (1974). A specific probe for A-T regions of duplex deoxyribonucleic acid, *J. Biol. Chem.*, **249**, 6719–6731.

71. Narayanaswamy, N., Narra, S., Nair, R., Saini, D. K., Kondaiah, P. and Govindaraju, T. (2016). Stimuli-responsive colorimetric and NIR fluorescence combination probe for selective reporting of cellular hydrogen peroxide, *Chem. Sci.*, **7**, 2832–2841.

72. Komeda, S., Moulaei, T., Woods, K. K., Chikuma, M., Farrell, N. P. and Williams, L. D. (2006). A third mode of DNA binding: phosphate clamps by a polynuclear platinum complex, *J. Am. Chem. Soc.*, **128**, 16092–16103.

73. Paolantoni, D., Rubio-Magnieto, J., Cantel, S., Martinez, J., Dumy, P., Surin, M. and Ulrich, S. (2014). Probing the importance of π-stacking interactions in DNA-templated self-assembly of bisfunctionalized guanidinium compounds, *Chem. Commun.*, **50**, 14257–14260.

74. Roy, B., Noguchi, T., Yoshihara, D., Tsuchiya, Y., Dawn, A. and Shinkai, S. (2014). Nucleotide sensing with a perylene-based molecular receptor *via* amplified fluorescence quenching, *Org. Biomol. Chem.*, **12**, 561–565.

75. Kocsis, I., Rotaru, A., Legrand, Y.-M., Grosu I. and Barboiu, M. (2016). Supramolecular rulers enabling selective detection of pure short ssDNA *via* chiral self-assembly, *Chem. Commun.*, **52**, 386–389.

76. Kim, Y., Li, H., He, Y., Chen, X., Ma X. and Lee, M. (2017). Collective helicity switching of a DNA–coat assembly, *Nat. Nanotechnol.*, **12**, 551–558.

77. Ensslen, P. and Wagenknecht, H.-A. (2015). 1D multichromophor arrays based on DNA – from self-assembly to light-harvesting, *Acc. Chem. Res.*, **48**, 2724–2733.

78. Teo, Y. N. and Kool, E. T. (2012). DNA-multichromophore systems, *Chem. Rev.*, **112**, 4221–4245.

79. Clever, G. H., Kaul, C. and Carell, T. (2007). DNA–metal base pairs, *Angew. Chem. Int. Ed.*, **46**, 6226–6236.

80. Polonius, F.-A. and Müller. J. (2007). An artificial base pair, mediated by hydrogen bonding and metal-ion binding, *Angew. Chem. Int. Ed.*, **46**, 5602–5604.

81. Katz, S. (1952). The reversible reaction of sodium thymonucleate and mercuric chloride, *J. Am. Chem. Soc.*, **74**, 2238–2245.

82. Thomas, C. A. (1954). The interaction of HgCl2 with sodium thymonucleate, *J. Am. Chem. Soc.*, **76**, 6032–6034.

83. Clever, G. H. and Shionoya, M. (2010). Metal-base pairing in DNA, *Coord. Chem. Rev.*, **254**, 2391–2402.

84. Clever, G. H., Kaul, C. and Carell, T. (2007). DNA–metal base pairs. *Angew. Chem. Int. Ed.*, **46**, 6226–6236.

85. Tanaka, K., Yamada, Y. and Shionoya, M. (2002). Formation of silver(I)-mediated DNA duplex and triplex through an alternative base pair of pyridine nucleobases, *J. Am. Chem. Soc.*, **124**, 8802–8803.

86. Heuberger, B. D., Shin, D. and Switzer, C. (2008). Two Watson–Crick-like metallo base-pairs, *Org. Lett.*, **10**, 1091–1094.

87. Miyake, Y., Togashi, H., Tashiro, M., Yamaguchi, H., Oda, S., Kudo, M., Tanaka, Y., Kondo, Y., Sawa, R., Fujimoto, T., Machinami, T. and Ono, A. (2006). MercuryII-mediated formation of thymine–HgII–thymine base pairs in DNA duplexes, *J. Am. Chem. Soc.*, **128**, 2172–2173.

88. Shionoya, M., Kimura, E. and Shiro, M. (1993). A new ternary zinc(II) complex with [12]aneN4 (=1,4,7,10-tetraazacyclododecane) and AZT (=3'-azido-3'-deoxythymidine). Highly selective recognition of thymidine and its related nucleosides by a zinc(II) macrocyclic tetraamine complex with novel complementary associations, *J. Am. Chem. Soc.*, **115**, 6730–6737.

89. Yamane, T. and Davidson, N. (1961). On the complexing of desoxyribonucleic acid (DNA) by mercuric ion, *J. Am. Chem. Soc.*, **83**, 2599–2607.

90. Kosturko, L. D., Folzer, C. and Stewart, R. F. (1974). Crystal and molecular structure of a 2:1 complex of 1-methylthymine-mercury(II), *Biochemistry*, **13**, 3949–3952.

91. Huang, C. C., Yang, Z., Lee, K. H. and Chang, H. T. (2007). Synthesis of highly fluorescent gold nanoparticles for sensing mercury(II), *Angew. Chem. Int. Ed.*, **46**, 6824–6828.

92. Lee, J. S., Han, M. S. and Mirkin, C. A. (2007). Colorimetric detection of mercuric ion (Hg2+) in aqueous media using DNA-functionalized gold nanoparticles, *Angew. Chem. Int. Ed.*, **46**, 4093–4096.

93. Li, D., Wieckowska, A. and Willner, I. (2008). Optical analysis of Hg2+ ions by oligonucleotide-gold-nanoparticle hybrids and DNA-based machines, *Angew. Chem. Int. Ed.*, **47**, 3927–3931.

94. Zhu, Y., Xu, L., Ma, W., Xu, Z., Kuang, H., Wang, L. and Xu, C. A (2012). One-step homogeneous plasmonic circular dichroism detection of aqueous

mercury ions using nucleic acid functionalized gold nanorods, *Chem. Commun.*, **48**, 11889–11891.

95. Zhang, L., Li, T., Li, B., Li, J. and Wang, E. (2010). Carbon nanotube-DNA hybrid fluorescent sensor for sensitive and selective detection of mercury(II) ion, *Chem. Commun.*, **46**, 1476–1478.

96. Helwa, Y., Dave, N., Froidevaux, R., Samadi, A. and Liu, J. (2012). Aptamer-functionalized hydrogel microparticles for fast visual detection of mercury(II) and adenosine, *ACS Appl. Mater. Interfaces*, **4**, 2228–2233.

97. Xu, H., Zhu, X., Ye, H., Yu, L., Liu, X. and Chen, G. (2011). A simple "molecular beacon"-based fluorescent sensing strategy for sensitive and selective detection of mercury (II), *Chem. Commun.*, **47**, 12158–12160.

98. Nakamura, M., Okaue, T., Takada, T. and Yamana, K. (2012). DNA-templated assembly of naphthalenediimide arrays, *Chem. Eur. J.*, **18**, 196–201.

99. Tsuto, K., Nakamura, M., Takada, T. and Yamana, K. (2014). Diketopyrrolopyrrole J-aggregates formed by self-organization with DNA *Chem. Asian J.*, **9**, 1618–1622.

100. Nakamura, M., Tsuto, K., Jomura, A., Takada, T. and Yamana, K. (2015). Donor–acceptor heterojunction configurations based on DNA–multichromophore arrays, *Chem. Eur. J.*, **21**, 11788–11792.

101. Guo, S., Wu, W., Guo, H. and Zhao, J. (2012). Room-temperature long-lived triplet excited states of naphthalenediimides and their applications as organic triplet photosensitizers for photooxidation and triplet–triplet annihilation upconversions, *J. Org. Chem.*, **77**, 3933–3943.

102. Favereau, L., Warnan, J., Pellegrin, Y., Blart, E., Boujtita, M., Jacquemin, D. and Odobel, F. (2013). Diketopyrrolopyrrole derivatives for efficient NiO-based dye-sensitized solar cells, *Chem. Commun.*, **49**, 8018–8020.

Chapter 4

Functional DNA Amphiphiles

Murali Golla, Hari Veera Prasad Thelu, Shine K. Albert, Nithiyanandan Krishnan, Siriki Atchimnaidu, Devanathan Perumal, Sai Praveen Thaddi, and Reji Varghese

School of Chemistry, Indian Institute of Science Education and Research Thiruvananthapuram, Thiruvananthapuram, Kerala, 695551, India

reji@iisertvm.ac.in

4.1 Introduction

Bottom-up molecular self-assembly through noncovalent interactions is a powerful, efficient, and convenient approach for the crafting of 1D, 2D, and 3D functional nanostructures [1]. During the past five or six decades, researchers have made significant progress in understanding the natural self-assembly process, and by exploring the knowledge of natural self-assembly, numerous examples of artificial nanostructures were designed and constructed. Self-assembly of amphiphilic molecules is a unique case of the bottom-up approach for the crafting of self-assembled nanostructures [2].

Templated DNA Nanotechnology: Functional DNA Nanoarchitectonics
Edited by Thimmaiah Govindaraju

ISBN 978-981-4800-21-1 (Hardcover), 978-0-429-42866-1 (eBook)
www.panstanford.com

Amphiphiles are molecular systems consisting of hydrophobic and hydrophilic moieties joined together either covalently or noncovalently and are capable of spontaneously assembling into diverse nanostructures, mainly through noncovalent interactions. Self-assembly of an amphiphilic lipid in an aqueous medium is a classic example of natural amphiphilic self-assembly, which is an integral part of all our cellular systems. A remarkable feature of amphiphilicity-driven self-assembly is the morphology tunability of the nanostructures, which is routinely achieved by controlling the volume ratio between hydrophilic and hydrophobic segments. This approach has been successfully applied in several classes of organic systems for the crafting of nanoarchitectures with interesting optical and electronic properties [3].

The discovery of the molecular structure of DNA by Watson and Crick is one of the breakthroughs during the last several decades and revealed several unique structural characteristics of DNA [4]. These include the following: (i) there is spontaneous self-assembly of single-stranded oligonucleotides into a duplex helical structure as encoded by the DNA base sequence, (ii) the base pair distance along the helical axis is 3.4 Å for B-DNA, (iii) the B-DNA has a diameter of 2 nm and a pitch of 3.4 nm, and (iv) the persistence length of double-stranded DNA (dsDNA) is 50 nm, which allows dsDNA to serve as a rigid scaffold for short nucleic acid segments (Fig. 4.1). Furthermore, the automated oligonucleotide chemistry makes DNA available in any desired base sequence and also in sufficient quantities. These remarkable structural features of DNA and ease of DNA synthesis have greatly attracted researchers to use DNA as a building block in the design of DNA-based nanostructures [5]. This gave birth to the emerging field of DNA nanotechnology, which is a bottom-up nanofabrication method based on Watson–Crick DNA base pairing, which permits the crafting of extremely complex 1D, 2D, and 3D nanostructures [6]. The most unique feature of DNA nanostructures is their inherent ability to act as soft nanoscaffolds for the organization of other functional molecules through sequence-specific DNA hybridization. Several DNA superstructures have been designed using the principles of structural DNA nanotechnology [7] and DNA origami [8] approaches and been successfully applied for the organization of various nanoscale objects, such as metal nanoparticles (NPs) [9], quantum dots [10], carbon nanotubes [11], small molecules [12], and enzymes [13], for diverse material and technological applications. The basic double-helical structure of DNA has also been effectively employed for the helical organization of small molecules into optically active nanoassemblies [14].

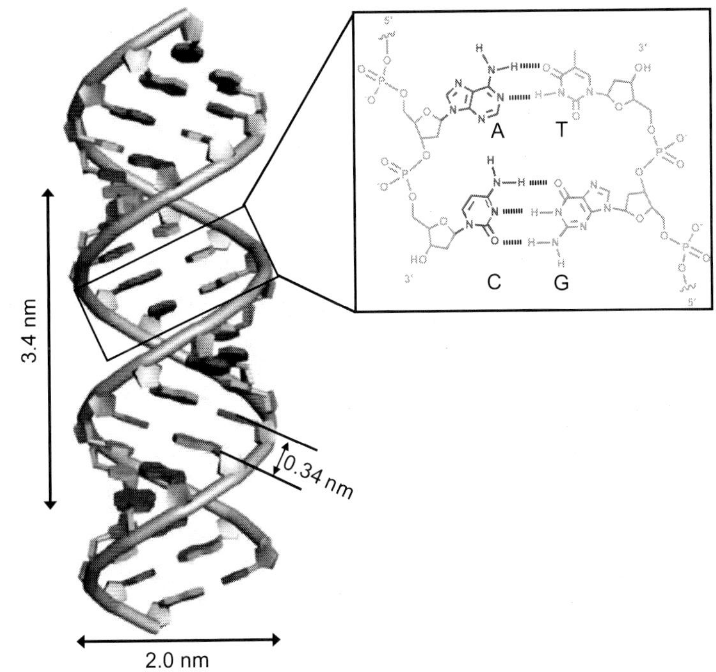

Figure 4.1 Chemical structure of dsDNA (B-DNA). The inset shows the hydrogen bonding of nucleobases. The important structural dimensions of dsDNA are also shown in the structure.

Recently, DNA has been applied as the hydrophilic segment in the design of amphiphilic systems known as DNA amphiphiles [15]. The most striking feature of the self-assembly of DNA amphiphiles in an aqueous medium is the formation of a nanostructure with a corona made of hydrophilic DNA. Another notable feature of DNA nanostructures derived from the self-assembly of a DNA amphiphile is the dense display of single-stranded DNA (ssDNA) on the surface of the nanostructure, which is extremely hard to achieve using the principles of structural DNA nanotechnology and DNA origami approaches. Hence, the self-assembly of DNA amphiphiles is preferred when dense surface decoration of nanostructures with ssDNA is required. The display of ssDNA on the surface of DNA nanostructures potentially allows the reversible integration of other functional molecules onto their surface through sequence-specific DNA hybridization, which has been exploited for many technological and biomedical applications. These include the self-assembly of NPs, carbon nanotubes, enzymes, cell-targeting ligands, and other functional molecules.

In this chapter, we provide a very detailed description of the design, synthesis, and self-assembly of DNA amphiphiles and the applications of self-assembled DNA nanostructures. One of the major challenges in this research area is the synthesis of DNA amphiphiles, which is mainly due to the incompatibility (polarity) of the hydrophilic and hydrophobic segments. A detailed discussion is done of the various approaches reported in the literature for the synthesis of DNA amphiphile. The synthetic section is divided into various approaches depending upon the coupling strategy. Self-assembly of DNA amphiphiles into various supramolecular nanostructures, with a special emphasis on the factors affecting the self-assembly, is provided, and the discussion will be restricted only to assembly in an aqueous medium. Utilization of the DNA nanostructure as a soft nanoscaffold for the reversible assembly of different functional molecules and nanomaterials for various applications is provided. Finally, we give an outlook on the challenges and future directions for this emerging research area.

4.2 Self-Assembly of Amphiphiles

Amphiphiles are molecules having hydrophilic (water loving) and hydrophobic (water hating) segments joined together either covalently or noncovalently. Classical amphiphiles can be very broadly classified into ionic and nonionic amphiphiles. An ionic amphiphile typically consists of a polar charged head group, either cationic or anionic, attached to a long hydrocarbon chain, whereas a nonionic amphiphile has polyethylene glycol (PEG) or a similar polar chain as the hydrophilic group. The thermodynamic properties of amphiphiles in solution are primarily controlled by the strong tendency of hydrophobic tails to avoid any contact with water.

This unfavorable interaction can significantly be diminished by the aggregation of amphiphilic molecules in such a way that hydrophilic domains are exposed to water and hydrophobic parts are completely shielded from the aqueous environment. It is also to be noted that the hydrophobic association of amphiphile is entropically favorable. The hydrophilic parts interact with polar solvent molecules (water) through various noncovalent interactions, such as hydrogen boding and electrostatic interactions, for enthalpic gain. The morphology of the resulting nanostructures depends on the molecular geometry of the amphiphile, hydrophobic-to-hydrophilic ratio, and self-assembly conditions. By appropriate modification of these parameters, nanostructures with different morphologies can be achieved. For a classical amphiphile self-assembly, the shape of the nanostructure can be predicted using the analysis of the packing factor (*P*), which is defined as $P = V/A.l$, where *V* is the effective volume of the hydrophobic domain, *A* is the effective area of the hydrophilic segment at the aggregate-solution interface, and *l* is effective chain length of the hydrophobic domain (Fig. 4.2). The morphology of the nanostructure transforms from a spherical micelle at $P < 1/3$ to a cylindrical micelle at $1/3 < P < 1/2$, a vesicle at $1/2 < P < 1$, and a sheet at $P = 1$. Inverse micelles are formed at a packing parameter larger than 1 ($P > 1$) [16].

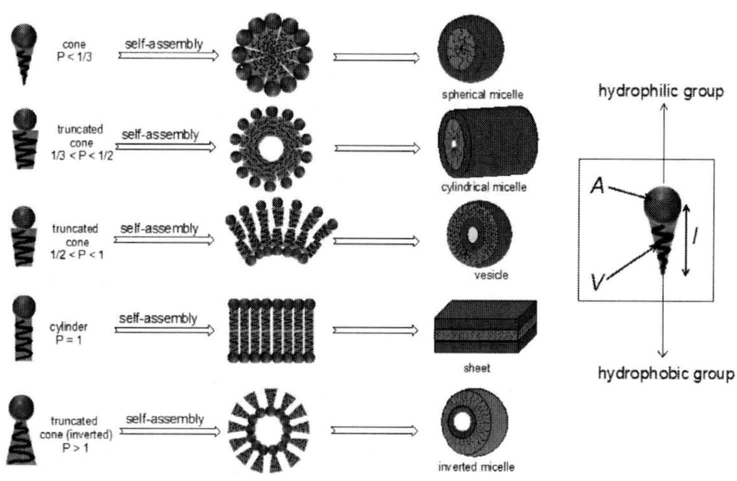

Figure 4.2 Schematic representation illustrating the prediction of self-assembly pathways of amphiphiles depending upon the shape and packing parameter (*P*) value.

4.3 DNA-Based Amphiphiles

The characteristic structural features of DNA, such as predictable secondary structure, nanoscale dimension, ease of synthesis, and molecular recognition properties, have greatly attracted researchers to use DNA as a building block in the design of DNA-based nanostructures. Very recently, DNA has also been applied as the hydrophilic segment in the design of DNA-based amphiphiles. The hydrophilic nature of DNA is due to its negatively charged anionic phosphate backbone. The conjugation of DNA to a hydrophobic moiety of interest covalently or noncovalently results in DNA-based amphiphiles, commonly known as DNA amphiphiles, capable of undergoing amphiphilicity-driven self-assembly in an aqueous medium into various nanostructures (Fig. 4.3). The morphology of the nanostructure can be predicted in most of the cases using the analysis of the packing parameter (P) as described for the self-assembly of classical amphiphiles. The most attractive feature of the self-assembly of DNA amphiphiles in an aqueous medium is the formation of a nanostructure with a corona made of hydrophilic DNA whereas the hydrophobic part of the amphiphile is completely shielded from the polar environment. DNA-based surface addressability of such DNA nanostructures allows the integration of other functional molecules onto their surface through sequence-specific DNA hybridization.

Hence, DNA nanostructures of this kind can act as soft nanoscaffolds for the organization of other functional molecules and nanomaterials in a reversible manner. Furthermore, a DNA nanostructure consists of a hydrophobic domain and a hydrophilic DNA shell. Undoubtedly, these nanostructures could be potential nanocarriers for drug delivery applications. This is because the hydrophobic domain of the nanostructure allows efficient encapsulation of a hydrophobic drug during self-assembly while the DNA shell provides excellent biocompatibility with the system. More importantly, the DNA shell of such a nanostructure offers a unique opportunity for targeted drug delivery by allowing the incorporation of specific-cell-targeting ligands onto the surface of the nanostructure through DNA hybridization or replacement of the random DNA sequence with a DNA or RNA aptamer for a specific target cell. Furthermore, DNA nanostructures exhibit excellent cell permeability and also show enhanced resistance toward enzymatic degradation. In this chapter, the discussion is restricted only to the self-assembly of DNA amphiphiles in an aqueous medium, and the reader is referred to other reviews for self-assembly of DNA amphiphiles in nonpolar solvents.

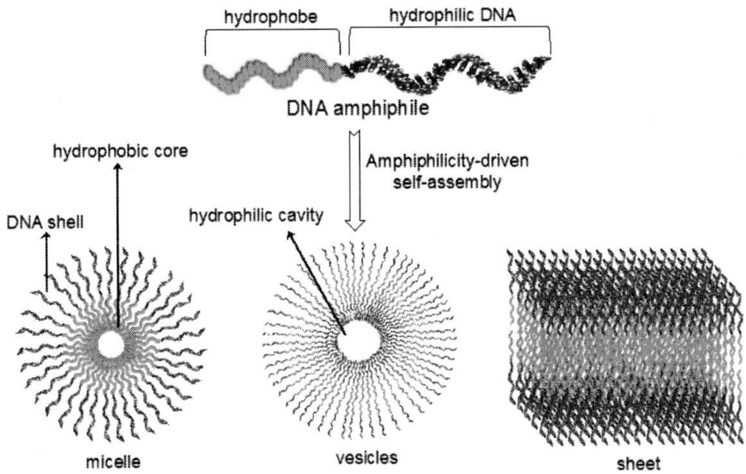

Figure 4.3 Schematic representation of the general structure of a DNA amphiphile and its self-assembly into different types of DNA-based nanostructures.

4.4 Different Types of DNA Amphiphiles

DNA amphiphiles are generally classified into linear and brush-type amphiphiles. In a linear-type amphiphile, the hydrophilic and hydrophobic segments are connected in an end-to-end fashion, whereas the brush-type structure consists of a linear backbone onto which multiple oligomer or polymer segments are grafted. There are different classes of linear and brush-type DNA amphiphiles reported. In a linear DNA amphiphile, the hydrophobe is attached to the hydrophilic DNA either through the 3′-end or the 5′-end in a linear fashion (Fig. 4.4a). On the other hand, a brush-type amphiphile consists of DNA grafted on a long hydrophobic polymer backbone (Fig. 4.4b) and a diblock copolymer consists of hydrophilic DNA grafted on a hydrophobic polymer backbone (Fig. 4.4c) [17]. Supramolecular interactions were also explored for the design and synthesis of a DNA-grafted bottlebrush-type polymer (Fig. 4.4d) [18].

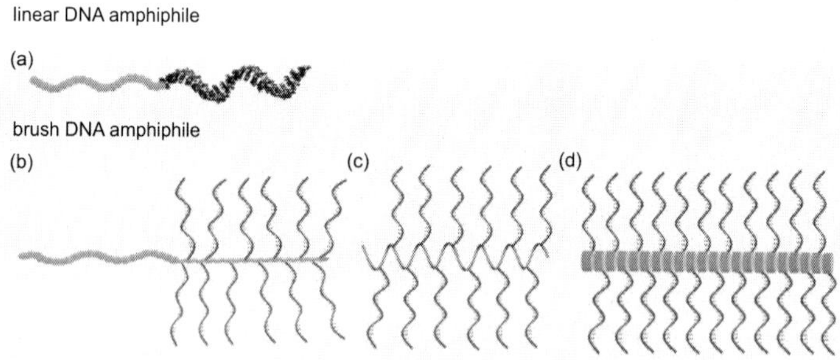

Figure 4.4 Pictorial representation of different types of DNA amphiphiles: (a) linear and (b–d) brush-type DNA amphiphiles.

4.5 Synthesis of DNA Amphiphiles

For the synthesis of a linear DNA amphiphile, one terminus of the oligonucleotide (3′ or 5′ terminus) needs to be conjugated to a hydrophobic segment either covalently or noncovalently. Oligonucleotides with up to 100 bases can easily be synthesized in a high yield by iterative chemical synthesis on solid supports using activated phosphoramidite building blocks on a DNA synthesizer. The four-step cycle in the phosphoramidite approach is (i) deblocking, (ii) coupling, (iii) capping, and (iv) oxidation. The solid phase technique not only allows the synthesis of an oligonucleotide of any sequence but also permits the incorporation of the desired functional group at any place during DNA synthesis, and the functional group–incorporated oligonucleotides are still present on the solid support. In the subsequent step, the chemically modified oligonucleotide grown on the support can either react on the solid phase directly for the conjugation of the hydrophobe or be initially deprotected from the support and then carry out reaction in the solution phase. Details of both approaches are discussed with relevant literature examples.

4.5.1 Solution Phase Synthesis

Due to the solvent incompatibilities of hydrophilic DNA and a hydrophobic segment, solution phase synthesis often fails or results in very poor yields on the conjugation of highly hydrophobic moieties to oligonucleotides, and hence solution phase synthesis is typically applied only for the conjugation of hydrophobic segments that are relatively polar in nature. Different coupling strategies have been demonstrated for the solution phase synthesis of DNA amphiphiles (Fig. 4.5). These include the reaction between acid and amine, the reaction between thiol groups, thiol-Michael addition, and standard phosphoramidite chemistry. In the case of acid-amine coupling, typically 5′-primary amine (–NH_2)–functionalized oligonucleotides couple with an activated acid (ester)–functionalized hydrophobic moiety at basic pH. Activation of the acid is generally achieved using common activating groups, such as *N*-hydroxysuccinimide ester (NHS), *N*-(3-dimethylaminopropyl)-*N*′-ethylcarbodiimide (EDC), *N,N*′-dicyclohexylcarbodiimide (DCC), or sulfo-N-hydroxysuccinimide ester (sulfo-NHS). Hoffman et al. have employed the acid-amine coupling strategy for the synthesis of DNA-*b*-poly(*N*-isopropylacrylamide) amphiphiles (**DNA1**) [19]. Similarly, Park et al. have demonstrated the synthesis of DNA-*b*-poly(D,L-lactic-co-glycolic acid) (**DNA2**) [20] and DNA-PEG (**DNA3**) [21] amphiphiles using an acid-amine coupling strategy. Coupling between a thiol-modified oligonucleotide and a thiol-functionalized hydrophobic segment produces a disulfide linkage (–S–S–) between the oligonucleotide and the hydrophobic segment. Since the thiol group easily undergoes air oxidation to disulfide, the thiol group needs to be protected or converted into another functional group to avoid air oxidation. Nagasaki et al. have used this strategy for the synthesis of PEG-*b*-DNA block copolymers in which the DNA and PEG segments are connected through a disulfide bond (**DNA4**) [22]. In their approach, the thiol group of PEG was converted to the corresponding 2-pyridyl-disulfide group to prevent the oxidation of the thiol group to the corresponding disulfide of PEG. The thiol-functionalized oligonucleotide was initially treated with dithiothreitol (DTT) to cleave any possible formation of disulfide of the oligonucleotides. The coupling of DTT-treated thiol-functionalized oligonucleotide and 2-pyridyl-disulfide-modified PEG in a phosphate buffer yielded

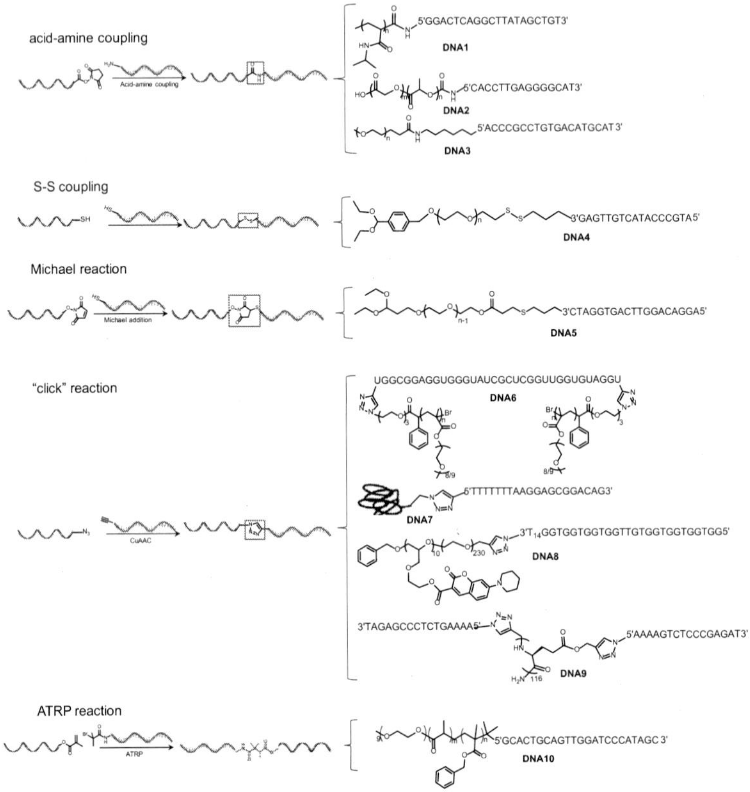

Figure 4.5 Different strategies for the synthesis of DNA amphiphiles in the solution phase.

the PEG-*b*-DNA conjugate connected through a disulfide bond. It is important that this coupling reaction be performed at basic pH. Alternatively, Michael addition was also very successful for the synthesis of amphiphilic DNA block copolymers. In the case of Michael addition, the nucleophile (commonly $-NH_2$ or –SH) functionalized at the 5′-end of the oligonucleotide reacts with an α,β-unsaturated carbonyl containing a hydrophobe at neutral pH to yield the corresponding DNA amphiphile. This approach generally results in an excellent yield of water-soluble hydrophobes such as PEG and poly(*N*-isopropylacrylamide). For example, Nagasaki et al. explored Michael addition for the synthesis of DNA-*b*-PEG amphiphiles (**DNA5**) [23], which was achieved by the reaction of thiol-functionalized oligonucleotide and acrylate-modified PEG at neutral pH. In this case, the hydrophobic moiety and the oligonucleotide were connected through an acid-labile ester linkage. Very recently, copper-catalyzed alkyne-azide cycloaddition (CuAAC) reaction, commonly known as "click" reaction, has emerged as a highly versatile approach for the biorthogonal functionalization of DNA and proteins, and this approach has been successfully applied for the synthesis of a DNA amphiphile as well. In this method, an oligonucleotide functionalized with alkyne or azide at the 5′-terminus of the oligonucleotide reacts with the azide- or alkyne-modified hydrophobe, respectively, to give the corresponding DNA-hydrophobe conjugate. This strategy yields DNA and hydrophobe connected through a very stable triazol linkage with exclusive formation of 1,4-regioisomer. Matyjaszewski et al. demonstrated the click chemistry approach for the synthesis of a series of DNA-based bolamphiphiles. The reaction between bis-alkyne, modified at the 3′-end and 5′-end of the oligonucleotide, and the azide-modified polymers yielded the corresponding polymer-*b*-DNA-*b*-polymer bolamphiphile (**DNA6**) [24]. The click chemistry approach has also been applied for the synthesis of DNA-silk conjugates (**DNA7**) by Humenik and coworkers [25]. Recently, Hawker et al. reported improved click chemistry by using a tricarboxylate-based copper-stabilizing ligand. In comparison to conventional CuAAC ligands, the use of a tricarboxylate-based ligand does not cause any chemical damage to the aptamers, leads to a higher yield, and offers excellent water solubility compared to traditional ligands for CuAAC. The reaction between 3′-azide-modified aptamer DNA and alkyne-modified ethylene glycol–based hydrophobic polymer under the click chemistry reaction conditions in the presence of a tricarboxylate-based ligand yielded the corresponding DNA amphiphile in good amounts (**DNA8**) [26]. Synthesis of bottlebrush-type DNA amphiphiles was also achieved using the click chemistry approach by Liu and coworkers (**DNA9**) [17]. In another interesting approach, Matyjaszewski et al. reported the solution phase synthesis of amphiphilic DNA-polymer block copolymer using atom-transfer radical polymerization (ATRP) as the key step of the synthesis. This approach involves the incorporation of the ATRP initiator (α-bromoisobutyryl) group to the 5′-terminus of the oligonucleotide during the solid phase DNA synthesis, and the subsequent deprotection of the oligonucleotide from the solid support yielded the corresponding ATRP initiator–incorporated oligonucleotide. The polymer was then grown from the DNA-bound initiator, both in solution, after the DNA initiator was released from the solid support or directly on the solid support itself (**DNA10**) [27].

4.5.2 Solid Phase Synthesis

Solvent incompatibility of solution phase synthesis can be overcome to a great extent by solid phase synthesis. The most attractive feature of this approach is that it allows the coupling of an oligonucleotide to a hydrophobe in an organic solvent. In general, this approach involves the synthesis of an appropriately functionalized oligonucleotide on a solid support such as controlled pore glass (CPG) and the subsequent reaction of the CPG-bound oligonucleotide with the hydrophobe of interest in a nonpolar or polar solvent. The deprotection of the hydrophobe-conjugated oligonucleotide from the solid support yields the corresponding hydrophobe-oligonucleotide amphiphile. Different chemical reactions were employed for the solid phase coupling reaction (Fig. 4.6). One of the most widely used reactions is phosphoramidite chemistry, which involves the conversion of a hydroxyl-terminated hydrophobe to the corresponding phosphoramidite upon reaction with phosphoramidite chloride. The phosphoramidite derivative of the hydrophobe is then coupled to the detritylated 5′ hydroxyl end of the oligonucleotide on the solid support using a DNA synthesizer. After deprotection from the solid support and removal of the protecting groups on DNA bases, the yield of DNA amphiphiles is good. In this case, the oligonucleotide and the hydrophobe are connected through a phosphate linkage. Mirkin et al. have demonstrated the synthesis of DNA-*b*-polystyrene block copolymers using the solid phase phosphoramidite chemistry [28]. In their approach, alcohol-terminated polystyrene was converted into the corresponding phosphoramidite and incorporated into the 5′-end of an oligonucleotide on a DNA synthesizer, followed by the deprotection of the oligonucleotide from the solid support, giving an excellent yield of the DNA-*b*-polystyrene amphiphile (**DNA11**) [28]. This strategy has later been explored by several research groups for the synthesis of different types of DNA amphiphiles. For example, Herrmann et al. reported the synthesis of DNA-*b*-polypropyleneoxide

(**DNA12**) [29] and DNA-*b*-polyfluorenes (**DNA13**) [30] amphiphiles. Liu et al. also explored this strategy for the synthesis of a series of DNA-poly(benzylether) dendron (**DNA14**)-based amphiphiles [31]. Using the commercially available hexaethylene glycol phosphoramidites, Sleiman et al. reported the synthesis of DNA amphiphiles containing dendritic hexaethyleneglycols as the hydrophobic chains (**DNA15**) [32]. Furthermore, Sleiman et al. demonstrated the potential of the solid phase approach for the synthesis of DNA amphiphiles having sequence-defined hydrophobic segments [33]. Recently, our group has shown the synthesis of a series of DNA-hexabenzocoronene (HBC)-based amphiphiles (**DNA16**) using the phosphoramidite approach [34]. Another chemical reaction that was widely used for the solid phase synthesis of DNA amphiphiles is the acid-amine coupling. In this approach, a CPG-bound amine-modified oligonucleotide reacts with an acid-modified hydrophobic polymer, followed by deprotection from the solid support, yielding the corresponding polymer-attached oligonucleotide. As in the solution phase coupling, the hydrophobe is attached to the oligonucleotide through an amide linkage. Gianneschi and coworkers used this approach for the synthesis of a good yield of DNA-3,4-di(octadecyloxy)benzene-based amphiphiles (**DNA17**) [35]. In a similar approach, Park et al. demonstrated the synthesis of amphiphilic DNA-*b*-poly(*N*-isopropylacrylamide) (**DNA18**) and DNA-*b*-poly(*N*-isopropylacrylamide)-*b*-poly(methylacrylate) (**DNA19**) di- and triblock copolymers, respectively [36]. A click reaction has also been shown to be an efficient approach for the synthesis of DNA amphiphiles. Our group has reported the synthesis of a DNA-oligo(phenylene ethynylene) hybrid amphiphile (**DNA20**) using the solid phase click chemistry [37]. In our approach, an alkyne functionality was incorporated into the DNA at the 5′-end using a commercially available alkyne-modified phosphoramidite. The reaction between the CPG-bound alkyne-functionalized DNA with an azide-functionalized alkyl chain–tethered oligo(phenylene ethynylene) in organic solvents (tetrahydrofuran [THF]) followed by the deprotection from CPG yielded DNA-oligo(phenylene ethynylene) hybrid amphiphiles. Later, we have applied this synthetic strategy for the synthesis of DNA-porphyrin (**DNA21**)- and DNA-merocyanine (**DNA22**)-based amphiphiles as well [38].

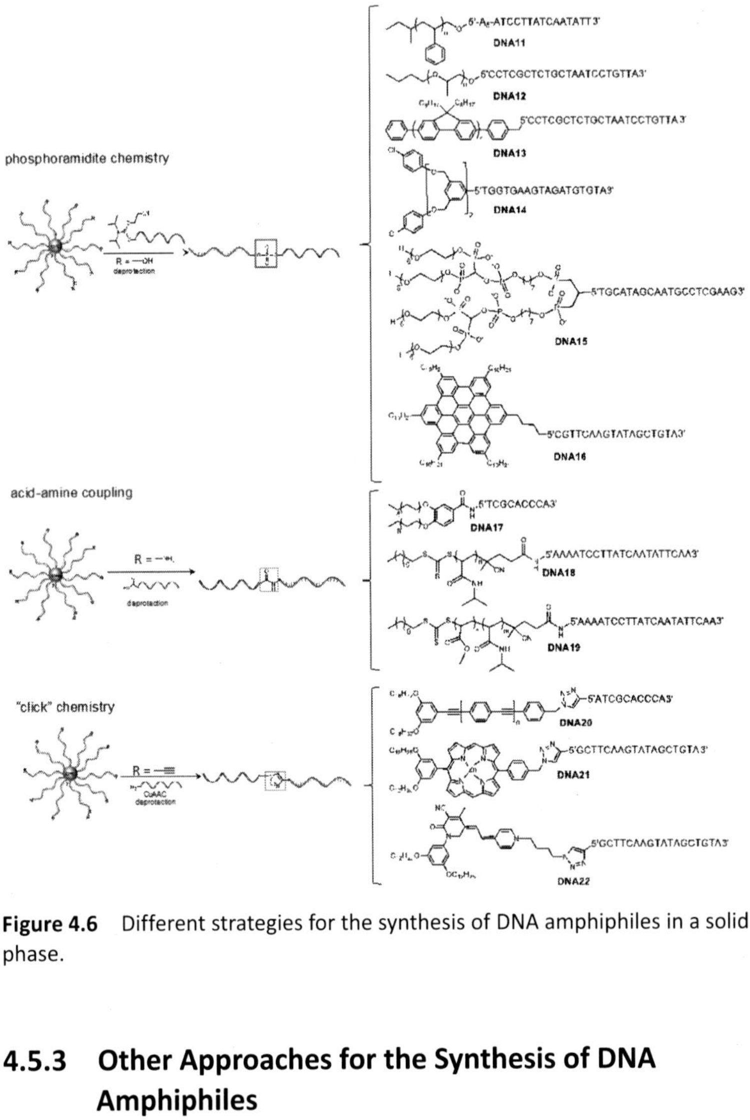

Figure 4.6 Different strategies for the synthesis of DNA amphiphiles in a solid phase.

4.5.3 Other Approaches for the Synthesis of DNA Amphiphiles

Synthesis of oligonucleotides of more than 100 bases is hard to achieve using a conventional DNA synthesizer. Hence, the synthesis of DNA amphiphiles with DNA segments larger than 100 bases as the hydrophilic segments is extremely difficult using the conventional solution or solid phase approaches. Herrmann et al. overcame this drawback by employing polymerase chain reaction (PCR) and demonstrated the syntheses of multiblock copolymers with prolonged DNA segments [39]. This approach involves the use of an ssDNA amphiphile, which was synthesized using the traditional solution or solid phase approach, as the primer for the PCR reaction. The utilization of one ssDNA-polymer amphiphile and a conventional oligonucleotide primer yielded a dsDNA diblock copolymer amphiphile with an extended nucleic acid block as the hydrophilic segment. Similarly, using two ssDNA-polymer amphiphiles with the same polymer segment as the forward and backward primers yielded polymer (X)-*b*-dsDNA-*b*-polymer (X) triblock amphiphiles having the same hydrophobic polymer (X) as the hydrophobic domains. On the other hand, the use of DNA amphiphiles having different polymers as primers yielded polymer (X)-*b*-dsDNA-*b*-polymer (Y) amphiphiles. Even more complex amphiphilic block polymers, such as pentablock amphiphiles, were also achieved by using ssDNA-*b*-polymer (X)-*b*-ssDNA and ssDNA-*b*-polymer (X or Y) as primers. DNA blocks having lengths even up to 1500 nucleobases were easily made by the appropriate annealing conditions during the PCR reaction. Authors have also investigated the PCR tolerance of different hydrophobic segments and found that PEG, polypropyleneoxide, polystyrene, and poly(*N*-isopropylacrylamide) are tolerable under PCR conditions. It is also worth noting that the block copolymers synthesized using the PCR approach were monodisperse and high molecular weight and had defined DNA sequences (Fig. 4.7) [39].

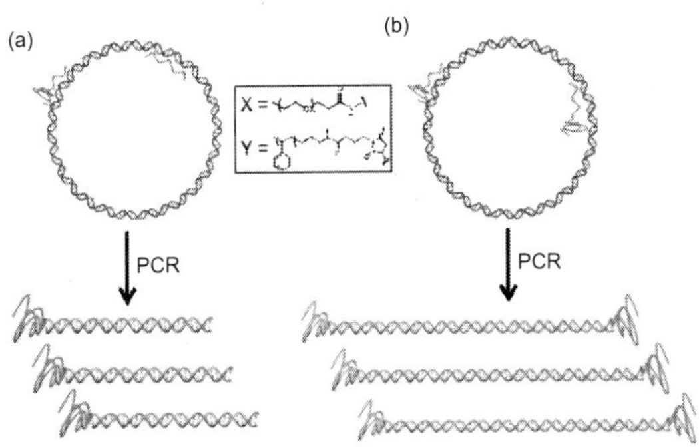

Figure 4.7 Schematic representation of the PCR-based synthesis of (a) DNA diblock and (b) DNA triblock copolymers. Reprinted with permission from Ref. [39]. Copyright 2007, John Wiley and Sons.

Though the solution phase and solid phase approaches allow the incorporation of highly hydrophobic segments into an oligonucleotide through an appropriate chemical reaction, the yield of the reaction dramatically decreases with the increase in hydrophobicity. Though the PCR-based approach is an alternative for solution and solid phase syntheses, not all hydrophobic segments are tolerable under PCR reaction conditions and hence the PCR-based approach cannot be considered as a very general approach for the synthesis of DNA amphiphiles. The incorporation of a highly hydrophobic segment into an oligonucleotide with a reasonable yield is extremely challenging. Sleiman et al. reported an efficient micellar-templated approach for the incorporation of highly hydrophobic molecules into oligonucleotides with a high yield under mild reaction conditions [40]. This approach involves the initial synthesis of a DNA amphiphile using one of the conventional approaches, which self-assembles into a micellar nanostructure. The hydrophobic core of the micelle is then used as a reaction chamber. The hybridization of a primary amine functionalized oligonucleotide with the DNA on the surface of the micelle brings the functional group (NH_2) oriented toward the hydrophobic core of the micelle. Subsequently, an activated acid–functionalized hydrophobe that is to be incorporated into the DNA is added into the micelle, which is efficiently encapsulated inside the hydrophobic cavity due to hydrophobic interaction. Thus, the DNA micelle brings the reactant in a very high concentration into the hydrophobic cavity and increases the rate of the reaction significantly. The authors have demonstrated the efficient conjugation of different types of hydrophobes, such as long alkyl chains, branched alkyl chains, pyrene, and polystyrene, to oligonucleotides (Fig. 4.8) [40].

We have recently reported a noncovalent approach for the synthesis of DNA amphiphiles, where the hydrophilic DNA and the hydrophobic moiety are connected through a noncovalent bond instead of a covalent bond. We have explored the host-guest interaction between β- cyclodextrin (β-CD) and adamantane for the noncovalent synthesis of a series of DNA amphiphiles. The host-guest interaction between β-CD and adamantane has been explored extensively in supramolecular chemistry due to their efficient molecular recognition with a high association constant (K_a = 10^4–10^5 M^{-1}). Our approach involves the separate synthesis of β-CD-functionalized DNA (host, **DNA-β-CD**) and adamantane functionalized hydrophobe (guests, **1** and **2**), followed by the self-assembly of host and guest in an aqueous medium. It is to be noted that the host DNA is water soluble whereas the guest hydrophobe is insoluble in water. However, molecule recognition between host and guest in water resulted in the spontaneous formation of a supramolecular DNA amphiphile. The modular nature of the synthesis allows the efficient coupling (~77%) of DNA with any hydrophobic segment of interest through self-assembly and avoids the time-consuming chemical synthesis to some extent. Since the hydrophilic DNA and the hydrophobic segment are synthesized separately and conjugate through self-assembly, solvent incompatibility can be completely avoided (Fig. 4.9) [41].

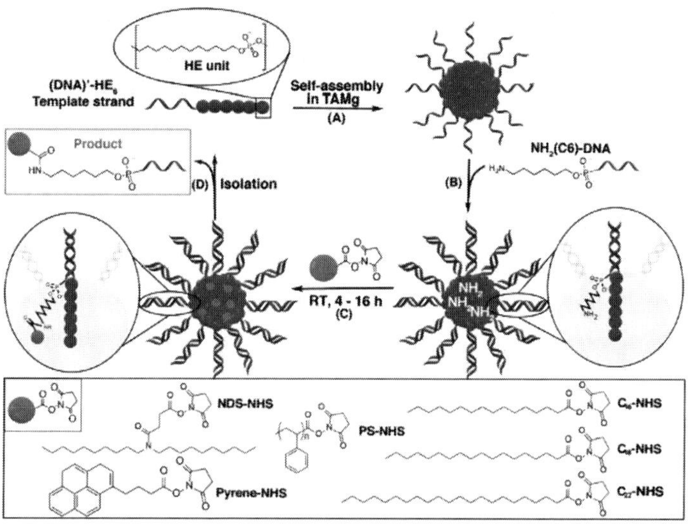

Figure 4.8 DNA micelle–templated synthesis of DNA amphiphiles. Step 1: Self-assembly of a DNA amphiphile into micelles. Step 2: Hybridization of amine-modified DNA, with DNA micelle bringing the amine functional group into the hydrophobic core. Step 3: Incubation of activated hydrophobe. The DNA micelle allows efficient coupling of the hydrophobe with amine-modified DNA, yielding the DNA amphiphile. Reproduced from Ref. [40] with permission of The Royal Society of Chemistry.

Herrmann et al. reported another coupling strategy based on DNA-surfactant interaction for the efficient conjugation of DNA with different hydrophobe molecules in an organic medium. This approach involves the separate synthesis of DNA-surfactant complexes by electrostatic complexation of 3′-amino-modified oligonucleotides with the cationic surfactant didodecyldimethylammonium bromide (DDAB) in the aqueous phase. Complex formation leads to the precipitation of the DNA from the aqueous solution and allows the recovery of the surfactant-coated DNA by centrifugation, and the DNA-DDAB complex is soluble in organic solvents (*N*,*N*-dimethylformamide [DMF], dimethylsulfoxide [DMSO], THF, and $CHCl_3$). Separately, activated ester derivatives of hydrophobic molecules were prepared by the reaction of a carboxylate group with NHS. The solution phase acid-amine coupling reaction between DNA-DDAB and the hydrophobic moiety was carried out in an organic solvent, after which the products were recovered by means of highly concentrated salt solution treatment to release the DNA from the surfactant shell. This is a very general approach for the synthesis of a high yield of DNA amphiphiles. This strategy suggests that the synthesis of DNA amphiphiles can be achieved without a DNA synthesizer (Fig. 4.10) [42].

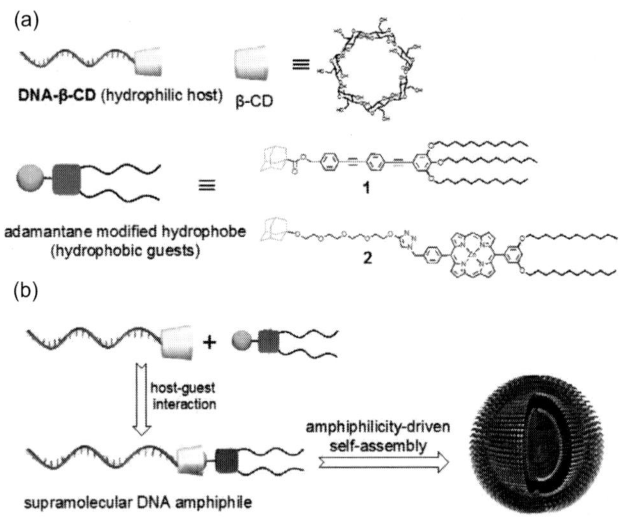

Figure 4.9 Noncovalent synthesis of DNA amphiphiles. (a) Structures of DNA-based β-CD host (**DNA-β-CD**) and adamantane-based guest molecules (**1** and **2**). (b) Noncovalent synthesis of a DNA amphiphile through host-guest interaction between β-CD and adamantane and its self-assembly into a DNA-based vesicle. Reprinted from Ref. [41] with permission of The Royal Society of Chemistry.

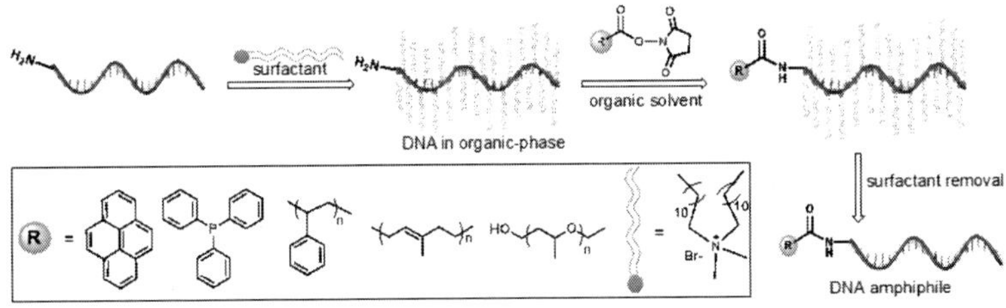

Figure 4.10 Synthesis of DNA amphiphiles in an organic medium. Amine-modified DNA is precipitated out of the aqueous phase through complexation with cationic surfactants. The insoluble complex is then extracted and redissolved in an organic solvent. The activated ester form of the hydrophobe is then coupled to the terminal amine of the DNA in the homogeneous organic phase.

4.6 Self-Assembly of DNA Amphiphiles

One of the unique features of DNA amphiphiles is their spontaneous self-assembly in an aqueous medium into diverse nanostructures. The main driving force behind the self-assembly is the microphase separation of the hydrophobic segment of the amphiphile to reduce its exposure to water medium. The morphology of the nanostructures depends upon the self-assembly conditions and the hydrophilic-to-hydrophobic ratio of the amphiphile. Nanostructures such as spherical micelles, cylindrical micelles, vesicles, tapes, twisted tapes, tubes, and sheets were achieved by the appropriate design of the DNA amphiphile. The most remarkable feature of this class of nanostructures is the external stimuli responsive nature with respect to temperature, pH, enzymatic action, slat concentration, etc. A very detailed discussion of different nanostructures reported from the self-assembly of DNA amphiphiles is provided.

4.6.1 Micellar (Spherical and Cylindrical) Nanostructures

A DNA amphiphile with a flexible polymer as the hydrophobic chain with a packing parameter $P < 1/3$ typically self-assembles into spherical micellar nanostructures. Micelle formation was achieved either through the spontaneous assembly of the amphiphile in water (or in a buffer) or with the slow addition of water into the DMF solution of the amphiphile. Gel electrophoresis, dynamic light scattering (DLS), transmission electron microscopy (TEM), atomic force microscopy (AFM), and small-angle X-ray scattering (SAXS) analyses were usually used for the characterization of DNA nanostructures. Spherical micelles were obtained from the self-assembly of a series of linear DNA block copolymer amphiphiles, including **DNA2** [20], **DNA11** [28], DNA-*b*-poly(*N*-isopropylacrylamide) (**DNA18**) [36], and **DNA12** [19]. In all these cases, the hydrophobic polymer segment forms the core of the micelle, with the shell made of hydrophilic DNA (Fig. 4.11).

Spherical micelle formation was observed for the self-assembly of DNA-brush copolymer amphiphiles as well. In an interesting report, the spherical micelle formed from the self-assembly of the DNA-brush copolymer (**DNA23**) was shown to transform into a cylindrical micelle upon enzymatic reaction in a reversible way. The design features of DNA-brush block copolymer include an RNA base as an enzymatic cleavage site and two ethylene glycol moieties that increase the steric bulk of the hydrophilic block. To facilitate a spherical-to-cylinder micellar transition, the spherical micelles were mixed with a DNA-based phosphodiesterase (DNAzyme), which was designed to recognize a given DNA sequence and cut at an RNA base. This process resulted in the complete, rapid catalytic turnover of the DNA substrate that formed the bulk of the hydrophilic block and generated a truncated ssDNA sequence. A subsequent spherical-to-cylinder phase transition occurs in the new amphiphile pack in a different way. Furthermore, the reversible cylinder-to-spherical micellar transition was achieved with the addition of a 19-base input DNA (**In1**). This sequence was designed to form a 9-base duplex with the truncated DNA in the cylinder shell. Subsequent cylindrical-to-spherical micellar transition occurs as the bulky, extended duplex is better accommodated in the spherical micelle phase. Therefore, the new structure contains a partial DNA duplex that can be used for the spherical-to-cylindrical phase transition utilizing the DNA strand replacement approach. This was achieved by the addition of a 19-base ssDNA, which is completely complementary to **In1**, and the more thermodynamically stable 19-base-long duplex departs, leaving the truncated ssDNA amphiphile to reassemble into the cylindrical micelle (Fig. 4.12a) [43].

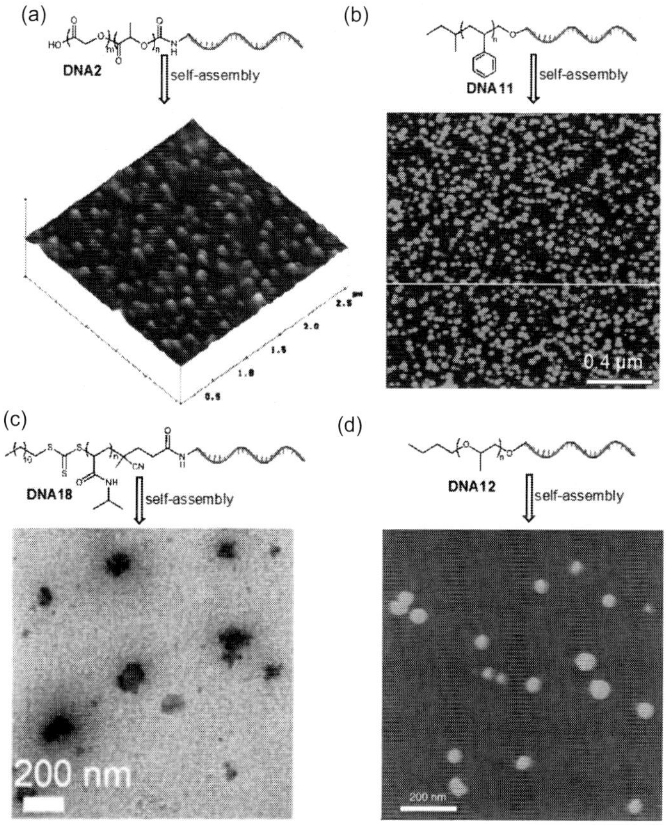

Figure 4.11 Micelles derived from the self-assembly of (a) DNA-*b*-poly(D,L-lactic-co-glycolic acid) (**DNA2**), (b) DNA-*b*-polystyrene (**DNA11**), (c) DNA-*b*-poly(*N*-isopropylacrylamide) (**DNA18**), and (d) DNA-*b*-polypropyleneoxide (**DNA12**). (a) Reprinted with permission from Ref. [20]. Copyright (2001) American Chemical Society. (b) Reprinted with permission from Ref. [28]. Copyright (2004) American Chemical Society. (c) Reprinted with permission from Ref. [36]. Copyright (2016) American Chemical Society. (d) Reprinted with permission from Ref. [29]. Copyright 2006, John Wiley and Sons.

In another report, the structural transformation associated with C-rich DNA with respect to pH was used for the spherical micelle–to–cylindrical micelle transformation. The DNA-*b*-PPO amphiphile (**DNA24**) contains a pH-responsive C-rich DNA as the hydrophilic segment, which can form an intermolecular "i-motif" structure through cytosine–protonated cytosine ($C:C^+$) base pairing at acidic

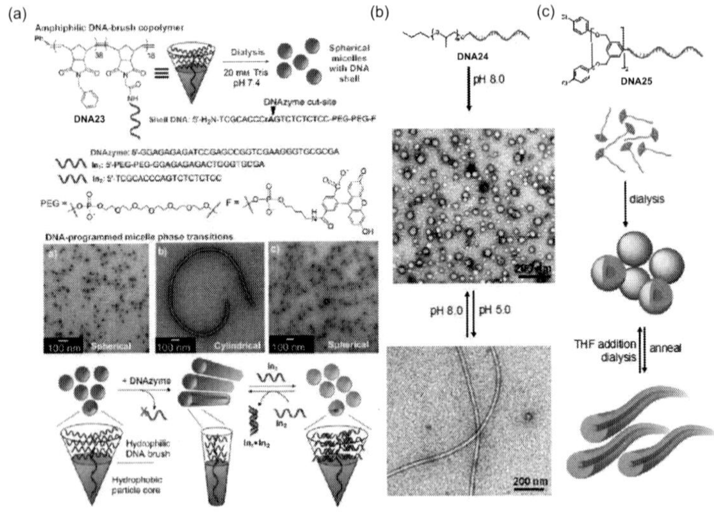

Figure 4.12 (a) Self-assembly of DNA-brush copolymers (**DNA23**) into micelles with spherical or cylindrical morphologies. TEM images of spherical micelles assembled from initial DNA-brush copolymers (bottom left); cylindrical morphology formed following DNAzyme addition to spheres (bottom center); spherical micelles (bottom right) formed after addition of **In1** to cylindrical micelle. (b) pH-controlled morphological transformation of a cytosine-rich DNA amphiphile. (c) Self-assembly condition–induced reversible morphological transformation of a DNA amphiphile. (a) Reprinted with permission from Ref. [43]. Copyright 2010, John Wiley and Sons. (b) Reprinted from Ref. [44] with permission of The Royal Society of Chemistry. (c) Reprinted from Ref. [45] with permission of The Royal Society of Chemistry.

pH (pH ≤ 6.3) and dissociate into random coils with increasing pH to a slightly basic condition in a completely reversible manner. Accordingly, the DNA-*b*-PPO behaves as a diblock amphiphile at basic pH and undergoes amphiphilicity-driven self-assembly into spherical micelles. On the other hand, at acidic pH C-rich DNA undergoes structural transformation through intermolecular i-motif formation into a triblock amphiphile, which subsequently self-assembles into a cylindrical micellar nanostructure. These structural transformations are found to be completely reversible with respect to the change in pH of the solution for several cycles (Fig. 4.12b) [44]. A spherical micelle–to–cylindrical micelle transformation was also achieved by changing the solvent polarity in the self-assembly of DNA-dendron hybrid amphiphiles (**DNA25**). Dissolving **DNA25** in a 1:1 mixture of

THF and water and the slow addition of water to 90%, followed by dialysis to remove the residual THF, yielded spherical micelles. On the other hand, annealing of the spherical micelle solution at 90°C for 30 min. in the presence of 5 vol% dichloromethane (DCM) yielded cylindrical micelles. The morphological transformations were found to be reversible for several cycles with the change in solvent polarity (Fig. 4.12c) [45].

4.6.2 Vesicular Nanostructures

With the increase in the hydrophobic volume of the amphiphile ($P = 1/2–1/3$), generally vesicular nanostructures are formed. The main driving force behind the assembly, and also to reduce the exposure of the hydrophobic segment to the water medium, is the hydrophobic interactions. A vesicular nanostructure consists of a hydrophobic membrane formed from the lamellar organization of the amphiphile and a hydrophilic DNA shell, which is exposed to the polar medium on either side of the hydrophobic membrane. Hence, a vesicle has a hydrophilic internal cavity and a hydrophobic membrane (unilamellar vesicle). Vesicular nanostructures are typically characterized through a TEM experiment, which generally provided significant contrast difference for the membrane and the inner part of the vesicle. Furthermore, matching of the experimental membrane thickness from the TEM with the calculated value confirms the vesicular nanostructure. In most of the cases a perfect match is observed, and the deviations observed in exceptional cases were attributed to the flexibility of the molecular components. The micellar or vesicular nature of the spherical particle can easily be distinguished by comparing the experimental diameter of the particle and the calculated bilayer packing distance, which is equal to twice the molecular length of the amphiphile. For the micellar assemblies, the particle size would be equal to the bilayer packing distance, whereas in the case of vesicular assemblies, the particle size would be significantly larger than the bilayer packing distance, because the bilayer distance in the case of vesicles represents only the wall thickness of the vesicles. The first example of a DNA-based vesicular nanostructure was reported for the self-assembly of a cytidine$_{12}$-*b*-poly(butadiene) hybrid amphiphile (**DNA26**), which self-assembled into vesicle of ~100 nm in diameter (Fig. 4.13a) [46].

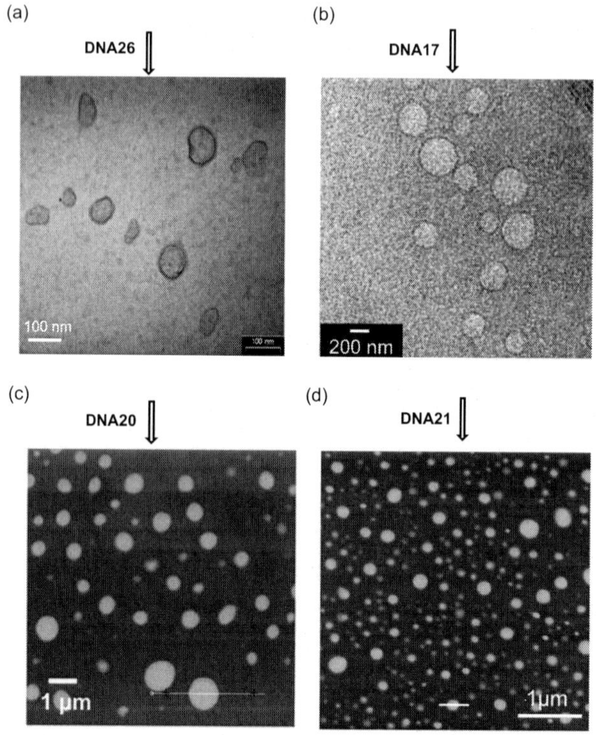

Figure 4.13 Self-assembly of a DNA amphiphile into vesicles. DNA vesicle derived from the self-assembly of (a) polybutadine, (b) dialkyloxybenzene, (c) alky chain–tethered phenylene ethynylene, and (d) alkyl chain–tethered porphyrin-based DNA amphiphiles. (a) Reprinted from Ref. [46] with permission of The Royal Society of Chemistry. (b) Reprinted with permission from Ref. [35]. Copyright (2010) American Chemical Society. (c) Reprinted with permission from Ref. [37]. Copyright 2014, John Wiley and Sons. (d) Reprinted from Ref. [38] with permission of The Royal Society of Chemistry.

Later, several examples of DNA-based vesicles were reported from different research groups. For example, **DNA17** self-assembles into unilamellar vesicles with diameters in the range of ~500 nm, as evident from various microscopic and light scattering experiments (Fig. 4.13b) [35]. As observed in most of the cases, though exceptions are there, DNA amphiphiles having flexible hydrophobic chains typically self-assemble into DNA micelles. In support of this, we have demonstrated that the incorporation of π-surface as part of the hydrophobic domain can facilitate the self-assembly through strong π-π stacking, which directs the self-assembly in a lamellar fashion to vesicle or 2D sheets (lamellar nanostructures). Self-assembly of a series of DNA amphiphiles containing a π-conjugated backbone as part of the hydrophobic segment such as **DNA20** [37] and **DNA21** [38] amphiphiles was studied, and it was found that this class of amphiphiles self-assembles into vesicular nanostructures, as evident from DLS, AFM, and TEM analyses (Fig. 4.13c,d).

4.6.3 Other DNA Nanostructure

Other morphologies that were achieved from the self-assembly of DNA amphiphiles include fibers, tapes, twisted ribbons, tubes, and 2D sheets (Fig. 4.14). Several micrometer-long fibers were observed from the self-assembly of **DNA14**, in which the periphery of the dendron is functionalized with multiple chlorine atoms (Fig. 4.14a). The fibers are several micrometers in length, with a uniform diameter of ~10 nm. The fiber formation for this DNA amphiphile was attributed to the cooperative assembly of the amphiphile through several noncovalent interactions, such as hydrophobic interaction, π-stacking, and halogen bonding. The proposed molecular-level structure of the fibers consists of a hydrophobic core made of a self-assembled dendron segment, whereas the hydrophilic DNA is exposed to the aqueous medium [31]. In another report, self-assembly of DNA aptamer-dihexadecyl lipid (C_{16})-based amphiphiles (**DNA27** and **DNA28**) with varying spacer lengths was systematically investigated. The amphiphile with the shorter, hydrophilic (PEG based) linker (**DNA27**) self-assembles into spherical micelles (Fig. 4.14b). Interestingly, flat, ribbon-like tapes and twisted tapes were obtained when the linker connecting the aptamer DNA and dilakyl C_{16} chains was replaced with longer, hydrophobic chains (**DNA28**) (C_{12} and C_{24} alkyl chains) [47]. The observed nanotape structures had lengths in the range of 100–1000 nm and widths between 25 and 100 nm. The thickness of the tapes was found to be ~15 nm (Fig. 4.14c). The influence of the sequence of hydrophilic DNA on the self-assembly behavior of DNA amphiphiles was also studied (Fig. 4.14d). A series of DNA with different sequences was synthesized keeping the hydrophobic segment the same for this purpose, and their self-assembly was investigated. This study revealed that a

DNA amphiphile having a DNA sequence lacking a guanine base and containing a C_{12} hydrophobic linker (**DNA29**) self-assembled into twisted nanotapes as the major structure. However, detailed TEM analyses have shown that helical nanotapes and nanotubes were also formed. Each of these structures is formed from the bilayers of amphiphiles, with the hydrophobic tails organized into an interior core and the ssDNA head groups forming an outside shell. Transition from a twisted to a helical nanotape morphology requires a change in the membrane curvature from saddle-like to cylindrical, which is attributed to a rearrangement of the individual amphiphiles into a molecular organization that is more ordered or crystalline. The forces that are often identified as causing the order or crystallinity are hydrogen bonding and π–stacking between individual amphiphiles, although electrostatic and hydrophobic forces are also likely important. The C_{12} spacer may also be helping to ensure that the amphiphiles can organize into crystalline or well-ordered bilayers by extending the large ssDNA head groups away from the interface and relieving some of the electrostatic or steric constraints that could impede close and ordered packing of the amphiphiles. This may be especially important in the case of the head groups without guanine bases that do not appear to form significant interactions with each other. However, the self-assembly pathway may be completely different when the head group DNA has nucleobases that can interact with each other through specific noncovalent interaction, such as hydrogen bonding. For example, guanine bases can form G-quadraplex structures through hydrogen-bonding interactions. Hence, the incorporation of G-rich DNA as the hydrophilic head group offers the opportunity for the formation of parallel G-quadraplex structures. Amphiphiles having $(GGGT)_3$-modified DNA as the head group form only tapes and twisted tapes, and helical nanotapes or nanotubes were not observed [48]. Helical ribbons were also obtained from the self-assembly of the DNA-oligo(pyrene) amphiphile (**DNA30**) [18] in an aqueous medium, and the hydrophobic interactions and π-stacking of pyrene are the main driving force for the ribbon formation. In this case, the ribbons were ~500 nm in length and had a helical pitch of ~50 nm, and the dimensions of the ribbon depend heavily on the ionic strength of the medium (Fig. 4.14e). Very recently, we have reported the fabrication of extremely high-aspect-ratio 2D crystalline DNA sheets from the self-assembly of **DNA16**. The formation of the sheet structure is attributed to the strong π-stacking of HBC in an aqueous medium. A sheet structure consists of a DNA-based hydrophilic surface and a functional core that results from the self-assembly of the alkylated HBC segment through π-stacking and van der Waals interactions (Fig. 4.14f) [34].

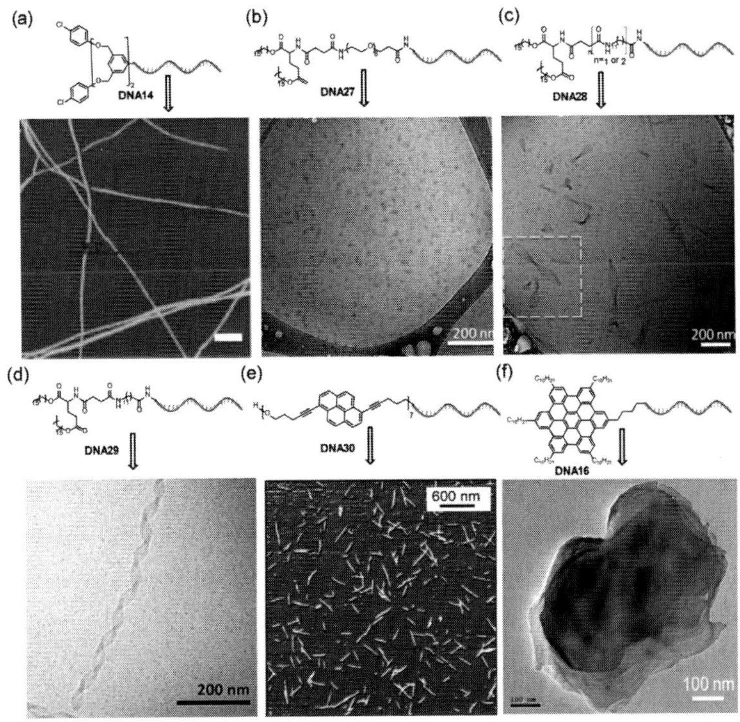

Figure 4.14 (a) Self-assembly of a DNA-dendron amphiphile into fibrous nanostructures. (b–d) Self-assembly of a DNA amphiphile into different nanostructures depending upon the spacer length/hydrophobicity and the DNA sequence. (e) Self-assembly of a DNA-oligo(pyrene) and DNA-hexabenzocoronene amphiphile into a twisted ribbon. (f) Self-assembly of a DNA-oligo(pyrene) and DNA-hexabenzocoronene amphiphile into a 2D sheet. (a) Reprinted from Ref. [31] with permission of The Royal Society of Chemistry. (b–d) Reprinted from Refs. [47, 48] with permission of The Royal Society of Chemistry. (e) Reprinted with permission from Ref. [18]. Copyright 2014, John Wiley and Sons. (f) Reprinted with permission from Ref. [34]. Copyright (2017) American Chemical Society.

4.7 Applications of DNA Nanostructures

The unique structural features of DNA nanostructures formed from the self-assembly of DNA amphiphiles include the following: (i) there is an extremely dense display of ssDNA on the surface of the nanostructures, which allows the reversible integration of other functional molecules/nanomaterials onto the surface of the nanostructures, (ii) it consists of a hydrophobic core/membrane that allows efficient encapsulation of hydrophobic guest molecules, (iii) surface decoration of the nanostructures with ssDNA provides excellent biocompatibility, and (iv) the structures exhibit resistance against enzymatic degradation due to the extremely dense array of ssDNA. Due to these remarkable structural properties, DNA nanostructures of this kind have found potential applications in material science, nanotechnology, and medicine.

4.7.1 Applications in Medicine

Since the micelle/vesicle derived from the self-assembly of DNA amphiphiles consists of a hydrophobic membrane and a hydrophilic DNA shell, undoubtedly, these nanostructures could be potential nanocarriers for drug delivery applications. This is because the hydrophobic membrane of the vesicle allows efficient encapsulation of hydrophobic drugs during self-assembly while the DNA shell provides excellent biocompatibility with the system. More importantly, the DNA shell of such a nanostructure offers a unique opportunity for targeted drug delivery by allowing the incorporation of specific-cell-targeting ligands onto the surface of the vesicle through DNA hybridization or replacement if the random DNA sequence with a DNA or RNA aptamer for a specific target. It has also been reported that DNA micelles are resistant to nuclease digestion due to the dense packing of DNA on the surface of the nanostructures [55]. Recent years have witnessed the emergence of different DNA nanostructures as carriers in several biomedical applications (Fig. 4.15). For example, DNA micelles obtained from the self-assembly of **DNA12** were demonstrated as drug carriers for the targeted killing of Caco-2 cancerous cell (Fig. 4.15a). The Caco-2 cell line is a folate-receptor-bearing cell line, and the folate residue was used as the targeting ligand. For this purpose, micelles were initially decorated with folate through DNA hybridization. Subsequently, doxorubicin (anticancer drug) was loaded into the hydrophobic core of the micelle through hydrophobic interactions and the drug-loaded micelles were found to efficiently deliver the anticancer drug selectively into the Caco-2 cell. The mechanism of cell penetration of the drug-loaded micelle was believed to be through receptor-mediated endocytosis [49]. In another interesting approach, the DNA aptamer specific for the Ramos cell line (B-cell lymphoma cell line) was used as the hydrophilic segment for the design of the DNA amphiphile, and hence the self-assembled micelle has an aptamer decoration on its surface. In this approach, the DNA aptamer segment acts not only as the hydrophilic domain but also as a targeting agent for cancer therapy. The advantages of aptamer-based micelles include enhanced binding affinity to the target cancer cell, rapid targeting ability, and high sensitivity [50]. In another report, DNA micelles were explored as a soft template for the assembly of cowpea chlorotic mottle virus (CCMV) capsids with $T = 1$ and 2 geometries at neutral pH. The resulting nanostructures were also found to be stable against dilution. Furthermore, both hydrophilic- and hydrophobic-molecule-loaded micelles acted as the template for the assembly of the CCMV capsid. This report represents a general and facile supramolecular loading strategy for hydrophobic or hydrophilic small molecules within the protein nanocontainer, an essential process for the application of these nano-objects in biomedicine [51]. DNA micellar self-assembly was also demonstrated for the encapsulation and controlled release of a hydrophobic guest molecule in a DNA cage (Fig. 4.15b). Self-assembly of the DNA-dendritic amphiphile (**DNA31**) below the critical micellar concentration (CMC) of the amphiphile in the presence of an addressable DNA cage resulted in the positioning of dendritic alkyl chains at specific sites on the cage with a predefined number of alkyl chains. Depending upon the number and position of the DNA amphiphiles present on the cage surface, intermolecular or intramolecular assembly of the alkyl chain occurs through hydrophobic interactions. For example, four amphiphiles assembled on a DNA cube face selectively undergo intermolecular association, resulting in the dimerization of two cubes. On the other hand, eight amphiphiles assembled on opposite faces of a cube assemble together intramolecularly to form a spherical micellar environment inside the DNA cage. These structures differ from conventional micelles in that they are monodisperse and stabilized by encapsulating inside the DNA cage. More importantly, these scaffolded micelles allow the encapsulation and release of the hydrophobic payload, as the core-forming units can be erased from the scaffold by DNA strand displacement. This unique class of micelles can potentially act as smart nanocarriers for drug delivery applications [52].

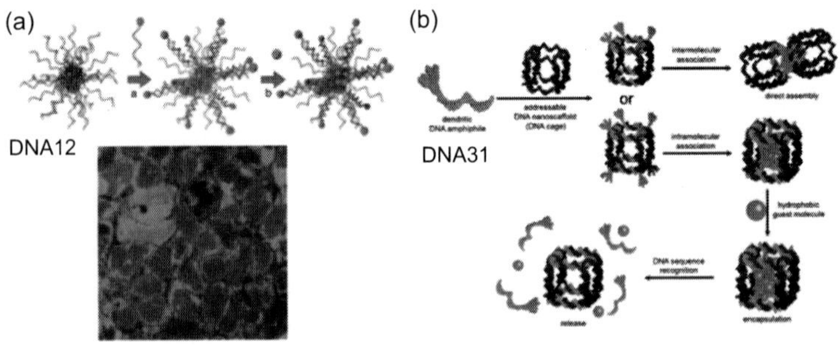

Figure 4.15 (a) Schematic representation of the drug delivery system based on DNA block copolymers. Targeting units (red dots) that are connected to the complementary sequence of the micelles are hybridized to decorate the micellar surface with cancer cell–targeting ligands. The anticancer drug (green dots) is loaded into the core of the micelles through hydrophobic interactions. (b) Positioning of DNA amphiphiles at defined sites on a DNA cage and their assembly into a micellar nanostructure inside the DNA cage. The encapsulation of the hydrophobic guest and its release using DNA hybridization as the trigger is also shown. (a) Reprinted with permission from Ref. [49]. Copyright 2008, John Wiley and Sons. (b) Reprinted by permission from Springer Customer Service Centre GmbH: Springer Nature, *Nature Chemistry*, Ref. [52]. Copyright (2013).

4.7.2 Material Science

Typically, the dense display of ssDNA on the surface of the nanostructures has been mainly explored for the application of DNA nanostructures in material science (Fig. 4.16). For example, cylindrical micelles formed from the self-assembly of **DNA23** were successfully applied as a template for the synthesis of inorganic nanowire (Fig. 4.16a). The key structural feature of the cylindrical micelle for this purpose is the incorporation of cyclic disulfide bonds in the DNA sequence. Hence, the protruded ssDNA on the surface of the cylindrical micelle allows for the efficient binding of Au NPs onto the surface of the micelle through Au-S bonding. Accordingly, the reaction between cylindrical DNA micelles and 2 nm citrate-stabilized Au NPs resulted in the formation of crystalline gold nanowires [53]. Interestingly, the template effect was not observed for spherical micellar nanostructures. The protruding ssDNA on the surface of the nanostructures can also be used for the assembly of other functional molecules through sequence-specific DNA hybridization. For example, vesicles derived from the self-assembly of DNA-oligo(phenylene ethynylene) hybrid amphiphiles (**DNA20**) were used for the assembly of Au NPs of different sizes (Fig. 4.16b). This was achieved by the self-assembly of DNA vesicles and Au NPs (1:1 molar ratio) that are surface-modified with DNAs completely complementary to the DNA on the surface of the vesicles. Hybridization of the DNAs led to the decoration of the vesicular surface with Au NPs. Extremely high loading of NPs onto the surface of the vesicles was observed due to the dense display of ssDNA on the vesicular surface. Dense packing of NPs resulted in weak interparticle surface plasmon coupling, as evident from the red-shift in the absorption maximum of the assembled particles when compared with the bare NP. However, no aggregation was observed for the NPs due to the repulsion between the negatively charged DNAs [37]. In a similar approach, NPs were assembled on the faces of DNA nanosheets derived from the self-assembly of **DNA16** [34]. DNA-based surface addressability of the vesicle was also explored for the design of a thermally reversible light-harvesting antenna (Fig. 4.16c). Vesicles obtained from the self-assembly of the **DNA20** amphiphile was used as the donor scaffold for this design, and ethynylpyrene was selected as the acceptor, considering the good spectral overlap between the emission spectrum of DNA vesicle (donor) and the absorption spectrum of ethynylpyrene (acceptor). Initially the acceptor molecule was conjugated to a DNA that is completely complementary to the DNA on the surface of the donor vesicle. Accordingly, the self-assembly of a 1:1 molar solution of the vesicle and the acceptor DNA places the acceptor molecule at a distance of ~5 nm from the chromophore stacks. Interestingly, efficient fluorescence resonance energy transfer (FRET) was observed upon selective excitation of the donor molecule. More importantly, since the donor and acceptor molecules are assembled through DNA hybridization, FRET was found to be thermally reversible with the assembly and disassembly

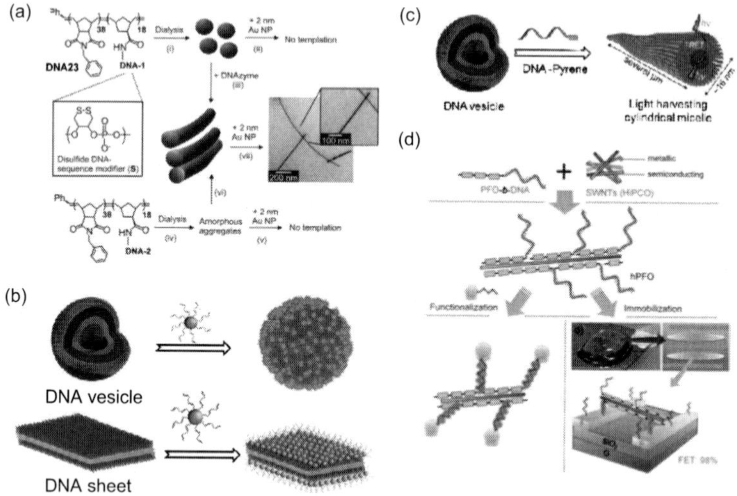

Figure 4.16 (a) Templated synthesis of crystalline nanowires using cylindrical DNA micelles as the template. (b) Schematic representation depicting the decoration of Au NPs on the surface of a DNA vesicle (top) and a DNA sheet (bottom) through sequence-specific DNA hybridization. (c) Design of a light-harvesting system using a DNA vesicle. (d) Schematic representation of programmed functionalization of SWCNTs and their sequence-specific assembly in nanoelectronic devices. (a) Reprinted with permission from Ref. [53]. Copyright 2011, John Wiley and Sons. (b, top) Reprinted with permission from Ref. [37]. Copyright 2014, John Wiley and Sons. (b, bottom) Reprinted with permission from Ref. [34]. Copyright (2017) American Chemical Society. (c) Reprinted with permission from Ref. [37]. Copyright 2014, John Wiley and Sons. (d) Reprinted with permission from Ref. [30]. Copyright 2011, John Wiley and Sons.

of the DNA. This was the first report demonstrating the design of a thermally reversible light-harvesting system using DNA-based amphiphilic self-assembly [37]. In another report, the amphiphilic DNA was applied as an all-in-one solution for single-walled carbon nanotube (SWCNT) technology applications (Fig. 4.16d). This study was done with the DNA-*b*-poly(9,9-di-n-octylfluorenyl-2,7-diyl) (**DNA13**) amphiphile. The hydrophobic segment of this amphiphile has a large π-surface, which can very efficiently interact with the SWCNT surface through π-π interaction in an aqueous medium. The interaction of the **DNA13** amphiphile with SWCNT resulted in the diameter-selective self-assembly of **DNA13** with SWCNT, which resulted in the selective dispersion of semiconducting SWCNT. Furthermore, decoration of the SWCNT with the **DNA13** amphiphile not only makes the nanotube water soluble but also provides DNA-based surface addressability for SWCNT, and the latter property was explored for the alignment of the metallic Au NPs along the SWCNT though sequence-specific DNA hybridization. Similarly, DNA hybridization allowed the attachment of DNA-decorated SWCNT onto a complementary DNA (cDNA)-functionalized surface. This was used for the surface immobilization of SWCNTs onto lithographically prepared electrodes, which resulted in 98% fabrication efficiency for SWCNT field-effect transistors (FETs). This represents a unique example of a hybrid system that exhibits all the advantageous properties of DNA, SWCNTs, and the individual polymer. [30].

4.7.3 Sensor Applications

The sequence-specific hybridization of the DNA present on the surface of the DNA nanostructures has been explored for the detection of different biologically important molecules/events. Self-assembly of DNA-polymer micelles of **DNA11** and DNA-modified Au NPs was used for single base mismatch detection (Fig. 4.17a). For example, self-assembly of DNA micelles and Au NPs modified with completely complementary DNA strands resulted in the formation of DNA micelle–Au NP hybrid assembly through sequence-specific DNA hybridization. Accordingly, the hybrid system exhibited a sharp melting temperature of 57.8°C, which corresponds to the disassembly of duplex DNA connecting the micelle and Au NPs. On the other hand, the hybrid assembly of the DNA micelle and Au NP modified with DNA strands that possess a single base mismatch in the DNA sequence shows a lower melting temperature, of 55.2°C. This difference in the melting temperature was attributed to the single base mismatch in the DNA sequence and thus provides a simple yet efficient strategy for the detection of a single base mismatch in a given DNA sequence [28]. DNA micelles were also used for the detection of mRNA in living cells using a molecular beacon (MB)-based DNA micelle (Fig. 4.17b). In this approach, a MB was used as the hydrophilic segment of the amphiphile design (**DNA32**); hence the self-assembly of the amphiphile resulted in the formation of

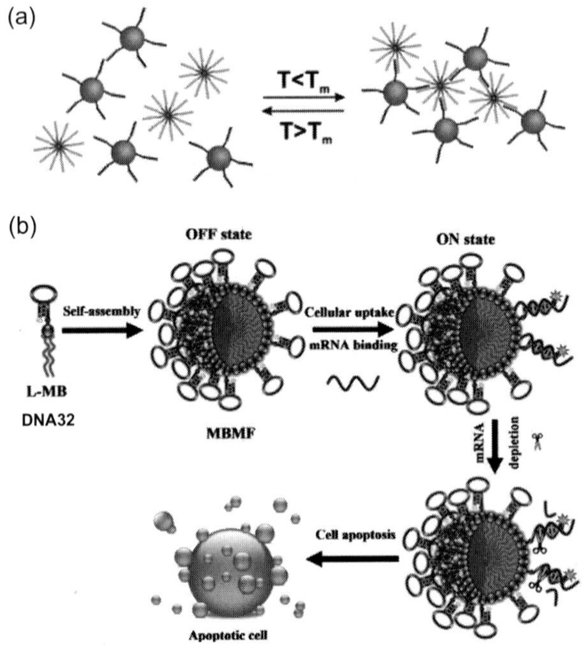

Figure 4.17 (a) Schematic representation of the reversible self-assembly of DNA micelles and DNA-modified gold nanoparticles through sequence-specific DNA hybridization. Reprinted with permission from Ref. [30]. Copyright 2004 American Chemical Society. (b) Schematic representation of a molecular beacon micelle for intracellular mRNA detection and gene therapy. Reprinted with permission from Ref. [54]. Copyright 2013, John Wiley and Sons.

DNA micelles with MBs in the folded state exposed onto the surface of the micelle. The folded conformation of the MB is a fluorescent "OFF" state due to the close proximity of the fluorophore and the quencher. The loop region of the MB was designed to be completely complementary to c-raf-1 mRNA, a cancer biomarker and antisense therapeutic target of great significance in cancer diagnostics and theranostics. DNA micelles exhibited excellent cell permeability toward the A549 cell line, and very interestingly, the binding of the loop region of the MB with the target mRNA resulted in the opening up of the MB to a linear duplex DNA. This conformational change associated upon target binding separates the fluorophore and the quencher, which leads to a large enhancement in fluorescence emission (fluorescence "ON" state) and offers a potential strategy for mRNA detection in living cells. This approach offers several advantages compared to the other mRNA detection approaches, including easy probe synthesis, efficient cellular uptake, enhanced enzymatic stability, high S/B ratio, excellent target selectivity, and superior biocompatibility [54].

4.8 Conclusions

Self-assembly of DNA amphiphiles is an emerging research area with great promise in material science, nanotechnology, and medicine. Self-assembled nanostructures of DNA-hydrophobe conjugates exhibit astonishing functional properties and structural features that are not achievable with DNA or hydrophobe alone. Diverse synthetic approaches have been developed during the last couple of decades, and the synthesis of DNA amphiphiles with any hydrophobe of interest is now easily achievable; the recent years have witnessed tremendous growth in this research field. However, increasing the yield of the coupling reaction and synthesizing a DNA amphiphile having a longer oligonucleotide segment remain two of the major hurdles in this field. Furthermore, synthesis of DNA block copolymers with multiple blocks needs to be addressed as DNA block copolymers are potential candidates in material science. The recent examples of self-assembly of DNA amphiphiles have revealed that one can mostly predict the self-assembly pathways of DNA amphiphiles through the careful analyses of self-assembly conditions and the hydrophobic-to-hydrophilic ratio. However, we are yet to understand the molecular-level packing of the self-assembly, which includes how the densely packed DNA is assembled in the shell of DNA nanostructures. Taken together the excellent biocompatibility, cell permeability, high drug (hydrophobic/hydrophilic) encapsulation ability, cell-targeting opportunity, and high stability of DNA nanostructures under enzymatic conditions suggest that DNA nanostructures derived from the self-assembly of DNA amphiphiles would be ideal nanocarriers for the delivery of biologically important molecules and drugs. Preliminary results have shown that self-assembled nanostructures of DNA amphiphiles have great potential as nanocarriers for targeted drug delivery applications. Future direction should be focused on the design of

DNA nanocarriers for the controlled release of encapsulated drug molecules at the targeted site with high efficiency. To conclude, DNA amphiphiles are unique molecular building blocks for bottom-up self-assembly, with great promise in several fields of science, and exciting applications are expected in the near future.

Acknowledgments

Financial support from DBT (BT/PR7030/NNT/28/63/2012) and KSCSTE-KSYSA is gratefully acknowledged. We thank UGC and CSIR for research fellowships.

References

1. Hoeben, F. J. M., Jonkheijm, P., Meijer, E. W. and Schenning, A. P. H. J. (2005). About supramolecular assemblies of π-conjugated systems, *Chem. Rev.*, **105**, 1491–1546.
2. (a) Shimizu, T., Masuda, M. and Minamikawa, H. (2005). Supramolecular nanotube architectures based on amphiphilic molecules, *Chem. Rev.*, **105**, 1401–1443. (b) Wang, C., Wang, Z. and Zhang, X. (2012). Amphiphilic building blocks for self-assembly: from amphiphiles to supra-amphiphiles, *Acc. Chem. Res.*, **45**, 608–618.
3. Barclay, T. G., Constantopoulos, K. and Matisons, J. (2014). Nanotubes self-assembled from amphiphilic molecules via helical intermediates, *Chem. Rev.*, **114**, 10217–10291.
4. Watson, J. D. and Crick, F. H. C. (1953). Molecular structure of nucleic acids, *Nature*, **171**, 737–738.
5. Seeman, N. C. (2010). Nucleic acid junctions and lattices, *J. Theor. Biol.*, **99**, 237–247.
6. Pinheiro, A. V., Han, D., Shih, W. M. and Yan, H. (2011). Challenges and opportunities for structural DNA nanotechnology, *Nat. Nanotechnol.*, **6**, 763–772.
7. Seeman, N. C. (2010). Structural DNA nanotechnology: growing along with nano letters, *Nano Lett.*, **10**, 1971–1978.
8. Rothemund, P. W. K. (2006). Folding DNA to create nanoscale shapes and patterns, *Nature*, **440**, 297–302.
9. (a) Schreiber, R., Do, J., Roller, E.-M., Zhang, T., Schüller, V. J., Nickels, P. C., Feldmann, J. and Liedl, T. (2014). Hierarchical assembly of metal nanoparticles, quantum dots and organic dyes using DNA origami

scaffolds, *Nat. Nanotechnol.*, **9**, 74–78. (b) Kuzyk, A., Schreiber, R., Fan, Z., Pardatscher, G., Roller, E.-M., Högele, A., Simmel, F. C., Govorov, A. O. and Liedl, T. (2012). DNA-based self-assembly of chiral plasmonic nanostructures with tailored optical response, *Nature*, **483**, 311–314.

10. Sharma, J., Ke, Y., Lin, C., Chhabra, R., Wang, Q., Nangreave, J., Liu, Y. and Yan, H. (2008). DNA-tile-directed self-assembly of quantum dots into two-dimensional nanopatterns, *Angew. Chem. Int. Ed.*, **47**, 5157 -5159.
11. Mangalum, A., Rahman, M. and Norton, M. L. (2013). Site-specific immobilization of single-walled carbon nanotubes onto single and one-dimensional DNA origami, *J. Am. Chem. Soc.*, **135**, 2451–2454.
12. Voigt, N. V., Tørring, T., Rotaru, A., Jacobsen, M. F., Ravnsbæk, J. B., Subramani, R., Mamdouh, W., Kjems, J., Mokhir, A., Besenbacher, F. and Gothelf, K. V. (2010). Single-molecule chemical reactions on DNA origami, *Nat. Nanotechnol.*, **5**, 200–203.
13. Park, S. H., Yin, P., Liu, Y., Reif, J. H., LaBean, T. H. and Yan H. (2005). Programmable DNA self-assemblies for nanoscale organization of ligands and proteins, *Nano Lett.*, **5**, 729–733.
14. Varghese, R. and Wagenknecht, H.-A. (2009). DNA as a supramolecular framework for the helical arrangements of chromophores: towards photoactive DNA-based nanomaterials, *Chem. Commun.*, 2615–2624.
15. Kwak, M. and Herrmann, A. (2011). Nucleic acid amphiphiles: synthesis and self-assembled nanostructures, *Chem. Soc. Rev.*, **40**, 5745–5755.
16. Lombardo, D., Kiselev, M. A., Magazù, S. and Calandra, P. (2015). Amphiphiles self-assembly: basic concepts and future perspectives of supramolecular approaches, *Adv. Condens. Matter Phys.*, **2015**, 1–22.
17. Chen, P., Li, C., Liu, D. and Li, Z. (2012). DNA-grafted polypeptide molecular bottlebrush prepared via ring-opening polymerization and click chemistry, *Macromolecules*, **45**, 9579–9584.
18. Vyborna, Y., Vybornyi, M., Rudnev, A. V. and Häner, R. (2015). DNA-grafted supramolecular polymers: helical ribbon structures formed by self-assembly of pyrene–DNA chimeric oligomers, *Angew. Chem. Int. Ed.*, **54**, 7934–7938.
19. Fong, R. B., Ding, Z., Long, C. J., Hoffman, A. S. and Stayton, P. S. (1999). Thermoprecipitation of streptavidin via oligonucleotide-mediated self-assembly with poly(*N*-isopropylacrylamide), *Bioconjugate Chem.*, **10**, 720–725.
20. Jeong, J. H. and Park, T. G. (2001). Novel polymer-DNA hybrid polymeric micelles composed of hydrophobic poly(D,L-lactic-*co*-glycolic acid) and hydrophilic oligonucleotides, *Bioconjugate Chem.*, **12**, 917–923.

21. Jeong, J. H., Kim, S. H., Kim, S. W. and Park, T. G. (2005). Polyelectrolyte complex micelles composed of *c-raf* antisense oligodeoxynucleotide-poly(ethylene glycol) conjugate and poly(ethylenimine): effect of systemic administration on tumor growth, *Bioconjugate Chem.*, **16**, 1034–1037.

22. Oishi, M., Hayama, T., Akiyama, Y., Takae, S., Harada, A., Yamasaki, Y., Nagatsugi, F., Sasaki, S., Nagasaki, Y. and Kataoka K. (2005). Supramolecular assemblies for the cytoplasmic delivery of antisense oligodeoxynucleotide: polyion complex (PIC) micelles based on poly(ethyleneglycol)-SS-oligodeoxynucleotide conjugate, *Biomacromolecules*, **6**, 2449–2454.

23. Oishi, M., Sasaki, S., Nagasaki, Y. and Kataoka K. (2003). pH-responsive oligodeoxynucleotide (ODN)-poly(ethylene glycol) conjugate through acid-labile β-thiopropionate linkage: preparation and polyion complex micelle formation, *Biomacromolecules*, **4**, 1426–1432.

24. Averick, S. E., Paredes, E., Dey, S. K., Snyder, K. M., Tapinos, N., Matyjaszewski, K. and Das, S. R. (2013). Autotransfecting short interfering RNA through facile covalent polymer escorts, *J. Am. Chem. Soc.*, **135**, 12508–12511.

25. Humenik, M. and Scheibel, T. (2014). Nanomaterial building blocks based on spider silk–oligonucleotide conjugates, *ACS Nano*, **8**, 1342–1349.

26. Oh, S. S., Lee, B. F., Leibfarth, F. A., Eisenstein, M., Robb, M. J., Lynd, N. A., Hawker, C. J. and Soh, H. T. (2014). Synthetic aptamer-polymer hybrid constructs for programmed drug delivery into specific target cells, *J. Am. Chem. Soc.*, **136**, 15010–15015.

27. Averick, S. E., Dey, S. K., Grahacharya, D., Matyjaszewski, K. and Das, S. R. (2014). Solid-phase incorporation of an ATRP initiator for polymer–DNA biohybrids, *Angew. Chem. Int. Ed.*, **53**, 2739–2744.

28. Li, Z., Zhang, Y., Fullhart, P. and Mirkin, C. A. (2004). Reversible and chemically programmable micelle assembly with DNA block-copolymer amphiphiles, *Nano Lett.*, **4**, 1055–1058.

29. Alemdaroglu, F. E., Ding, K., Berger, R. and Herrmann, A. (2006). DNA-templated synthesis in three dimensions: introducing a micellar scaffold for organic reactions, *Angew. Chem. Int. Ed.*, **45**, 4206–4210.

30. Kwak, M., Gao, J., Prusty, D. K., Musser, A. J., Markov, V. A., Tombros, N., Stuart, M. C. A., Browne, W. R., Boekema, E. J., ten Brinke, G., Jonkman, H. T., van Wees, B. J., Loi, M. A. and Herrmann, A. (2011). DNA block copolymer doing it all: from selection to self-assembly of

semiconducting carbon nanotubes, *Angew. Chem. Int. Ed.*, **50**, 3206–3210.

31. Wang, L., Feng, Y., Sun, Y., Li, Z., Yang, Z., He, Y.-M., Fan, Q.-H. and Liu, D. (2011). Amphiphilic DNA-dendron hybrid: a new building block for functional assemblies, *Soft Matter*, **7**, 7187–7190.
32. Carneiro, K. M. M., Aldaye, F. A. and Sleiman, H. F. (2010). Long-range assembly of DNA into nanofibers and highly ordered networks using a block copolymer approach, *J. Am. Chem. Soc.*, **132**, 675–685.
33. Edwardson, T. G. W., Carneiro, K. M. M., Serpell, C. J. and Sleiman, H. F. (2014). An efficient and modular route to sequence-defined polymers appended to DNA, *Angew. Chem. Int. Ed.*, **53**, 4567–4571.
34. Albert, S. K., Sivakumar, I., Golla, M., Thelu, H. V. P., Krishnan, N., Libin, J. K. L., Ashish, and Varghese, R. (2017). DNA-decorated two-dimensional crystalline nanosheets, *J. Am. Chem. Soc.*, **139**, 17799–17802.
35. Thompson, M.-P., Chien, M.-P., Ku, T.-H., Rush, A. M. and Gianneschi, N. C. (2010). Smart lipids for programmable nanomaterials, *Nano Lett.*, **10**, 2690–2693.
36. Kim, C.-J., Hu, X. and Park, S.-J. (2016). Multimodal shape transformation of dual-responsive DNA block copolymers, *J. Am. Chem. Soc.*, **138**, 14941–14947.
37. Albert, S. K., Thelu, H. V. P., Golla, M., Krishnan, N., Chaudhary, S. and Varghese, R. (2014). Self-Assembly of DNA–oligo(p-phenylene-ethynylene) hybrid amphiphiles into surface engineered vesicles with enhanced emission, *Angew. Chem. Int. Ed.*, **53**, 8352–8357.
38. Albert, S. K., Golla, M., Thelu, H. V. P., Krishnan, N., Deepak, P. and Varghese, R. (2016). Synthesis and self-assembly of DNA–chromophore hybrid amphiphiles, *Org. Biomol. Chem.*, **14**, 6960–6969.
39. Safak, M., Alemdaroglu, F. E., Li, Y., Ergen, E. and Herrmann, A. (2007). Polymerase chain reaction as an efficient tool for the preparation of block copolymers, *Adv. Mater.*, **19**, 1499–1505.
40. Trinh, T., Chidchob, P., Bazzib, H. S. and Sleiman, H. F. (2016). DNA micelles as nanoreactors: efficient DNA functionalization with hydrophobic organic molecules, *Chem. Commun.*, **52**, 10914–10917.
41. Albert, S. K., Thelu, H. V. P., Golla, M., Krishnan, N. and Varghese, R. (2017). Modular synthesis of supramolecular DNA amphiphiles through host–guest interactions and their self-assembly into DNA-decorated nanovesicles, *Nanoscale*, **9**, 5425–5432.
42. Liu, K., Zheng, L., Liu, Q., Willem de Vries, J., Gerasimov, J. Y. and Herrmann, A. (2014). Nucleic acid chemistry in the organic phase:

from functionalized oligonucleotides to DNA side chain polymers, *J. Am. Chem. Soc.*, **136**, 14255–14262.

43. Chien, M.-P., Rush, A. M., Thompson, M. P. and Gianneschi, N. C. (2010). Programmable shape-shifting micelles, *Angew. Chem. Int. Ed.*, **49**, 5076–5080.

44. Zhao, Z., Wang, L., Liu, Y., Yang, Z., He, Y.-M., Li, Z., Fan, Q.-H. and Liu, D. (2012). pH-induced morphology-shifting of DNA-b-poly(propylene oxide) assemblies, *Chem. Commun.*, **48**, 9753–9755.

45. Wang, L., Feng, Y., Yang, Z., He, Y.-M., Fan, Q.-F. and Liu, D. (2012). Reversibly controlled morphology transformation of an amphiphilic DNA–dendron hybrid, *Chem. Commun.*, **48**, 3715–3717.

46. Teixeira Jr., F., Rigler, P. and Vebert-Nardin, C. (2007). Nucleo-copolymers: oligonucleotide-based amphiphilic diblock copolymers, *Chem. Commun.*, 1130–1132.

47. Pearce, T. R., Waybrant, B. and Kokkoli, E. (2014). The role of spacers on the self-assembly of DNA aptamer-amphiphiles into micelles and nanotapes, *Chem. Commun.*, **50**, 210–212.

48. Pearce, T. R. and Kokkoli, E. (2015). DNA nanotubes and helical nanotapes via self-assembly of ssDNA-amphiphiles, *Soft Matter*, **11**, 109–117.

49. Alemdaroglu, F. E., Alemdaroglu, N. C., Langguth, P. and Herrmann, A. (2008). DNA block copolymer micelles – a combinatorial tool for cancer nanotechnology, *Adv. Mater.*, **20**, 899–902.

50. Wu, Y., Sefah, K., Liu, H., Wang, R. and Tan, W. (2010). DNA aptamer–micelle as an efficient detection/delivery vehicle toward cancer cells, *Proc. Natl. Acad. Sci. U.S.A.*, **107**, 5–10.

51. Kwak, M., Minten, I. J., Anaya, D.-M. Musser, A. J., Brasch, M., Nolte, R. J. M., Müllen, K., Cornelissen, J. J. L. M. and Herrmann, A. (2010). Virus-like particles templated by DNA micelles: a general method for loading virus nanocarriers, *J. Am. Chem. Soc.*, **132**, 7834–7835.

52. Edwardson, T. G. W., Carneiro, K. M. M., McLaughlin, C. K., Serpell, C. J. and Sleiman, H. F. (2013). Site-specific positioning of dendritic alkyl chains on DNA cages enables their geometry-dependent self-assembly, *Nat. Chem.*, **5**, 868–875.

53. Chien, M.-P. and Gianneschi, N. C. (2011). A morphology-dependent bio-organic template for inorganic nanowire synthesis, *Small*, **7**, 2041–2046.

54. Chen, T., Wu, C. S., Jimenez, E., Zhu, Z., Dajac, J. G., You, M., Han, D., Zhang, X. and Tan, W. (2013). DNA micelle flares for intracellular mRNA imaging and gene therapy, *Angew. Chem. Int. Ed.*, **52**, 2012–2016.

55. Rush, A. M., Thompson, M. P., Tatro, E. T. and Gianneschi, N. C. (2013). Nuclease-resistant DNA via high-density packing in polymeric micellar nanoparticle coronas, *ACS Nano*, **7**, 1379–1387.

Chapter 5

Nucleoside Lipid–Based Soft Materials

Alexandra Gaubert, Laurent Latxague, and Philippe Barthélémy
ARNA, ChemBioPharm, INSERM U1212, CNRS UMR 5320, Université de Bordeaux, 146 rue Léo Saignat, F-33076 Bordeaux, France
philippe.barthelemy@inserm.fr

Soft matters such as supramolecular gels are of great interest for several biomedical applications, including tissue engineering, regenerative medicine, and drug delivery. Hybrid amphiphiles featuring DNA moieties possess unbelievable self-assembly properties leading to various supramolecular systems. Interestingly, these behaviors occurring at the nanometer scale can be translated into advanced soft materials at the macroscopic level.

5.1 Introduction

Gels are soft materials displaying a broad range of structures and properties according to the nature of the gelling agent(s) called gelator(s). These molecules form networks, which entrap the

Templated DNA Nanotechnology: Functional DNA Nanoarchitectonics
Edited by Thimmaiah Govindaraju

ISBN 978-981-4800-21-1 (Hardcover), 978-0-429-42866-1 (eBook)
www.panstanford.com

surrounding fluid. As a result, depending on the nature of the fluid, different gels are observed at the macroscopic level, including organogel (organic solvents), oleogel (oil), and hydrogel (water) [1–2]. Owing to their biocompatibility, hydrogels are of special importance, and among them, nucleoside-based hydrogels, more particularly, will be introduced in this chapter.

Hydrogels can be defined as molecular networks able to immobilize large amounts of water (up to more than 99% w/w), giving rise to a soft, jelly-like material [3]. Such material can be made of polymeric chains or from the self-assembly of small molecules (a.k.a low-molecular-weight gelators [LMWGs]), a process driven by weak molecular interactions (Table 5.1) [4]. Note that the connectivity between the gelator molecules, including polymers or small molecules, can be of two types: permanent with covalent bonds or reversible (resulting from hydrogen bonding, van der Waals interactions, etc.). As a consequence, it is possible to differentiate two main families of gels, chemical and physical gels, resulting from either covalent or reversible connectivities. Low-molecular-weight hydrogels (LMWHs), which are stabilized by small molecules such as nucleoside-based gelators, belong to the family of physical gels.

It is commonly admitted that gelation in the case of physical gels is a stepwise process, starting from the molecular level to the macroscopic structure. First, noncovalent interactions (H bonding, π-π stacking, hydrophobic forces) must take place between gelling molecules, leading to their nanostructuration in a 1D fashion (fiber formation). Then, the growing fibers form a continuous 3D entangled network in water under the same noncovalent set of forces in order to eventually give rise to a macroscopic soft material, that is, a gel. Recently, Zelzer et al. [5] carried out an in-depth study of the gelation mechanism of a nucleotide-based gel (namely *N*-4-octanoyl-2′-deoxycytidine) by combining a range of complementary experimental techniques as well as molecular dynamics simulations. It appears that gelator molecules subjected to solvophobic forces form oligomeric structures that rearrange into cylindrical fibers managing hydrophobic areas inside the fibers and hydrophilic areas outside. Aliphatic chains point to the inside of the fibers, while the ribose goes outside. The nucleobases are stacked together and the overall structure is stabilized by hydrogen bonding. From a practical point of view, the gelation process is not spontaneous, though, and is most commonly triggered by a heat–cool cycle. Therefore, LMWG solubility is achieved upon heating, and then self-organization (or unwanted precipitation/crystallization) takes place, while the system goes back to room temperature in a less soluble environment, suggesting a kinetically dependent process [6].

Table 5.1 Common noncovalent interactions involved in the gelification process

	(a) van der Waals	
Keesom force Interaction between permanent dipoles (polar molecules)	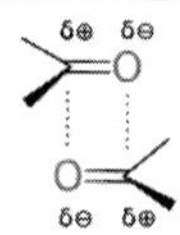	Strength = $f(1/r^6)$
Debye force Interaction between a permanent dipole and an induced dipole (polar/apolar molecules)	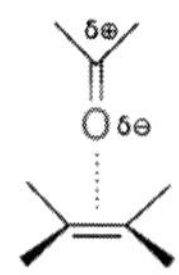	
London dispersion force Interaction between two instantaneous induced-dipoles (apolar molecules)	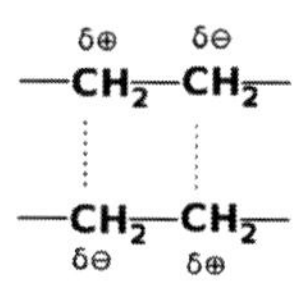	
π-π stacking Interaction between π-orbitals of two neighboring aromatic systems	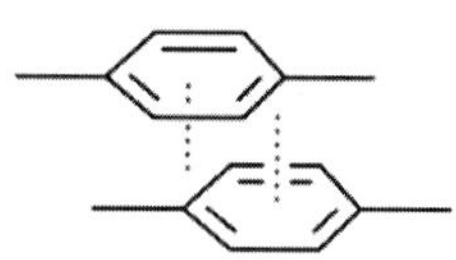	Electron-rich and electron-poor π-surfaces interact attractively
	(b) Hydrogen Bonding	
Interaction between a partially positive H and a partially negative heteroatom	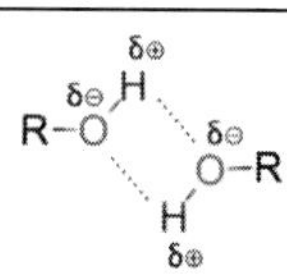	Strength drastically weakened in the presence of water
	(c) Hydrophobic Effects	
Tendency of hydrophobic moieties to aggregate in water, thanks to an entropic effect	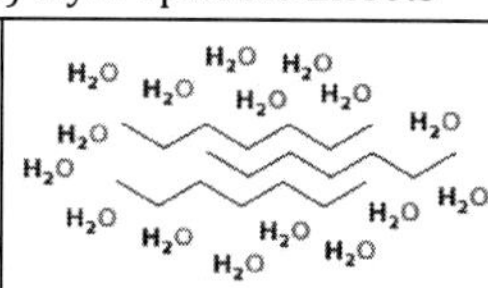	Strength = f(size of the nonpolar surface)

Nevertheless, the exact mechanism of self-organization of the gelling agents in water is still poorly understood from a thermodynamic point of view. In a recent work, Miravet et al. [7] carried out a series of experiments on bolaform amino acids to highlight the role of enthalpy and entropy during their self-assembly in water. It appears that hydrogel formation can be either enthalpy or entropy driven according to the hydrophobic character exhibited by the gelling agent. Similarly, van Bommel et al. [8] pointed out that the gel–sol transition temperature varies with the hydrophobic character and the hydrogen-bonding interactions in cyclohexane–amino acid-based hydrogelators.

Interestingly, Tuttle and coworkers [9] developed a simple packing model mimicking the fiber formation during self-assembly. The fiber is considered the stacking of discrete square units exhibiting hydrophobic and hydrophilic faces corrected with several weighting coefficients representing free-energy penalties. Use of this model demonstrated that for selected classes of LMWGs the fiber structure can represent the thermodynamic minimum.

5.2 Nucleoside Lipids as Bioinspired Materials

Quite a small numbers of building blocks are ubiquitous in the living world (carbohydrates, lipids, proteins, nucleic acids) and nature has successfully used their multiple combinations to build up complex organisms. Therefore, it is attractive for the scientists to follow its footsteps in order to rearrange for its own all these available biomolecules. With the double-helical DNA structure, nature has clearly provided one of the most sophisticated supramolecular organizations from which life originates. Starting from this simple fact, all kinds of variations around nucleic acids deserve particular attention and have already been reviewed in the past few years [10–12]. Herein, we will focus on the carbohydrate and/or lipid conjugates of nucleosides as charged or neutral LMWGs, which are currently under investigation in our laboratory.

5.2.1 Nucleolipids

The first nucleolipids, we were interested in, are strictly speaking nucleotide lipids [13] and pseudo nucleotide lipid derivatives (featuring a phosphate group on the 3′- and/or 5′-ribose position, respectively). Both classes of compounds give rise to charged amphiphiles (Table 5.2). In contrast, various nonionic nucleolipids were also synthesized [14].

Table 5.2 Various zwitterionic and cationic nucleolipids.

Zwitterionic	
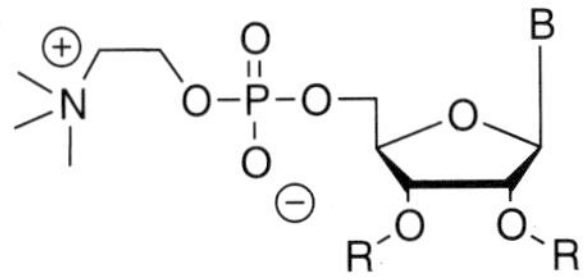 **1a-f** R= myristoyl, B= U: **DMUPC** R= palmytoyl, B= U: **DPUPC** R= stearoyl, B= U: **DSUPC** R= arachidonoyl, B= U: **DAUPC** R= oleoyl, B= U: **DOUPC** R= oleoyl, B= A: **DOAPC**	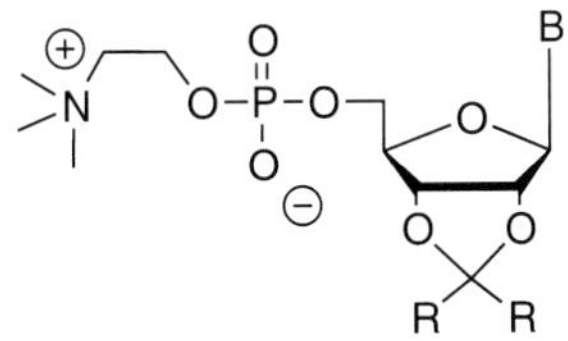 **2a-d** R= palmytyl, B= U: **PUPC** R= palmytyl, B= A: **PAPC** R= stearyl, B= U: **SUPC** R= stearyl, B= A: **SAPC**
Cationic	
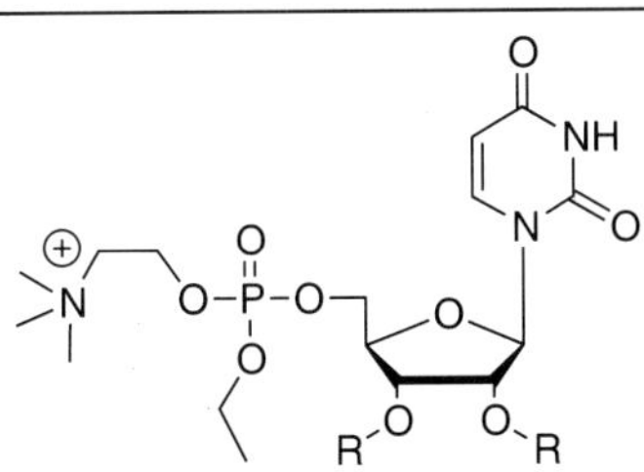 **3** R= oleoyl: **O-ethyl DOUPC**	

5.2.1.1 Cationic- and zwitterionic-based nucleolipids

Starting from the acetonide derivative of uridine or adenosine, zwitterionic phosphocholine nucleolipids have been synthesized in four steps with hydrophobic acid chain lengths from C14 to C20 (compounds **1a–f**, Table 5.2) attached on their 2′- and 3′-ribose positions [15–17]. A cationic counterpart was also similarly prepared with two oleyl chains [18] (compound **3**, Table 5.2).

Supramolecular assemblies were observed for all derivatives, but 1,2-dipalmitoyluridinophosphocholine (DPUPC; **1b**), 1,2-distearyluridinophosphocholine (DSUPC; **1c**), and 1,2-diarachidoyluridinophosphocholine (DAUPC; **1d**) provided hydrogels at the minimal concentration of 6% (w/w) for 1,2-dioleyluridinophosphocholine (DOUPC), for example [15]. In contrast, 1,2-dipalmitoylphosphocholine (DPPC), a glycerol phosphocholine, was unable to gel water, emphasizing the role played by the nucleobase in the self-assembly process to form a gel. Interestingly, DNA (at 0.0125% w/w) could be entrapped in a 5% (w/w) DSUPC hydrogel, showing the potential of such materials as nonviral gene delivery systems.

The combination of phosphocholine ketal nucleolipids (KNLs) was investigated. It was demonstrated that mixing complementary KNLs induced the formation of combined supramolecular systems (CSS), thanks to weak interactions. Indeed, ketal-based nucleoside phospholipids [19–20] PUPC **2a** and PAPC **2b** (Table 5.2) are an additional type of zwitterionic nucleolipids that individually form hydrogels above 4% (w/w). Remarkably, an equimolar mixture of both nucleosides exhibited a higher phase transition enthalpy than that of the individual ketals. This increased amount of energy needed for the main transition phase to occur was imputed to the extra stabilization resulting from complementary A:U recognition between the self-assembled nucleolipids. Moreover, when hydrated in close proximity to each other the two complementary nucleolipids spontaneously formed a new supramolecular structure at the interface, as evidenced by real-time imaging. This was confirmed by molecular modeling studies showing that (i) the bicyclic ribonucleoside was locked in the C-2′ endo conformation by the ketal functionality, which helped base pairing, meaning that hydrophobic domains are formed, preventing the competitive binding of water molecules, and (ii) the phosphate moieties also participate in the molecular recognition (Fig. 5.1).

A similar behavior was observed with a SUPC **2c** and SAPC **2d** mixture that forms a heterogeneous mixed hydrogel. When heated above each melting temperature (T_m), the gel irreversibly decomposes to give a white precipitate further analyzed by small-angle X-ray scattering (SAXS), transmission electron microscopy (TEM), and Fourier transform infrared (FTIR) as a multilamellar system with A–U interactions. Likely, the synthesis of combined supramolecular systems via nucleobase interaction could be of interest for the development of programmable soft matter in the field of material engineering [20].

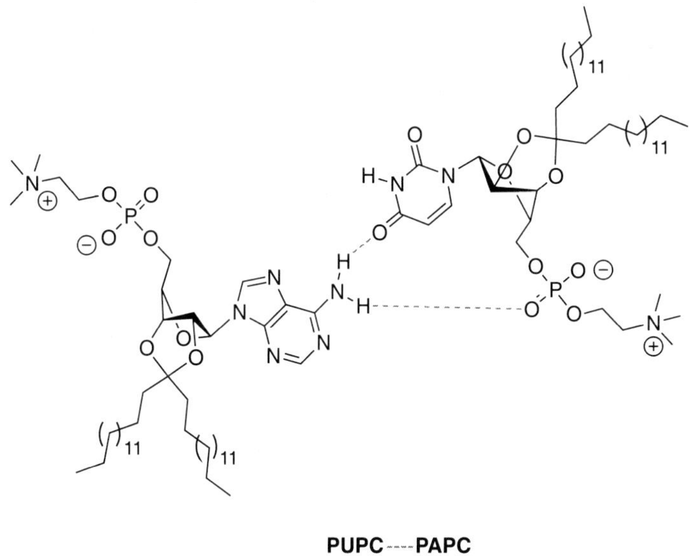

Figure 5.1 Proposed complementary association between PUPC and PAPC nucleolipids.

5.2.1.2 Anionic nucleoside lipids

Anionic nucleolipids can be synthesized as 3′-O-alkyl or di-acyl-nucleotides by using a phosphoramidite synthetic approach. In a first example, deoxythymidine and deoxyadenosine with a hexadecyl or eicosyl phosphate (compounds **4a–d**, Table 5.3) exhibited various nanostructured assemblies driven by the nature of the nucleobase [13].

In a second example, the possibility of using anionic nucleotide amphiphiles for gene delivery was examined. Then, a series of thymidine nucleotides possessing a 1,2-diacyl-*sn*-glycerol chain was synthesized (compounds **5a–c**, Table 5.3) [21]. Later, it was inadvertently discovered that the dipalmatoyl derivative $diC_{16}dT$ **5c** was able to efficiently gel water at 1.5% (w/w) in the presence of small cations (Li^+, Na^+, and K^+) [22]. The $diC_{16}dT/Na^+$ hydrogel was successfully evaluated in vitro as a reservoir and drug delivery system as well as in vivo: it is stable for several weeks after subcutaneous injection in mice, it remains at the original site of injection without inducing an immune response, and small entrapped molecules (e.g., the hydrophobic dye cyanine 5.5) can be gradually released with the natural gel degradation (Fig. 5.2).

Table 5.3 Various anionic nucleotide lipids

3′-Alkyl series	3′-Diacyl glycerol series
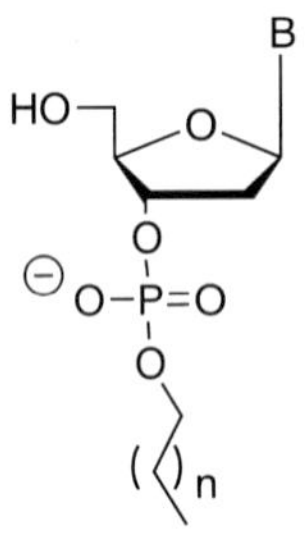	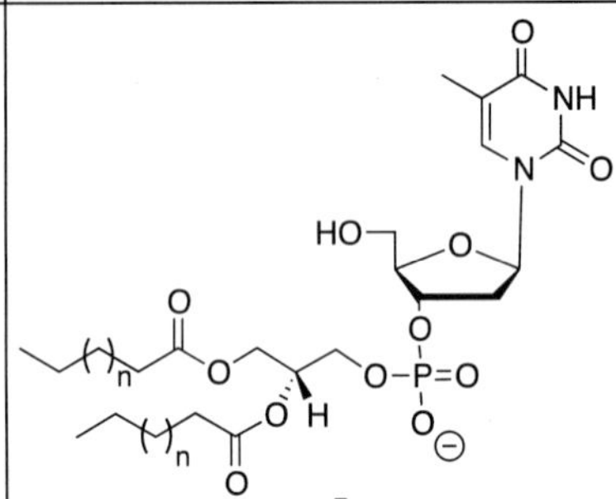
4a-d	**5a-c**
B= T, n= 14: **C16-3'TMP** B= T, n= 18: **C20-3'TMP** B= A, n= 14: **C16-3'AMP** B= A, n= 18: **C20-3'AMP**	n= 8: **diC$_{12}$dT** n= 10: **diC$_{14}$dT** n= 12: **diC$_{16}$dT**

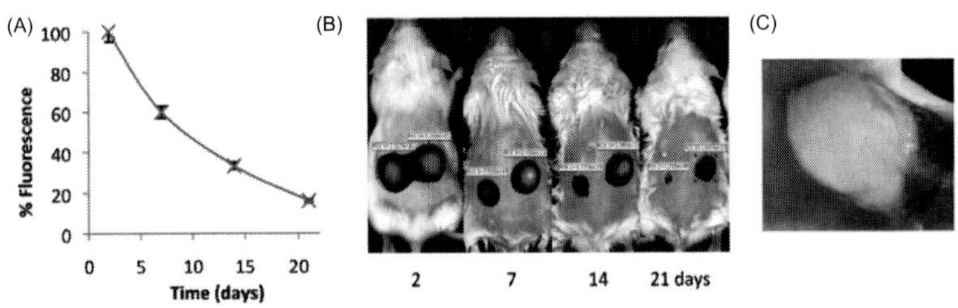

Figure 5.2 (A) Fluorescence of the gel-embedded dye, (B) fluorescence recorded with time after subcutaneous injection of the gel on each flank of one mouse, and (C) macroscopic aspect of the gel in the skin biopsy after sacrifice at 21 days.

This latter finding points out the potential of such hydrogels for long-term sustained release of active pharmaceutical ingredients (APIs).

5.2.2 Glyconucleolipids

In the design of nonionic nucleolipids, the replacement of the phosphate moiety by a neutral but hydrophilic moiety was necessary to maintain the amphiphilic properties of the molecules. Firstly, a polyethylene glycol chain was appended via DCC coupling to an uridine-based nucleolipid [23] and then a glucose moiety via a trichloroacetimidate coupling [24]. Later, the use of hydrophilic sugars was generalized simultaneously with a change in the synthetic strategy that exclusively relies now on the more versatile click chemistry. This newer family of nucleolipids is referred as glyconucleosidelipids (GNLs) [25].

The three building blocks (lipid, nucleoside, and glycosyl) are sequentially linked together by means of two click reaction key steps with their respective alkyne and azide derivatives (Fig. 5.3). The resulting Huisgen 1,3-dipolar addition provides a triazole spacer separating each block that adds its own contribution to the overall weak interactions governing the supramolecular assembly.

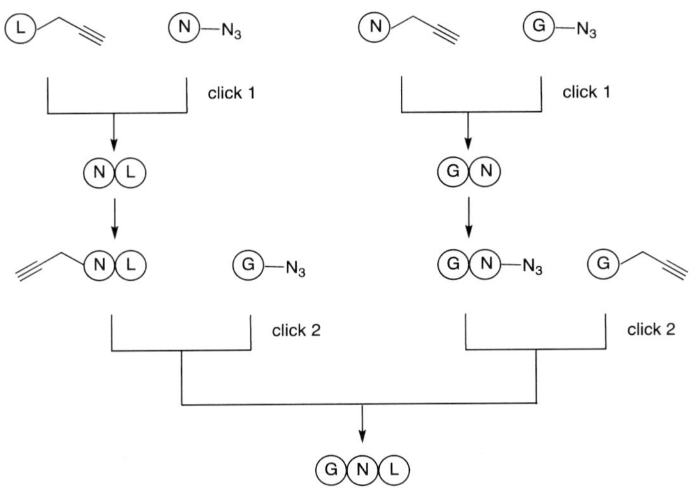

Figure 5.3 Schematic representation of the strategies used for GNL synthesis. G = glycosyl; N = nucleoside; L = lipid; NL = nucleolipid; and GN = glyconucleoside moieties.

A large series of GNLs has thus been synthesized (Table 5.4), and most of them give hydrogels or organogels at low concentration, whereas their nonnucleoside derivatives do not, demonstrating once again the key role played by the nucleobase in the gelation process.

Table 5.4 Minimum gelation concentrations in various solvents. CG = clear gel; G= gel; OG= opaque gel. Other GNLs unable to form a gel are not displayed here.

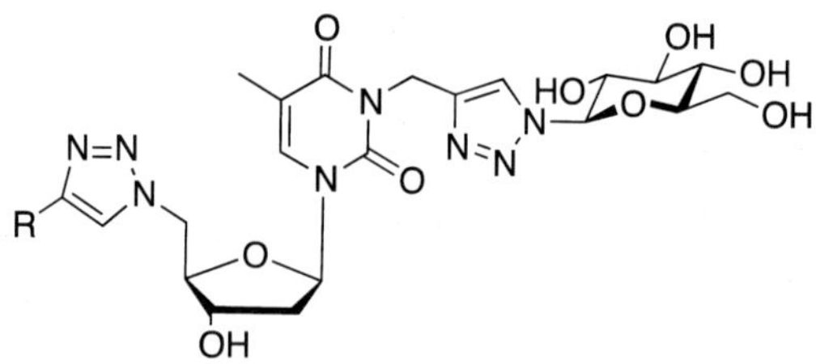

R	(a) Hydrogel (w/w)	Organogel (w/w, solvent)
6a oleylamide	CG (2.5%)	CG (2.5%, $CHCl_3$)
6b stearoylamide	G (0.1%)	G (8%, $CHCl_3$)
6c octadecyloxy	OG (3%)	CG (6%, toluene)
6d cholesteryloxy	–	OG (5%, $CHCl_3$) CG (9%, toluene) OG (9%, butan-1-ol)

For example, the hydrogel obtained from the oleoyl derivative **6a**, exhibited no cytotoxicity, and it has been evaluated for its capacity to trap a nucleic acid [26]. A fluorescein-labeled 17-mer oligonucleotide was then dissolved in an aqueous sample of the GNL. The resulting gel was characterized by a fiber network with small aggregates of oligonucleotides located at the fiber surface (Fig. 5.4).

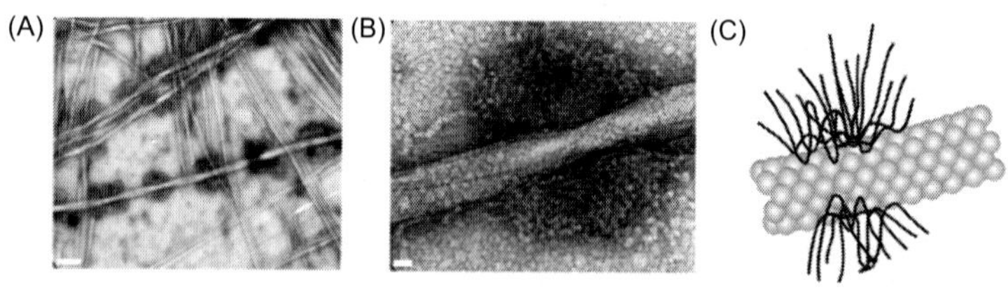

Figure 5.4 (A, B) TEM images showing the oligonucleotide clustering around a fiber. Scale bars: (A) 100 nm and (B) 20 nm. (C) Proposed model to account for the TEM observations.

More recently, this gel was also evaluated for its capacity to release a trapped bioactive substance (an immunoglobulin G anti–tumor necrosis factor alpha [TNFα]) under shear stress [27]. Submitted to a 10 Pa oscillatory shear stress, a relative release of 20% of the substance was measured compared to 7.5% without mechanical stress. This example of mechanoresponsiveness was attributed to the mild disruption of the hydrogel integrity during its stimulation.

Among GNLs, special attention needs to be paid to the subfamily of fluorinated glyconucleolipids (a.k.a. GNFs) [28] (Fig. 5.5). Our interest in per- or semifluorinated chains in designing nucleoamphiphiles lies in their unique properties brought by fluorine atoms. F-chains are bulkier than alkyl chains (fluorine van der Waals radius = 1.47 Å vs. 1.20 for H; single bond length = 1.35 Å for C-F vs. 1.09 for C-H) [29], and they are more hydrophobic than H-chains, although they are lipophobic and fluorophilic as well [30–32].

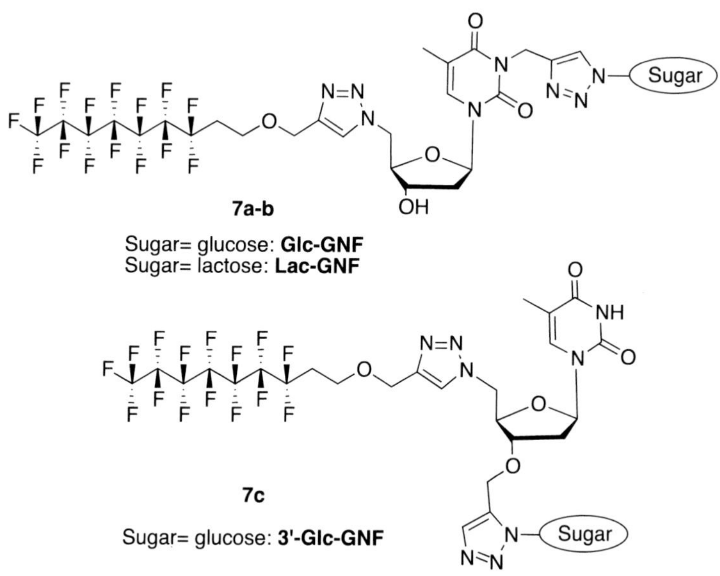

Figure 5.5 Subfamily of GNF derivatives.

Interestingly, compounds **7a** and **7b** form hydrogels at a concentration of 0.1% and 2.5% (w/w), respectively, whereas **7c** with the glucose moiety located on the 3′-ribose position does not

[33]. This distinctly different behavior shows that despite a very similar (if not identical) hydrophilic–lipophophilic balance (HLB), the molecular topology of the gelator may also play a role in the self-assembly process, a parameter poorly investigated beside HLB. In contrast with GNLs, it should be mentionned that gelation of GNFs is unaffected in the presence of Dulbecco's modified Eagle medium (DMEM). Moreover, **7a** was not cytotoxic, whereas its GNL counterpart becomes toxic above 100 μM against growing human cells, a phenomenon attributed to a detergent effect of H-chains on the cell membrane [26]. The above two properties of GNFs over GNLs make them especially usefull in biomedical applications. Thus, GNF hydrogel was used, for example, in tissue engineering as a scaffold for stem cell culture [34] or in the decontamination of nanoparticule waste from aqueous suspensions [35].

5.2.3 Bola-Glyconucleolipids

Inspired from the pioneering work of Shimizu et al. on nucleotide bola-amphiphiles [36, 37] and directly inspired from GNLs, we designed bola-glyconucleosidelipids (bola-GNLs) also referred to as glyconucleosidebola-amphiphiles (GNBAs), postulating that a higher symmetry in the gelling agent structure should favor afterward its self-assembly in solution [38, 39]. Generally, the most common GNBAs are composed of two hydrophilic head groups (not necessarily identical, though) covalently connected on both side of a hydrophobic spacer [40]. Here, click chemistry helped us again to link two glyconucleoside polar heads to a hydrocarbon chain of variable length, displaying either ether oxide, carbamate, urea, or amide functionalities (Fig. 5.6).

Interestingly, a higher success rate in terms of hydrogel formation was obtained with better rheological properties as well, compared to regular GNLs. Compound **8a**, the first GNBA synthesized of the series, is superior in many respects to the best hydrogelator obtained so far (GNF **7a**) with newer possibilities in tissue engineering applications: its gel is nontoxic, it exhibits thixotropic properties, which means that it is injectable (in vivo injectability was assessed on mice). More importantly, it allows the culture and proliferation of individual stem cells inside the material (3D scaffold) where GNF hydrogel was limited to surface cell culture, or enabled only cell clusters survival in 3D. As a potential scaffold for tissue regeneration, gel degradation in vivo is a key parameter apart with its absence of toxicity, of induced fibrosis, etc. The GNBA **8a** hydrogel, for example, was found unaltered after two weeks of implantation in mice. Following the same idea, gels made from **8c–e** after three weeks of implantation behave quite differently from each other [41]. A cyanine dye incorporated within each hydrogel was used to monitor the fluorescence signal decay (Fig. 5.7). It appears that GNBA hydrogels **8c** and **8e** degraded rapidly. Histological examination confirmed that GNBA **8d** provided the most stable gel, which was found integrated in the mouse skin. GNBA **8e** gel was found as small pieces, whereas GNBA **8c** was almost totally degraded.

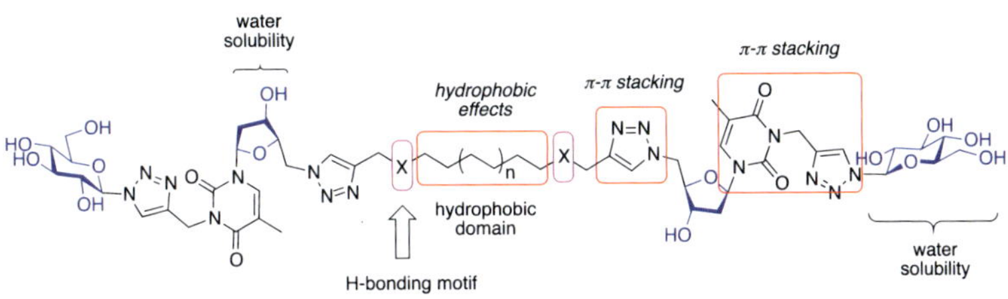

8a-e

X= O, n= 4	**ether-GNBA**
X= OC(O)NH, n= 4	**carbamate-GNBA**
X= NHC(O), n= 5	**amide-GNBA**
X= NHC(O)NH, n= 4	**urea1-GNBA**
X= $CH_2NHC(O)NH$, n= 4	**urea2-GNBA**

Figure 5.6 GNBAs' general structure. Hydrogen-bonding motifs X (purple) are surrounded by hydrophobic domains (red) and therefore shielded from water. Water solubility is provided by sugar motifs (ribose and glucose, in blue).

As evidenced by the preceding results, even a small variation at the molecular level of the gelling agent can affect its macroscopic behavior. However, gathering more data, especially with the help of rheological studies, will enable us to refine our global vision of this behavior to customize hydrogels to the desired biomedical application. In this regard, bola-amphiphiles of glyconucleosides have a great potential that needs yet to be explored.

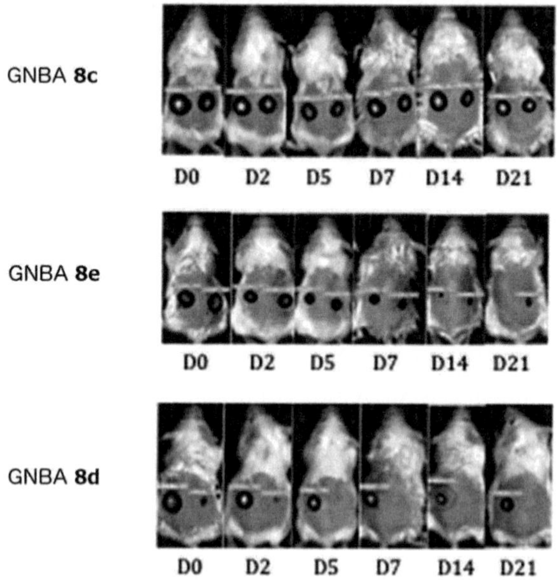

Figure 5.7 In vivo gel degradation monitoring by fluorescence imaging (4% w/w gel-embedded cyanine of GNBA **8c–e**).

5.3 Characterization of Nucleoside Lipid–Based Soft Materials

As presented before, a wide range of molecules are designed on the basis of a nucleobase conjugated with sugar/phosphate moieties and lipidic chains. The common feature between all these nucleoside lipids is their ability to self-assemble into nanostructured materials. This singular property opens a wide panel of potential applications, especially in the biomedical and health fields. Nevertheless, the process of self-assembling and the relationships existing between the molecular structures and the resulting nano-objects and/or materials is not fully understood yet. Many factors have been taken into account regarding the impact of nucleobases, solvent, temperature, concentration, conditions, etc., on the formation of supramolecular systems. Accordingly, the nanostructures obtained display various morphologies, including vesicles, fibers, lamellar structures, nanotubes, liposomes, and microspheres. Note that these structures organized at the nanometer scale induce the formation of soft materials at the macroscopic level, essentially as gels (hydrogels/organogels), dispersions, or colloidal suspensions. The main nucleolipid-based materials, their corresponding properties, and applications are summarized in Table 5.5.

Among the five nucleobases of the genetic code, uracil and thymine nucleolipid derivatives have been the most investigated so far. The three other ones have been less investigated likely due to their synthesis access, which involve additional steps of nucleobase protection and deprotection. Nevertheless, recent reports (see, for example, Ref. [42]) open the door to a wide range of original compounds with modifications on the nucleobase, the sugar, or the lipid chain. The supramolecular assemblies obtained, depending on the potential interaction, are presented in Table 5.1. Several physicochemical techniques are combined most of the time to characterize properly the new materials and propose supramolecular models and/or hypotheses.

Going from the macroscopic to the nanometric scale, the simplest way to characterize a system is by the visual aspect. To differentiate the liquid from the gel state in a formulation, the sample is turned upside down: if it is not flowing under its own weight, it is considered a gel. It is one characteristic for a global evaluation of the responsiveness hydrogels. Srivatsan et al. studied the impact of ions on their stimuli-sensitive systems by this technique [2]. The additions of Hg^{2+} or F^- led to the liquefaction of the gel, stressing the potential of such soft matters as ion sensors. In a similar way, Jial et al. developed a fluoride ion sensor [2]. The organogel in its original state exhibited a blue light further quenched following the addition of F^- anions. The visual investigation was made possible by the fluorescent nucleoside lipid–based molecules. Thus, the fluorescence decreased at the transition from the liquid state to the gel state, as presented by Zelzer et al. [5]. This transition is called the gel–sol transition and could be determined by different techniques such as differential scanning calorimetry (DSC) or rheology (addressed later in this chapter). The modification of the fluorescence intensity could result from new interactions occurring between the molecules [5] or with other components inserted in the sample (API, for example).

Table 5.5 Summary of nucleoside lipids and their nanostructures, applications, and characterization techniques

Base	Type of molecule	Solvent	Nanostructure	Application	Characterization	Ref.
Cytosine	Nucleolipid	Ethanol/water (20%/80% v/v)	Fibers		MDS	[5]
					CD	
				Drug delivery	Fluorescence	
				Tissue engineering	NMR	
					PXRD	
Adenine	Nucleolipid		Vesicles		Light microscopy	[16]
			Lamellar system		TEM	
			Small unilamellar vesicles by extrusion		QELS	
Uracil/ Adenine	Ketal-based nucleolipid	Aqueous media	Lamellar structure	($T° > T_m$)	TEM	[19]
			Gels	Hydrogels for each base (concentration > 4% and $T° < T_m$)	Monte Carlo simulation	
				Equimolar mixture → Base pairing		

Base	Type of molecule	Solvent	Nanostructure	Application	Characterization	Ref.
Uracil/ Adenine	Ketal-based nucleolipid	Aqueous media	Combined supramolecular systems (CSS)	Nonhomogeneous hydrogels	DSC	[20]
					SAXS	
					TEM	
			Single-ketal nucleolipids (SSS)	Hydrogels ($T° > T_m$)	AFM	
				Fluid lamellar phases ($T° > T_m$)	FTIR	
					Modeling	
Uracil	Nucleolipid	Organic	Fibers	Stimuli-sensitive organogels	NMR	[2]
			Helical ribbons	(Temperature/ultrasound/ chemical)	XRD	
			Nanotubes			
Uracil	Nucleolipid	Organic (Toluene/ petroleum ether)	Fibers	Fluoride ion sensors	Fluorescence	[2]

(*Continued*)

Table 5.5 *(Continued)*

Base	Type of molecule	Solvent	Nanostructure	Application	Characterization	Ref.
Uracil	Phosphocholine	Aqueous media	Fibers	Hydrogels (Cc > 6% w/w)	DSC	[15]
				Encapsulation of linear calf thymus DNA	SAXS	
			Vesicles/liposomes	Liposomes ($T° > T_m$)	UV-Vis	
					CD	
		Organic (cyclohexane)	Fibers	Organogels ($T° > T_m$)	TEM/SEM	
					TGA	
					XRD	
					Modeling	
Uracil	Phosphocholine	Aqueous media with actinide or lanthanide salts	Hollow microsphere		TEM / SEM	[17]
					SAXS	
					NMR	
					FTIR	
Uracil	Phosphocholine	Aqueous media	Vesicles (lamellar structure)	DNA entrapped	SAXS	[18]
		Organic	Fibers (helical wormlike structure)	«Lipoplexes» (larger multilamellar systems)		
				Transfection agent		
				Nontoxic/no inhibition of cell proliferation		

Base	Type of molecule	Solvent	Nanostructure	Application	Characterization	Ref.
Uracil	Nucleolipid	Aqueous media	Micelles	DNA–amphiphile supramolecular assemblies	AFM	[23]
			Toroid			
Uracil	GNL	Aqueous media	SUV (small unilamellar vesicle after extrusion)	DNA aggregation	UV-Vis	[24]
					QELS	
					TEM	
					Gel electrophoresis	
					NMR	
					FTIR	
					CD	
Thymine	Nucleolipid	Aqueous media	Ribbon-like structures	Colloidal dispersion	TEM/SEM	[13]
					Modeling	
Adenine			Small aggregates		FTIR	
					XRD	
Thymine/ Adenine			Cohabitation of the 2 systems			

(*Continued*)

Table 5.5 (*Continued*)

Base	Type of molecule	Solvent	Nanostructure	Application	Characterization	Ref.
Thymine	Nucleolipid	Organic (DMSO/DMF/MeCN/MeOH/CCl_4/Dioxane/Toluene)	Fibers	Metal ion–responsive organogels	NMR SEM XRD PXRD	[2]
Thymine	Nucleolipid	Aqueous media	Lipoplexes at room temperature	Transfection Nontoxic after 4 h of incubation	SAXS	[21]
Thymine	Nucleolipid	Aqueous media	Fibers (with Li^+/Na^+/K^+) Lamellar structure (with NH^{4+}/Et_3NH^+)	Drug delivery system Hydrogels Minimal inflammatory reaction	Rheology DSC	[22]
Thymine	GNL	Aqueous media	Liposomes	Interaction with ADSCs		[25]

Base	Type of molecule	Solvent	Nanostructure	Application	Characterization	Ref.
Thymine	GNL	Aqueous media	Fibers Nanotubes	Hydrogels Delivery of nucleic acid Nontoxic and noncytostatic	SAXS Microscopy	[26]
Thymine	GNF	Aqueous media	Fibers	Mechanoresponsive hydrogel with anti-TNFα antibody	TEM AFM Rheology Kinetic study	[27]
Thymine	GNF	Aqueous media	Fibers	Nontoxic for human cells		[28]
Thymine	GNL/GNF	Aqueous media			Surface tension DSC	[33]
Thymine	GNF	Aqueous media	Fibers	Tissue engineering	Rheology	[34]
Thymine	GNF	Aqueous media	Fibers	Decontamination of QDs and nanoparticles	TEM UV-Vis Fluorescence	[35]
Thymine	Bola-amphiphile	Aqueous media	Fibers (ribbon-like) Lamellar structure		EF-TEM FTIR UV-Vis	[36]

(*Continued*)

Table 5.5 (*Continued*)

Base	Type of molecule	Solvent	Nanostructure	Application	Characterization	Ref.
Thymine	Bola-amphiphile	Aqueous media	Fibers	DNA-like nanofibers Base pairing	ESI-FTICR-MS ATR-FTIR VT-CD	[37]
Thymine	Bola-amphiphile	Aqueous media	Fibers	Scaffold for stem cells No cytotoxicity	SAXS NMR Rheology TEM	[39]
Thymine	Bola-amphiphile	Aqueous media	Fibers	Regenerative medicine Tissue engineering Good cytocompatibility	Rheology DSC FTIR TEM	[41]

MDS, molecular dynamics simulation; CD, circular dichroism; NMR, nuclear magnetic resonance; PXRD, powder X-ray diffraction; TEM, transmission electron microscopy; QELS, quasi-elastic light scattering; T_m, melting temperature; DSC, differential scanning calorimetry; SAXS, small-angle X-ray scattering; AFM, atomic force microscopy; FTIR, Fourier transform infrared; XRD, X-ray diffraction; SEM, scanning electron microscopy; TGA, thermogravimetric analysis; Cc, concentration; DMSO, dimethylsulfoxide; DMF, dimethylformamide; MeCN, acetonitrile; MeOH, methanol; CCl_4, carbon tetrachloride; QD, quantum dot; EF, energy filtered; ESI-FTIRCR-MS, electrospray ionization Fourier transform ion cyclotron resonance mass spectrometry; ATR, attenuated total reflection; VT, variable temperature.

Among the spectral techniques commonly used for supramolecular structure characterization one can cite ultraviolet-visible (UV-Vis), fluorescence, FTIR, and nuclear magnetic resonance (NMR) spectroscopy. These techniques provide molecular and supramolecular insights such as base–base interaction involved in self-assembly, for example. These techniques, and more especially fluorescence or UV-Vis spectroscopy, enabled the evaluation of drug encapsulation [15], kinetic of drug release [22], and decontamination efficiency [35].

The morphology of the supramolecular structures obtained by nucleoside lipid–based molecules is routinely investigated by electron microscopies: TEM and scanning electron microscopy (SEM). In the first case, the sample is dried on a grid with subsequent alteration of its 3D arrangement, but in the latter, the native state of the sample is preserved, especially for cryo-SEM (Fig. 5.8).

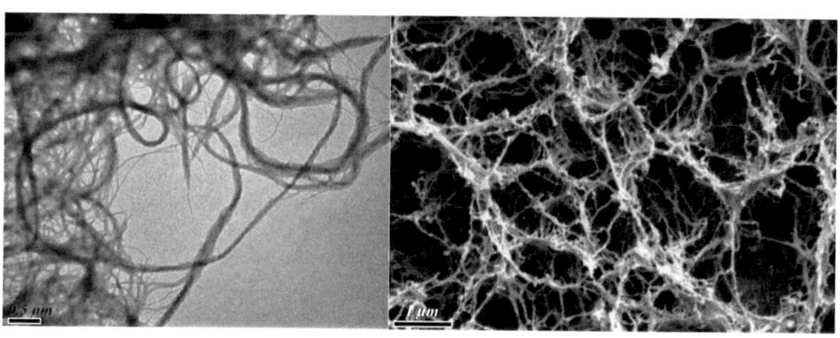

Figure 5.8 TEM (left) and cryo-SEM (right) images of GNBA bis urea. Scale bars = 0.5 μm (left) and 1 μm (right).

As an example, fibers presented in Fig. 5.8 are identified by TEM, whereas the scaffold and arrangement in 3D are highlighted with cryo-SEM. Freeze-drying of the water content of the gel without deterioration of the intertwining of fibers represents the main advantage of cryo-SEM. Pore diameters of the scaffold can also be determined, a parameter that is useful in the case of drug trapping in regenerative medicine [15]. Micelles, liposomes, lamellar structures, microspheres, or fibers can be visualized by electronic microscopy. This technique also provides details of supramolecular materials exhibiting helical, twisted, ribbon-like fibers. Barthélémy et al. stressed the cohabitation of two specific systems in an equimolar mixture of thymidine/adenine nucleolipids using TEM. Although duplex structures resulting from base pairing were expected, the specific structures of the nucleolipids alone along with a ribbon-like structure and small aggregates for thymidine and adenine nucleolipids, respectively, were observed in the mixture [13].

More information about the supramolecular structure and specifically the lengths of the assemblies were obtained by X-ray scattering. SAXS measures lengths in the range of ~10 Å to 1000 Å, Wide-angle X-ray scattering (WAXS) was used to gather information at the atomic scale. On the basis of the dimension of the nanoassembly, models were proposed for the 3D organization.

Another way to characterize soft matters and especially gels is rheology. This technique enables us to highlight the mechanical properties of a material: its strength, its linear viscoelastic region (LVER), its gel–sol transition, its self-healing ability, etc. The characteristics studied by oscillatory experiments are G^* (complex modulus), G' (elastic modulus), and G'' (viscous modulus). The evolution of G' and G'' depend of several parameters, including the oscillation frequencies, percentage of applied strain or deformation, time, and temperature. These parameters are specific to the viscoelastic properties of the biomaterial [43].

In the case of a gel (hydrogel, oleogel, or organogel), G' values are higher than G'' values and are independent of the frequency. G' defines the strength of the biomaterial, which can be essential for cell differentiation, for example [44]. To ensure the quality of the data obtained during the characterization, all the experiments need to be conducted within a specific region, the LVER, where the macroscopic organization of the hydrogel remains intact. Another essential parameter is the determination of the temperature corresponding to the transition between a solid state and a liquid state useful for the potential use of the biomaterial for in vitro and in vivo applications ($T_{\text{gel-sol}}$ > 37°C). Another interesting property exhibited by some supramolecular gels, such as those developed by Barthélémy's group, is their self-healing ability or thixotropy properties. To investigate this behavior, a time-dependent experiment is required. In that case, the hydrogel was submitted to a high mechanically stimulus, which induced the disruption of the 3D arrangement and a gel–sol transition. The withdrawing of this stimulus results in a sol–gel transition corresponding to the reassembly of the gel to its initial state. Note that the modifications take place only at the macroscopic scale without any alteration of the gelator molecule. Such biomaterial exhibiting this characteristic is called thixotropic. Hence, the biomaterial may be injected in the body through a syringe and gelified again [41]. All these viscoelastic properties are important in order to properly characterize the gel behavior. However, this part of the physicochemical investigation of soft matters is very often underestimated in the literature and needs to be properly developed for a better understanding of supramolecular assemblies.

Despite these studies, the phenomena responsible for self-assembly are not fully understood and remain very difficult to predict. A personalized design of new nucleoside lipid molecules function of the intended application is a complex approach. The development of theoretical chemistry to understand the mechanistic underlying self-assembly could be a new tool in soft-matter characterization.

5.4 Conclusion

Almost 30 years ago, the Nobel Prize in Physics was given to the French scientist Pierre Gilles de Gennes for his tremendous contribution in the field of soft matter. The past 30 years have seen new soft matters involving biomolecules, such as nucleic acids, peptides, lipids, and hybrid bioinspired molecules. Interestingly, the natural molecular species or moieties have been used as molecular building blocks, allowing the formation of supramolecular structures stabilizing soft materials at the macroscopic level. As an example, such an approach has been illustrated with the formation of low-molecular-weight gels using synthetic bioconjugates. Nevertheless, novel molecular structures remain to be explored in order to develop advanced biomaterials featuring behaviors adapted to a given specific application, such as new scaffolds for drug delivery, regenerative medicine, bioprinting, or tissue engineering, for example. In this context, polymer-free biomaterials that can overcome the weak viscoelastic properties and biocompatibility limitations are of great interest. Hybrid bioconjugates such as the nucleoside lipids described in this contribution represent a promising alternative for constructing biocompatible functional soft matter. Thanks to the genetic code inserted in the amphiphilic structures, one can design a new generation of functional and responsive materials. The future development of soft matter involving nucleoside lipids, in particular the synthesis of a large diversity of structures, will bring new opportunities to the field of biomedicine.

References

1. Du, X., Zhou, J., Shi, J. and Xu, B. (2015). *Chem. Rev.*, **115**, 13165–13307.
2. Basu, N., Chakraborty, A. and Ghosh, R. (2018). *Gels.*, **4**, 1–32.
3. Estroff, L. A. and Hamilton, A. D. (2004). *Chem. Rev.*, **104**, 1201–1217.
4. Krieg, E., Bastings, M. M. C., Besenius, P. and Rybtchinski, B. (2016). *Chem. Rev.*, **116**, 2414–2477.
5. Angelerou, M. G. F., Frederix, P. W. J. M., Wallace, M., Yang, B., Rodger, A., Adams, D. J., Marlow, M. and Zelzer, M. (2018). *Langmuir*, DOI:10.1021/acs.langmuir.8b00646.
6. Raeburn, J., Zamith Cardoso, A. and Adams, D. J. (2013). *Chem. Soc. Rev.*, **42**, 5143–5156.
7. Nebot, V. J., Armengol, J., Smets, J., Fernandez Prieto, S., Escuder, B. and Miravet, J. F. (2012). *Chem. Eur. J.*, **18**, 4063–4072.
8. van Bommel, K. J. C., van der Pol, C., Muizebelt, I., Friggeri, A., Heeres, A., Meetsma, A., Feringa, L. and van Esch, J. (2004). *Angew. Chem. Int. Ed.*, **43**, 1663–1667.
9. Ramos Sasselli, I., Halling, P., Ulijn, R. V. and Tuttle, T. (2016). *ACS Nano*, **10**, 2661–2668.
10. Berti, D. (2006). *Curr. Opin. Colloid Interface Sci.*, **11**, 74–78.
11. Araki, A. and Yoshikawa, I. (2005). *Top. Curr. Chem.*, **256**, 133–165.
12. Allain, V., Bourgaux, C. and Couvreur, P. (2011). *Nucleic Acids Res.*, **40**, 1891–1903.
13. Campins, N., Dieudonné, P., Grinstaff, M. W. and Barthélémy, P. (2008). *New J. Chem.*, **31**, 1928–1934.
14. Gissot, A., Camplo, M., Grinstaff, M. W. and Barthélémy, P. (2008). *Org. Biomol. Chem.*, **6**, 1324–1333.
15. Moreau, L., Barthélémy, P., El Maataoui, M. and Grinstaff, M. W. (2004). *J. Am. Chem. Soc.*, **126**, 7533–7539.
16. Moreau, L. Grinstaff, M. W. and Barthélémy, P. (2005). *Tetrahedron Lett.*, **46**, 1593–1596.
17. Moreau, L., Ziarelli, F., Grinstaff, M. W. and Barthélémy, P. (2006). *Chem. Commun.*, **15**, 1661–1663.

18. Moreau, L., Barthélémy, P., Li, Y., Luo, D., Prata, C. A. H. and Grinstaff, M. W. (2005). *Mol. BioSyst.*, **1**, 260–264.

19. Moreau, L., Camplo, M., Whatier, M., N. Taib, N., Laguerre, M., Bestel, I., Grinstaff, M. W. and Barthélémy, P. (2008). *J. Am. Chem. Soc.*, **130**, 14454–14455.

20. Taib, N., Aimé, A., Moreau, L., Camplo, M., Houmadi, S., Desbat, B., Laguerre, M., Grinstaff, M. W., Bestel, I. and Barthélémy, P. (2012). *J. Colloid Interface Sci.*, **377**, 122–130.

21. Khiati, S., Pierre, N., Andriamanarivo, S., Grinstaff, M. W., Arazam, N., Nallet, F., Navailles, L. and Barthélémy, P. (2009). *Bioconjugate Chem.*, **20**, 1765–1772.

22. Ramin, M. A., Sindhu, K. R., Appavoo, A., Oumzil, K., Grinstaff, M. W., Chassande, O. and Barthélémy, P. (2017). *Adv. Mater.*, **29**, 1605227.

23. Barthélémy, P., Prata, C. A. H., Filocamo, S. F., Immoos, C. E., Maynor, B. W., Nadeem Hashmi, S. A., Lee, S. J. and Grinstaff, M. W. (2005). *Chem. Commun.*, **10**, 1261–1263.

24. Arigon, J., Prata, C. A. H., Grinstaff, M. W. and Barthélémy, P. (2005). *Bioconjugate Chem.*, **16**, 864–872.

25. Latxague, L., Dalila, M.-J., Patwa, A., Ziane, S., Chassande, O., Kaplan, G., Barthélémy, P. (2012). *C.R. Chim.*, **15**, 29–36.

26. Godeau, G., Bernard, J., Staedel, C. and Barthélémy, P. (2009). *Chem. Commun.*, **34**, 5127–5129.

27. Kaplan, J. A., Barthélémy, P. and Grinstaff, M. W. (2016). *Chem. Commun.*, **52**, 5860–5863.

28. Godeau, G., Brun, C., Arnion, H., Staedel, C. and Barthélémy, P. (2010). *Tetrahedron Lett.*, **51**, 1012–1015.

29. O'Hagan, D. (2008). *Chem. Soc. Rev.*, **37**, 308–319.

30. Krafft, M.-P. (2001). *Adv. Drug Delivery Rev.*, **47**, 209–228.

31. Krafft, M.-P. and Goldmann, M. (2003). *Curr. Opin. Colloid Interface Sci.*, **8**, 243–250.

32. Riess, J. G. (2002). *Tetrahedron*, **58**, 4113–4131.

33. Latxague, L., Patwa, A., Amigues, E. and Barthélémy, P. (2013). *Molecules*, **18**, 12241–12263.

34. Ziane, S., Schlaubitz, S., Miraux, S., Patwa, A., Lalande, C., Bilem, I., Lepreux, S., Rousseau, B., Le Meins, J.-F., Latxague, L., Barthélémy, P. and Chassande, O. (2012). *Eur. Cell Mater.*, **23**, 147–160.

35. Patwa, A., Labille, J., Bottero, J.-Y., Thiéry, A. and Barthélémy, P. (2015). *Chem. Commun.*, **51**, 2547–2550.

36. Iwaura, R., Yoshida, K., Masuda, M., Yase, K. and Shimizu, T. (2002). *Chem. Mater.*, **14**, 3047–3053.

37. Iwaura, R., Yoshida, K., Masuda, M., Ohnishi-Kameyama, M., Yoshida, M. and Shimizu, T. (2003). *Angew. Chem. Int. Ed.*, **42**, 1009–1012.

38. Barthélémy, P., Ramin, M., Latxague, L., Appavoo, A., Chassande, O. and Ehret, C. (2014). *Glycosylated Nucleo Bolamphiphiles as New LMWG for Biomedical Applications*, EP1490302.0.

39. Latxague, L., Ramin, M. A., Appavoo, A., Berto, P., Maisani, M., Ehret, C., Chassande, O. and Barthélémy, P. (2015). *Angew. Chem. Int. Ed.*, **54**, 4517–4521.

40. Fuhrhop, J.-H. and Wang, T. (2004). *Chem. Rev.*, **104**, 2901–2937.

41. Ramin, M. A., Latxague, L., Sindhu, K. R., Chassande, O. and Barthélémy, P. (2017). *Biomaterials*, **145**, 72–80.

42. Alies, B., Ouelhazi, M. A., Patwa, A., Verget, J., Navailles, L., Desvergnes, V., Barthélémy, P. (2018). *Org Biomol. Chem.*, **16**, 4888–4894.

43. Latxague, L., Gaubert, A., Barthélémy, P. (2018). *Molecules,* **23**, 89.

44. Engler, A.J., Sen, S., Lee Sweeney, H., Discher, D.E. (2006). *Cell,* **126**, 677–689.

Chapter 6

Excited-State Dynamics in Chromophore-Appended Nucleic Acids

Abbey M. Philip, Vinayak Bhat, and Mahesh Hariharan
School of Chemistry, Indian Institute of Science Education and Research Thiruvananthapuram (IISER-TVM), Maruthamala P. O., Vithura, Thiruvananthapuram 695551, Kerala, India
mahesh@iisertvm.ac.in

6.1 Introduction

Photoinduced processes in nucleobases and DNA have attracted enormous attention owing to their implications in diverse fields, such as fundamental research in biochemistry and medicine, fluorescent probes, DNA nanotechnology, light harvesting, and bioorganic device applications [1]. The sequence and elementary structure, comprising heterocyclic nucleobases and a sugar-phosphate backbone, play a vital role in controlling the physical, biochemical, and photoexcited-state processes in DNA [2]. The aromatic nucleobases, that is (i) purine analogs (adenine [A] and guanine [G]) and (ii) pyrimidine analogs (thymine [T] and cytosine [C]), form the primary fortitude of

Templated DNA Nanotechnology: Functional DNA Nanoarchitectonics
Edited by Thimmaiah Govindaraju

ISBN 978-981-4800-21-1 (Hardcover), 978-0-429-42866-1 (eBook)
www.panstanford.com

the DNA structure and are also referred to as the alphabet of genetic materials (Fig. 6.1) [3, 4]. The self-sorted base pairing and the π-π interactions among the nucleobases lend a helical DNA skeleton with an ideal base stacking distance of 3.4–3.5 Å [3, 5, 6].

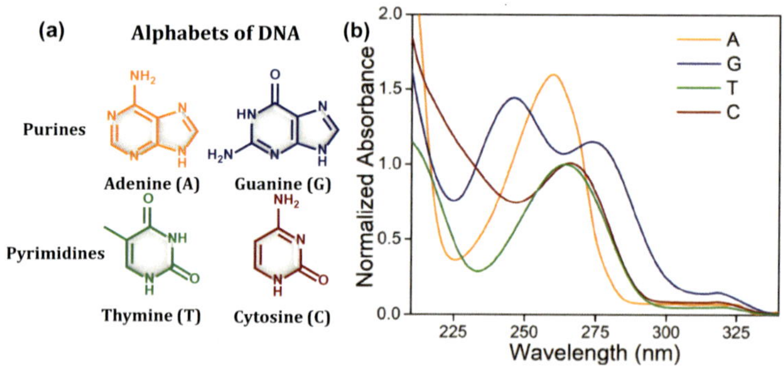

Figure 6.1 (a) Chemical structure and (b) UV absorption spectra of the DNA alphabet; adenine (orange), guanine (blue), thymine (green), and cytosine (maroon).

The helical stacked array renders a rigid electronically coupled column of aromatic nucleobases that can facilitate a "π-way" for the smooth conduction of charge across the DNA backbone [7]. Eley and Spivey, in 1962, gave the first notion that the π-stacked nucleobase array in DNA can facilitate charge migration [8]. Henceforth, various scientific reports debated charge migration in DNA and its function as a natural biopolymer for long-range charge transport [9]. Consequently, a myriad of supramolecular or covalently linked electron donor/acceptor–DNA assemblies were investigated for long-range charge transport across DNA [10].

On the other hand, DNA nucleobases possess intense absorption in the UV region of the solar spectrum and their excitation can deposit excess electronic energy, leading to highly reactive excited states (Fig. 6.1b) [11]. These electronic excited states can initiate destructive photochemical reactions and can have complex decay pathways, ultimately leading to mutation [12]. Yet, only a limited number of the excited DNA molecules undergo detrimental photochemical reactions owing to the rapid discharge of the excess electronic energy via efficient nonradiative processes and dissipation of the excess energy in the form of heat to the surroundings [11]. Thus, exploration of photoexcited-state processes in DNA can help understand the mechanism of light-induced lesions and can assist in deciphering the oxidative damage repair pathways in the biological system.

Alternatively, the self-sorting and favorable π-stacking characteristics highlight DNA as a suitable template for crafting a hierarchical chromophoric assembly [13]. A principal conundrum in the assembly of organic molecules is that it is dictated by the subtle balance between the competitive/synergistic noncovalent interactions between the chromophores [14–16]. Directional self-assembly of DNA can thus guide a predictable and reproducible assembly of the chromophores [17, 18]. Apparently, a controlled and predictable chromophore-appended DNA assembly can lend a better hold over the electronic communication, thereby modulating the photophysical characteristics of the chromophores for light-harvesting applications [13, 19]. As a result, several DNA-chromophore assemblies have been synthesized and explored for emergent photofunctional applications [13, 19]. Owing to numerous benefits, the design and study of excited-state properties in DNA-chromophore assemblies have grown tremendously over the last three decades.

6.2 Photophysical Properties of DNA: Excited-State Processes in Monomeric Nucleobases

Three-dimensional helical arrays of DNA oligonucleotides depend exclusively on the constituent heterocyclic nucleobase and the sugar–phosphate backbone [3]. Perceptibly, the photoinduced-excited-state properties of the integral aromatic nucleobases play a vital role in defining the excited-state properties of the DNA oligonucleotides. The photoexcited-state processes of single- and double-stranded oligonucleotides are different from that of the single nucleobases [11]. The photoexcited-state processes in constituent DNA nucleobases occur on an ultrafast timescale, and the consequent rapid nonradiative relaxation of the excited states via predominant internal conversion (IC) and thermal equilibration (vibrational cooling) with the surrounding medium safeguards the

DNA from deleterious effects [20]. Very low fluorescence ($\phi_f \sim 10^{-4}$) of the DNA nucleobases in solution state further infers ultrafast nonradiative relaxation of the excited states [5]. Consequently, exploration of the rapid electronic processes in DNA nucleobases entails an analogous ultrafast spectroscopic technique for mapping the short-lived transient intermediates.

Kohler and coworkers reported the first accurate measurement of the $^1\pi\pi^*$ lifetime in DNA nucleosides employing the femtosecond transient absorption (fTA) technique [21]. Upon UV excitation at ~260 nm, the aromatic nucleobases undergo a strongly allowed transition to populate the $^1\pi\pi^*$ energy level [5, 22]. The concept of rapid drifting of the wave packets along the conical intersection (CI) with the ground state is invoked here [21, 23]. Out-of-plane distortion along the double bonds activates the CIs and necessitates substantial molecular motions before the nucleobases reach the lowest-ground-state equilibrium structure [11]. During these processes, the initially populated lowest $^1\pi\pi^*$ state dumps energy ≥4 eV into the vibrational modes of the nucleobases via IC from $^1\pi\pi^*$ to the hot ground state (S_0), which survives a few picoseconds before decaying to the lowest ground state [11]. Subsequent release of the excess energy via vibrational energy transfer to the neighboring solvent molecules, also known as vibrational cooling, helps the hot ground state to decay to the lowest ground state [21]. The IC to the hot ground state is independent of the solvent polarity. However, the subsequent vibrational cooling to the lowest ground state is a solvent-polarity-dependent phenomenon [24].

Though most of the de-excitation in the nucleobases occurs via IC to the ground state, there can be an alternative relaxation pathway that can populate the $^1n\pi^*$ state [25, 26]. Despite being weak due to their forbidden nature, these transitions can be instigated by the lone pair of electrons in heterocyclic nucleobases [5, 22]. Apart from these transitions, there can be close triplet states ($^3\pi\pi$ and $^3n\pi^*$), which can be populated/activated via intersystem crossing (ISC) [11, 26]. The $^1n\pi^*$, $^3\pi\pi^*$, and $^3n\pi^*$ states are signified as the dark states, as the exact mechanism for the population of these excited states is not clear [4, 26]. Nevertheless, the population of these excited states occurs in low yields, and they evidently do not contribute much to the total spectrum. Though having a negligible quantum yield, the triplet excited state in the DNA nucleobases should not be underrated, as the long lifetime of these excited states can catalyze photodamaging reactions. Consequently, the study of ultrafast photoexcited processes in DNA nucleobases has grown over the last two decades, synonymous with the emergence of ultrafast spectroscopic techniques.

On the other hand, stacking and base pairing of the nucleobases can alter the photoexcited-state processes in DNA. The stacking of the aromatic nucleobases is known to cause a decrease in the 260 nm absorption and can help acknowledge base-pair-mediated DNA duplex formation. Moreover, duplex formation can affect the absorption of the left and right circularly polarized light, leading to distinct spectroscopic circular dichroism signatures for the single-stranded DNA, A, B, and Z forms of duplex DNA. An elaborate account of the excited-state properties in the single-/double-stranded oligonucleotides and DNA multimers is not discussed here but has been reported elsewhere [1, 5, 11].

6.3 Photophysical Properties of Functional Molecule-DNA Assemblies

The query whether DNA can act as a semiconductor or not has intrigued the scientific community across the globe since the discovery of the 3D helical structure of DNA by Watson and Crick (Fig. 6.2) [9]. Soon after the discovery (Fig. 6.2a), it was hypothesized that DNA could conduct charge owing to its similarity to the 1D aromatic crystals [8]. The key feature of the DNA double helix is the base-pairing-assisted stacking of the heterocyclic nucleobases at a distance of 3.4–3.5 Å. The π-stacked nucleobases ensure sufficient electronic coupling between the aromatic cores, thereby creating a columnar π-way for efficient migration of charge over nanoscale lengths (Fig. 6.2) [29]. Moreover, the electron-donating efficiency of the purine derivatives makes them an ideal candidate for the exploration of charge migration in electron donor/acceptor pendant DNA. Among the heterocyclic nucleobases, guanine has the lowest oxidation potential and hence can act as a hole carrier while mediating the electron transfer between donor-/acceptor-conjugated DNA (Table 6.1) [9, 27, 28]. In the absence of guanine, it has been observed that adenine can mediate hole transfer across DNA stacks [30, 31]. Likewise, pyrimidines exhibit facile reduction, which can facilitate electron transport in DNA [9, 27, 28]. Consequently, several covalent and noncovalent donor/acceptor pendant DNA assemblies emerged to examine stack-mediated conduction of the charge radical intermediates [32].

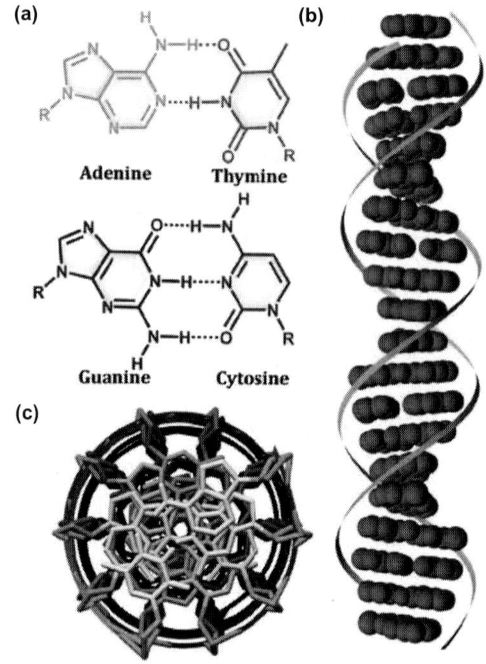

Figure 6.2 (a) Base pairing in DNA nucleobases; (b) side view and (c) top view of the base-pairing-induced stacking of the nucleobases in double-stranded DNA.

Table 6.1 Shows the redox potentials of the nucleoside derivatives of the DNA alphabet

Nucleoside	E_{ox}^{a} (V)	E_{red}^{b} (V)
G	1.24	−2.67
A	1.69	<−3.0
C	1.9	−2.36
T	1.9	−2.26

[a,b]Estimated values reported against the standard calomel electrode [9, 27, 28]

Mechanistically, charge transfer in DNA has been rationalized by two widely recognized models: (i) coherent superexchange (single step), prevalent for short distances, and (ii) incoherent hopping (multistep), applicable for longer distances (Fig. 6.3) [33, 34]. Both mechanisms depend on the match between the energy levels of the bridge in relation to that of the donor and the acceptor. The superexchange mechanism prevails when the energy levels of the bridge and the donor/acceptor do not match and the charge is tunneled across the donor and the acceptor in one coherent jump [9]. The kinetics of the charge transfer in the superexchange [34, 35] mechanism can be described as

$$k = Ae^{-\beta R}, \tag{6.1}$$

where β is the important factor defining the distance dependence of the charge transfer and depends on the energy match between the initial state and the bridge as well as the coupling between intervening bridges. A typical value of β for DNA-mediated charge transfer ranges from 0.1 to 1.5 Å^{-1} [33].

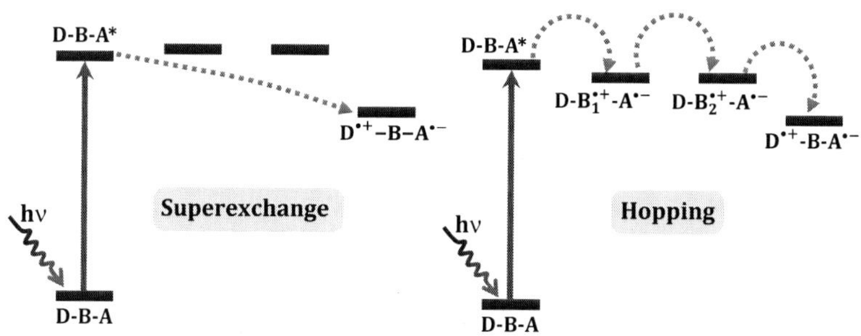

Figure 6.3 Schematic representation of superexchange (left) and hopping (right) mechanisms for DNA-mediated charge transfer.

Superexchange is distance dependent and occurs only over short distances. On the other hand, the hopping mechanism is feasible when the donor and the bridge states are nearly degenerate. In the hopping mechanism, the donor injects a charge onto the bridge, which hops across the bridge until it is accommodated by the distant charge acceptor [9]. Herein, the DNA nucleobase, especially guanine because of its lowest oxidation potential, can act as a bridge through which the charge can hop from donor to acceptor. The dynamics of the hopping mechanism can be explained by the following equation [34]:

$$k = k_0 N^{-\eta}, \tag{6.2}$$

where N is the number of hopping steps and η denotes the nature of the bridge and can have a value between 1 and 2. The hopping mechanism exhibits a shallow distance dependence and relies on the number of intervening hopping steps (N) [9].

6.3.1 Long-Range Charge Transfer in Chromophore-Appended DNA

Preliminary investigation of the DNA-stack-mediated charge transfer utilized self-assembly of donor/acceptor units in the DNA double helix. Fromherz and Rieger reported a colossal enhancement in the intermolecular electron transfer yields when ethidium (Et, donor) was intercalated in the DNA helix and methylviologen (MV^{2+}, acceptor) bound outside the helix [36]. Barton, Turro, and coworkers explored the charge transfer kinetics between the self-assembled ruthenium complex (donor) intercalated in the DNA helix and a suitable complementary metal complex (acceptor) bound to the major groove of the DNA double helix [37, 38].

Herein, charge-transfer-mediated quenching of the donor photoluminescence was enhanced when it was intercalated between the DNA stack, as compared to the mixture of donor and acceptor complexes in the solution [36–38]. An enhanced rate of photoluminescence quenching in the presence of DNA suggests the important role of the DNA π stack in mediating charge transfer between donor and acceptor units. Consequently, a myriad of metallointercalator-DNA assemblies emerged to modulate the π-stack-mediated charge transfer in DNA (Fig. 6.4) [38–44]. Though self-assembly-based intercalation of the donor to the DNA helix was confirmed via a multitude of experimental methods, possibilities of diffusion-mediated electron transfer between the nonintercalated donor and acceptor in the solution always exist.

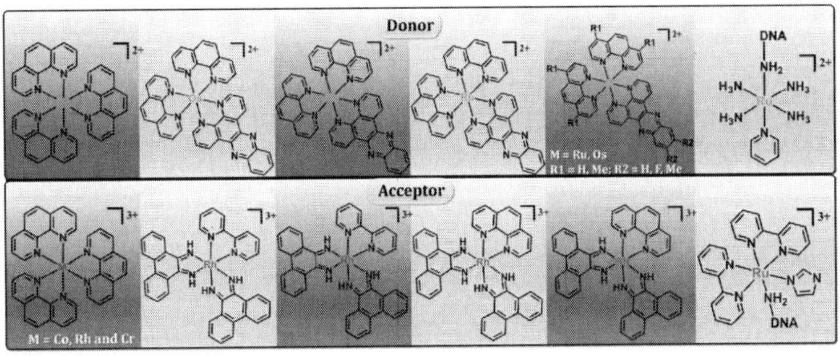

Figure 6.4 A few electron donor (top) and complementary acceptor (bottom) metallointercalators explored for DNA-π-stack-mediated charge transfer investigations.

In a subsequent pioneering work, Barton, Turro, and coworkers reported a covalent design strategy to connect donor and acceptor metallointercalators to the DNA strands via a flexible linker (Fig. 6.5) [7]. The covalent attachment of the donor and the acceptor units to the DNA helix can eliminate the complexities of diffusion-mediated processes and the ambiguities concerning the binding mode of the metallointercalators while self-assembling on DNA. Moreover, a precise distance dependence of the charge transfer dynamics in DNA can be accessed in the covalent attachment strategy [7, 45]. The designated donor and the acceptor complex were found to intercalate strongly into the DNA stack with affinities of $\geq 10^6$ M^{-1} and hence were perfect candidates to explore stack-mediated charge transfer in DNA [46, 47]. The donor (bis(phenanthroline(dipyrido phenazine)ruthenium(II),[Ru(phen)$_2$dppz]$^{2+}$) possesses negligible luminescence in an aqueous solution. However, it luminesces intensely when intercalated into the DNA stack. The lowest-excited-state transition in the donor corresponds to the metal-to-ligand charge transfer [48]. While the lowest-excited-state transition in the acceptor (bis(9,10-phenanthrenequinonediimine)(phenanthroline) rhodium(III), [Rh(phi)$_2$phen]$^{3+}$) has a ligand-to-metal charge transfer character, thereby guiding photoinitiated electron transfer from Ru(II) to Rh(III) center mediated by the DNA π stack.

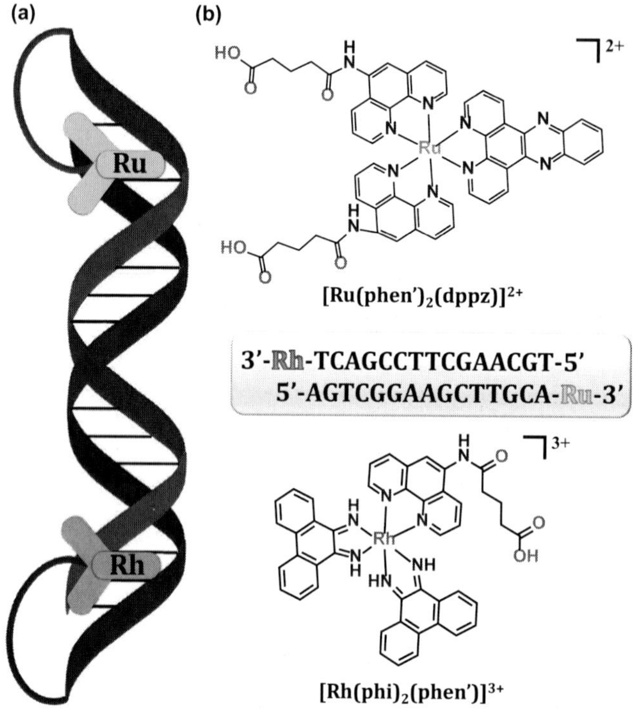

Figure 6.5 (a) Schematic representation depicting the Ru (donor) and Rh (acceptor) modified oligonucleotide and (b) Structure of the Ru (donor) and Rh (acceptor) complexes along with the DNA sequence.

The intercalation of the Ru(II) donor to the DNA stack was affirmed by the intense luminescence attained upon hybridizing the Ru(II)-appended DNA to its complementary unmetallated DNA strand. The hybridization of a complementary Rh(III)-modified oligonucleotide completely quenched the luminescence of the Ru(II) donor. The contribution from the intermolecular quenching was rebutted by experiments where Ru(II)-modified DNA and Rh(III)-modified DNA were mixed together and still 86% of the luminescence was retained. The possibility of Förster resonance energy transfer was ruled out as neither did the donor excited state have a singlet character nor did any spectral overlap between the photoexcited states of Ru(II) donor and Rh(III) acceptor complexes exist. The complete quenching of luminescence was accounted for intramolecular charge transfer among the Ru(II) and Rh(III) complexes facilitated by the DNA π

stack. Thus, a rapid photoinduced electron transfer over a distance of >40 Å was facilitated through the DNA π stack by tethering the donor and acceptor complexes to the 15–base pair DNA duplex. Similarly, Meade and Kayyem reported a rigid, nonintercalating donor and acceptor–modified 8-mer oligonucleotide wherein DNA-π-stack-mediated electron transfer occurred over a distance of 21 Å [43].

Barton and coworkers further explored the biological relevance of DNA-π-stack-mediated long-range charge transport in the oxidative damage and repair pathways of the genetic material [10, 49]. Herein, the DNA-tethered metallointercalator was used as a photo-oxidant to evaluate the oxidative damage occurring at a distance using biochemical approach [10]. The photo-oxidant was attached to a DNA strand having a guanine-rich sequence and that can act as an electron sink in DNA due to its low oxidation potential [50]. The presence of a neighboring guanine unit can further decrease the oxidation potential of DNA and hence guanine multiplets are an active site for photo-oxidation within DNA [49, 51]. Upon excitation of the metallointercalator and subsequent piperidine treatment, cleavage was observed at the 5′-end of the guanine-doublet-containing strand [10]. The photocleavage of the guanine-rich oligonucleotide at a distance from the metallointercalator suggests the role of DNA-π-stack-mediated ultrafast charge transfer in the photo-oxidation of DNA. Negligible photo-oxidation in the DNA-with an intervening bulge in the stack further emphasizes the significance of the DNA-π-way mediated long-range charge transport [52]. Consequently, shallow distance dependence of the DNA-mediated charge transfer was reported to occur over a distance of 200 Å.

The exploration of the charge transfer dynamics in DNA stacks was not limited to the donor and acceptor metallointercalators. A series of donor and acceptor organointercalators and modified nucleobases were also explored for probing the ultrafast DNA-mediated charge transfer (Fig. 6.6). Fromherz and Rieger showed one of the initial reports wherein the organointercalator, Et (donor), intercalated in the DNA helix efficiently donates electron to the MV^{2+} (acceptor) bound outside the helix [36]. Consequently, ethidium has been one of the highly sought organointercalator because of its exceptional intercalation and distinct luminescence signatures in the presence/absence of DNA and hence is successfully used as a tool for stack-mediated charge transfer in DNA [53–56]. Zewail and coworkers explored the femtosecond dynamics of the electron transfer between Et and a modified nucleobase 7-deazaguanine (7-dAzG) separated by a fixed distance [57]. The electron transfer was found to be distance independent. However, depending on the orientation of the donor and the acceptor, two different time constants were observed for the electron transfer [57]. Acridinium ions (Acr) [58–60], anthraquinone (AQ) [61–63], methylene blue, and daunomycin are other organointercalators that can strongly bind to the DNA and can be utilized for the exploration of DNA-mediated charge transfer [9, 64–66].

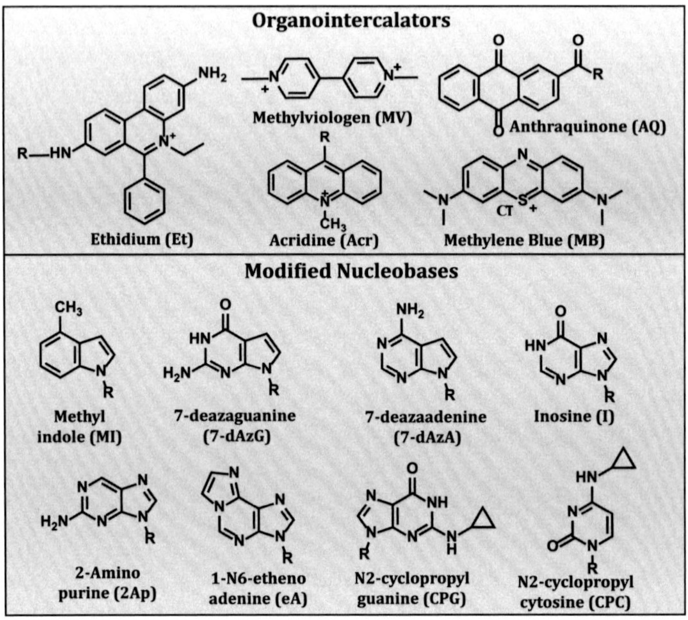

Figure 6.6 A few organointercalators and modified nucleobases explored for DNA-π-stacked-mediated charge transfer investigations.

On the other hand, introduction of modified nucleobases in the DNA strand causes minimal perturbation in the DNA base pairing/3D structure, while modifying the photophysics and the redox properties of the bases [9]. Methylindole (MI), 7-deaza guanine (7-dAzG), 7-deazaadenine (7-dAzA), inosine (I), 1-N6-ethenoadenine (eA), aminopurine (2Ap), and N2-cyclopropylamine-modified bases are some of the modified nucleobase derivatives explored for DNA-mediated charge transfer [9]. MI and 7-deaza purine derivatives possess a low oxidation potential compared to the native nucleobase (G and A) derivatives and hence can act as a deep hole trap [32, 57, 67, 68]. Barton and coworkers explored the oxidative charge transfer in DNA between MI (donor) and $[Ru(phen)_2(dppz)]^{2+}$ (acceptor) employing flash-quench experiments and electron paramagnetic resonance (EPR) techniques [67]. On the other hand, inosine, an analog of guanine, being harder to oxidize has been used as a control to compare the role of guanine in the charge transfer dynamics in DNA [62, 69–71].

The N2-cyclopropyl-substituted bases do not alter the redox potential; however, they accelerate the hole trapping, thereby allowing faster oxidative damage compared to native nucleobases [63, 72]. As a result, they have been used extensively to explore the charge transfer dynamics in DNA by probing the oxidative damage [9, 62, 63, 70, 73, 74]. 2Ap and eA are fluorescent-modified nucleobases and exhibit distinct fluorescence characteristics in the presence and absence of DNA and can further undergo charge transfer with native DNA bases, leading to the quenching of the fluorescence [9, 32, 69, 75, 76]. Consequently, 2Ap has been one of the most widely utilized fluorescent nucleobase analogs for the exploration of ultrafast dynamics in DNA [32, 62, 69, 70, 71, 76–79]. As a result, the exploration of the charge transfer dynamics in DNA using modified nucleobase derivatives as a photoactive redox gradient grew over the last two decades.

6.3.2 Ultrafast Charge Migration in Chromophore-Appended DNA Hairpins

While the insights into long-range charge transfer in DNA proposed by Barton and coworkers were debated [7], Lewis and coworkers explored the distance dependence and the mechanism of ultrafast charge migration in a family of chromophore-appended DNA hairpins [45]. Herein, stilbene (Sa)-linked DNA hairpins with a single G-C pair in a poly(A-T) domain were employed, while separating G and Sa with a varying number of intervening A_n (n = 0–4, Fig. 6.7) [45]. Stilbene-linked hairpins possess high thermal stability, and the incorporation of Sa into the DNA hairpin does not alter the structure of the B-form DNA, as is evident from the crystal structure, molecular modeling, and nuclear magnetic resonance (NMR) studies [33, 80].

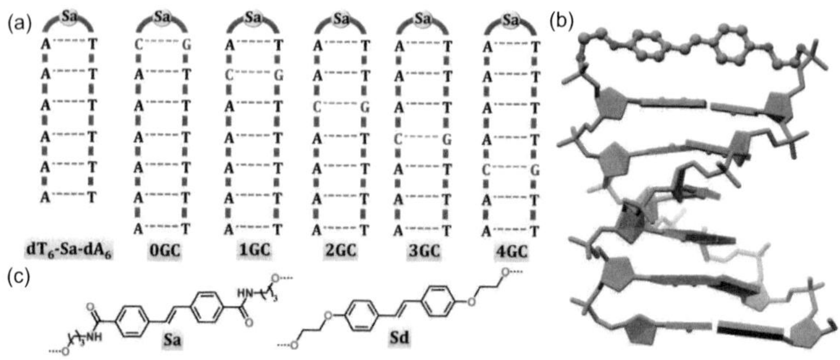

Figure 6.7 (a) A family of stilbene dicarboxamide (Sa)-linked hairpin with a varying A-T step between the G and Sa. (b) Crystal structure of the DNA hairpin with Sd linker (PDB accession code: 1PUY). (c) Structure of stilbene linkers stilbene dicarboxamide (Sa) and stilbenediether (Sd).

Photoinduced electron transfer (PET) in DNA hairpins was investigated by fluorescence and transient absorption spectroscopy. The parent DNA sequence (dT_6-Sa-dA_6) did not demonstrate any evident photoinduced electron transfer characteristics. However, introduction of a single G-C pair (0-4GC) resulted in distance-dependent fluorescence quenching and the consequent formation of a radical anion of stilbene ($Sa^{\bullet-}$). The rates for charge separation (k_{cs}) were calculated from the singlet decay times in the presence and absence of G, while the rates for the charge recombination (k_{cr}) were calculated from the decay of $Sa^{\bullet-}$ transient signature centered at 535 nm (Table 6.2) [45]. Linear plots of log(k_{cs}) or log(k_{cr}) versus R (distance between Sa and G) estimated the distance dependence of the PET as per Eq. 6.1. The values of β = 0.64 and 0.95 Å^{-1} obtained for the stilbene DNA hairpins suggest that the DNA is more efficient than protein (β = 1.0–1.4 Å^{-1}) but less than the fully conjugated bridges with wire-like distance dependence [45, 81].

Though PET in the Sa-linked hairpin was probed by monitoring the formation and decay of $Sa^{\bullet-}$, the dynamics of the hole acceptor ($G^{\bullet+}$) were difficult to perceive owing to its low extinction coefficient [33]. To monitor the dynamics of the hole acceptor, independently, Lewis, Wasielewski, and coworkers and Lewis, Fiebig, Shafirovich, and coworkers designed a hairpin wherein the hole donor (Sa) and acceptor (Sd) are separated by variable A-T base pairs (Fig. 6.8) [82, 83]. The end-capped hairpin sequences display the presence of $Sa^{\bullet -}$ and $Sa^{\bullet +}$ while revealing both charge injection and charge separation processes [82, 83]. When Sa (hole donor) and Sd (hole acceptor) are separated by one or two A-T base pair, the kinetics follow a single-step superexchange mechanism [82, 83]. A plot of $\log(k_{cr})$ versus R gives a β value of 0.97 Å^{-1} for $(AT)'_{1-4}$ sequences, indicating a larger distance dependence, similar to that observed for sequences containing Sa and G and exhibiting the superexchange mechanism [81].

Table 6.2 The rate constants for charge separation (k_{cs}) and charge recombination (k_{cr}) for DNA hairpins

Hairpin	k_{cr} (s^{-1})	k_{cr} (s^{-1})
0GC	1×10^{12}	3.8×10^{10}
1GC	2.0×10^{11}	7.1×10^{9}
2GC	7.8×10^{9}	2.6×10^{8}
3GC	5.1×10^{8}	4.1×10^{6}
4GC	2.1×10^{8}	NA
$(AT)_2$	1.3×10^{10}	1.2×10^{8}
$(AT)_3$	3.1×10^{9}	7.7×10^{6}
$(AT)_4$	1.3×10^{9}	1.7×10^{5}

NA = not available
Values obtained from fluorescence and transient absorption spectroscopy [45, 82]

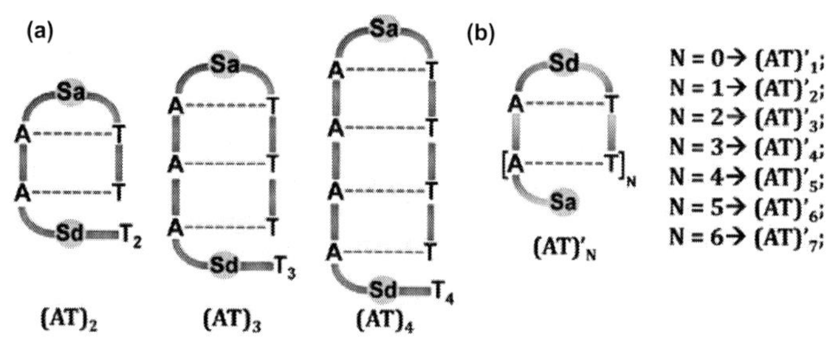

Figure 6.8 End-capped hairpins with (a) an electron acceptor (Sa) in the hairpin loop and (b) an electron donor (Sd) in the hairpin loop.

Long sequences wherein Sa and Sd are separated by two or more intervening A-T base pair exhibit a faster rate for charge injection as compared to charge separation [83]. This observation can be accounted for by a change in the mechanism from superexchange (for shorter strands) to charge injection followed by charge hopping across the DNA bases before the charge gets trapped by the acceptor [83]. The $(AT)'_{5-7}$ sequences exhibit weaker distance dependence, with a β value of 0.42 $Å^{-1}$, and are consistent with the results independently reported by Shafirovich and coworkers [79] and Majima and coworkers [84].

Burin and coworkers provided a kinetic model for charge separation in the stilbene end–capped hairpins (Fig. 6.9a) [85]. The model suggests an analogous insight as that of the experimental result, except that the initial hole injection leads to a $Sa^{\bullet-}A^{\bullet+}$ contact radical pair, or exciplex. The contact radical pair, or the exciplex, acts as an onset for both superexchange for short sequences $[(A\text{-}T)'_{1-3}]$ and the hopping mechanism for longer sequences $[(A\text{-}T)'_{4-7}]$. Ratner and coworkers provided a slightly different depiction of the charge transfer kinetics in the stilbene end–capped hairpins [86]. As per the model, the charge transfer over $(A\text{-}T)'_{1-3}$ follows the superexchange mechanism; however, the longer sequences $(A\text{-}T)'_{4-7}$ exhibit the charge delocalized over several bases (Fig. 6.9b) [86].

Thus, the mechanism of charge transfer in DNA has been highly debated and several mechanisms have been forwarded to explain the experimental results [9, 33, 87]. The results accumulated over more than two decades of research suggest that primary charge separation in DNA occurs via superexchange (exergonic processes) or exciplex/contact radical pair formation (endergonic processes), depending on the energetics of the chromophore and the adjacent nucleobase [87]. The consequent migration of the hole from a neighboring base to a remote trap has been proposed to occur via coherent superexchange [85, 88], incoherent hopping [89], and polaron-like [86, 90, 91] charge transport models [88]. One of the problems with the proposed schemes was that the observation of the intermediate G radical species in the DNA had not been possible to give unequivocal evidence for the projected charge transfer mechanisms. In this direction, Lewis, Wasielewski, and coworkers have recently reported direct observation of hole carriers in diphenylacetylenedicarboxamide (DPA)-linked DNA hairpins [88].

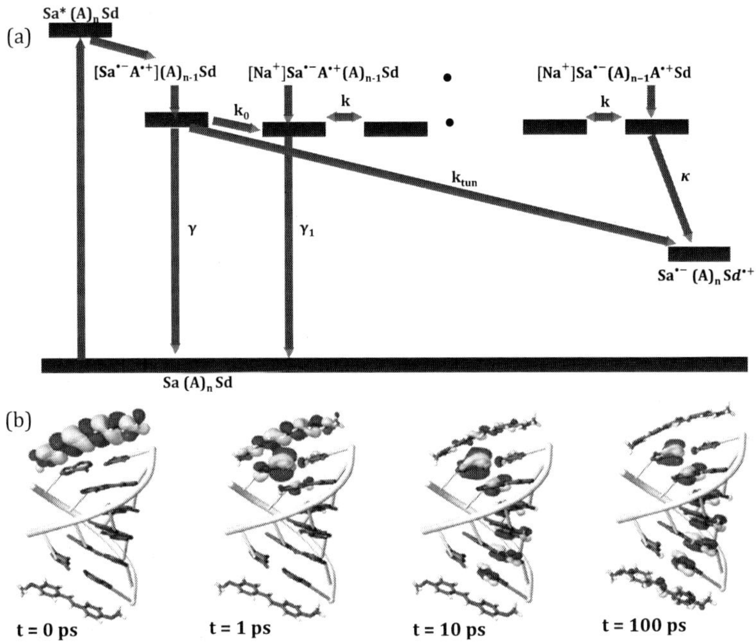

Figure 6.9 (a) Burin's mechanism. (b) Ratner's mechanism for end-capped hairpins with electron acceptor (Sa) and electron donor (Sd). (b) Reprinted with permission from Ref. [85]. Copyright (2013) American Chemical Society.

DPA-linked DNA hairpins consist of a single G-C base pair embedded in a poly(A-T) sequence at varying distances from the DPA chromophore located in the hairpin loop (Fig. 6.10a). For the hairpin wherein DPA and G-C are separated by zero and one (A-T) base pair, the formation of both DPA radical anion ($DPA^{\bullet-}$) and G radical cation ($G^{\bullet+}$) was perceived via fTA and femtosecond stimulated Raman spectroscopy (fSRS). The sequences wherein two or more (A-T) base pairs intervene in the G-C pair and DPA, $DPA^{\bullet-}$ and adenine polaron ($A_n^{\bullet+}$) were observed via fTA spectroscopy. These results illuminate the enduring ambiguity over the mechanism of hole transport following the charge injection process. In a succeeding contribution, Wasielewski, Young, Lewis, and coworkers developed a novel modified guanine nucleobase to envision the hole carrier ($G^{\bullet+}$) in the DNA hairpins. The modified base 8-(4′-phenylethynyl) deoxyguanosine (EG) possesses an oxidation potential similar to G but displays a unique transient signature at 460 nm and a distinct

Raman-active ethynyl stretch for the EG•+ radical species. The rational design of the hairpin embedded with EG at a specified location enabled straightforward tracking of the hole transport dynamics in DNA [92].

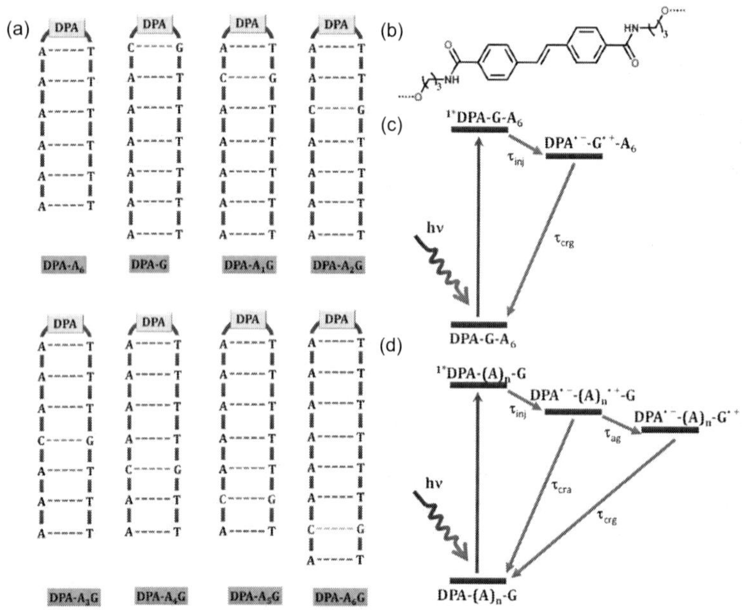

Figure 6.10 (a) A family of diphenylacetylenedicarboxamide (DPA)-linked hairpin with a varying A-T step between the G and DPA. (b) Structure of diphenylacetylenedicarboxamide (DPA). Mechanism of charge migration in (c) DPA-G-A_6- and (d) DPA-$(A)_n$-G-linked DNA hairpins.

While the intricacies of the reported charge transport mechanisms were explored, Grozema, Lewis, Wasielewski, and coworkers reported a novel mechanism, *deep-hole transfer*, accountable for ultrafast charge separation in naphthalenediimide (NDI)-linked DNA hairpins [93]. The structure of the DNA hairpin resembled that of stilbene end-capped hairpins [$(AT)'_{1-7}$], except that the Sa (hole donor) was switched with a more potent NDI moiety (Fig. 6.11) [93]. Depending on the energetics of the hole donor and the adjacent base, an alternate mechanism operated, wherein the low-lying orbitals were accessible by maneuvering the charge injection from chromophore to DNA. The charge transport rates in the NDI end-capped hairpins were found to be 2 orders faster and exhibited weak distance dependence compared to the corresponding hairpin with the Sa hole donor [(AT)′$_{1-7}$] [93]. For longer sequences, with five or more intervening bases, ultrafast charge separation was observed owing to the delocalization of the charge carrier across the base pair. The shorter sequences demonstrated a novel deep-hole transfer mechanism, which further increased the rate of charge separation by accessing the low-lying orbitals [93]. Contrarily, the Sa-linked hairpin can inject the charge only in the highest occupied molecular orbital (HOMO) of the adjacent base, resulting in slow charge injection and trapping at the first base, thereby unveiling a slow charge separation phenomenon. Thus, by manipulating the energetics of the hole donor and the adjacent base, charge migration dynamics in DNA can be regulated.

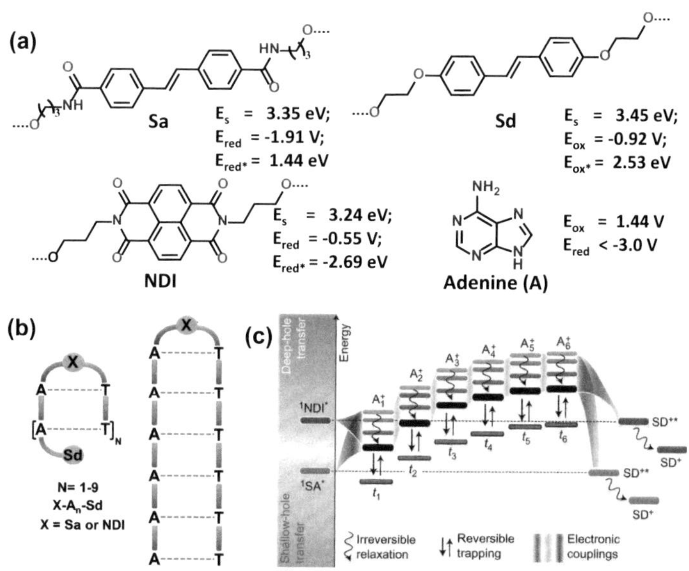

Figure 6.11 (a) Structures, singlet energies, and redox potentials of adenine and hairpin linkers, (b) end-capped hairpin with an electron acceptor (X) in the hairpin loop and an electron donor (Sd) in the hairpin loop and the model hairpin, and (c) mechanism of charge migration in NDI-linked DNA hairpins. (c) Reprinted with permission from Ref. [93]. Copyright © 2016, Springer Nature.

6.4 Nucleobase-Arene Assembly for Long-Lived Charge Separation and Light Harvesting

Pioneering work in the field of DNA charge transfer revolved around the mechanistic intricacies of charge migration in electron donor-/acceptor-conjugated DNA hairpins, or helices, however, with limited control over the electronic properties of the constituent nucleobases. Modified nucleobases unlock an avenue wherein the preprogrammed self-assembly traits of the DNA can be accessed while introducing a desired redox, electronic, or photophysical character [13]. Consequently, donor-/acceptor-conjugated DNA nucleobases have been used to create segregated hierarchical transport pathways for efficient energy and charge migration [17–19, 94, 95]. Some of the initial attempts to explore the electron transfer dynamics in arene-nucleobase hydrogen-bonded assemblies were reported via modification of the nucleobase with appropriate aromatic chromophores [96–102]. In one of the exemplary works, Sessler, Wiederrecht, and coworkers displayed a novel assembly wherein a dimethylaniline (donor)- and anthracene (acceptor)-modified guanosine:cytidine ensemble exhibited hydrogen-bond-mediated charge transfer [94]. Herein, the anthracene-dimethylaniline dyad was held together via noncovalent guanosine-cytidine Watson–Crick base pairing interactions (Fig. 6.12) [94]. The modified nucleobases **1** and **2** were synthesized via palladium-catalyzed cross-coupling reactions, and the formation of the molecular ensemble (1:1 stoichiometry) was confirmed by ^{1}H NMR titrations [94]. The driving force for the electron transfer exhibited an exergonic character (ΔG_{cs}= −0.41 eV and ΔG_{cr}= −2.5 eV), implying the feasibility of charge transfer when probed via steady-state emission, cyclic voltammetry, and ground-state absorption measurements.

The photoinduced electron transfer in the hydrogen-bonded ensemble was envisaged using fluorescence quenching and time-resolved spectroscopic measurements. Interestingly, the fluorescence experiments reveal significant quenching in ensemble I as compared to the binary anthracene:dimethylaniline mixture. Moreover, the lack of quenching in a masked guanosine (donor):cytidine (acceptor) ensemble where no Watson–Crick base pairing occurs suggests the vital role of hydrogen bond in mediating the electron transfer within ensemble I.

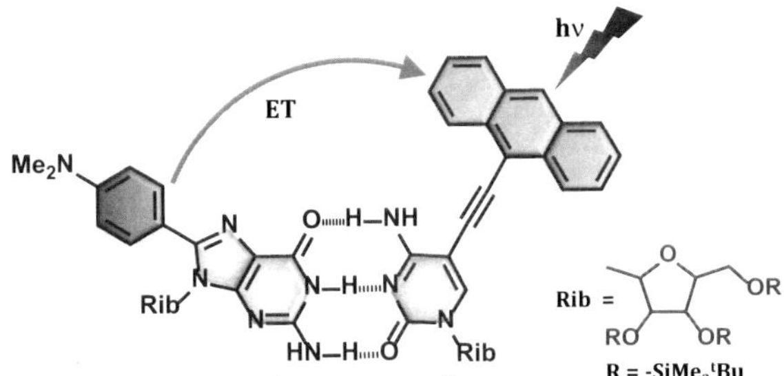

1 2

Ensemble I

Figure 6.12 The Watson–Crick-type noncovalent system investigated by Sessler, Wiederrecht, and coworkers for hydrogen-bond-mediated electron transfer. Dimethylaniline-coupled guanosine (1) and anthracene-coupled cytidine (2).

Furthermore, fTA and time-resolved fluorescence spectroscopy were employed to monitor the charge transfer dynamics of the system. The time-resolved fluorescence (λ_{ex} = 420 nm) measurements indicate a biexponential decay of the singlet excited state of anthracene while emission was monitored at 440 nm. The faster components of the biexponential decay were attributed to the fluorescence quenching arising from the hydrogen-bond-mediated electron transfer in the ensemble, whereas the slower component was attributed to emission from the uncomplexed chromophore. Additional evidence of the electron transfer in the ensemble was provided by fTA measurements. The fTA measurements reveal a strong absorption at 590 nm, corresponding to $S_1 \rightarrow S_n$ transition of anthracene, which decays (τ = 150 ps) with a concurrent rise of a radical cation of dimethylaniline (at 480 nm), leading to an overall k_{cs} = (3.5 ± 0.03) × 10^{-10} s^{-1}. These results demonstrate electron transfer in modified nucleobases and the effect of Watson–Crick hydrogen bonds in the electron transfer process [94].

Along similar lines, Wasielewski and coworkers employed the self-assembly of guanine to define a hierarchical D-A architecture in synthetic G-quadruplexes [17, 103]. Unlike guanine, which acts as a hole trap, G-quadruplexes/G-tetrads can function as a channel for hole migration [17]. Exploiting this unique property of the G-quadruplex, Wasielewski and coworkers devised a series of artificial arene-modified guanines that could self-sort and stack by virtue of hydrogen bonding and strong π-π stacking [17, 19, 95].

In one of the earlier efforts, Wasielewski and coworkers utilized a guanine-perylene-3,4:9,10-bis(dicarboximide) conjugate (GPDI) and explored the role of the G-quartet assembly in extending the lifetimes of the radical intermediates [17]. Upon addition of KF_6 in tetrahydrofuran, the GPDI units self-assembled in the form of a G-quartet, which further transformed into eclipsed G-quartet stacks, as confirmed via X-ray scattering, ^1H NMR, and steady-state ultraviolet-visible (UV-Vis) spectroscopy [17]. The photoexcitation of the G-quartet stacks resulted in photoinduced charge separation, leading to $G^{\bullet+}$– $PDA^{\bullet-}$, which exhibited a survival time >100 times that of the monomeric GPDI dyad. The delocalization of the charge across the G-quadruplex assembly reveals the utility of the nucleobase-arene hierarchical assembly in extending the survival time of the radical intermediates [17].

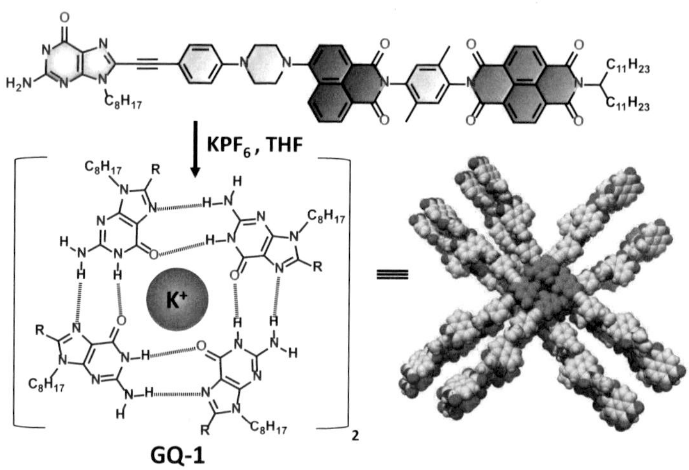

Figure 6.13 Scheme for the formation of the synthetic G-quadruplex GQ-1 from GEAn-ANI-NDI.

In continuation, Wasielewski and coworkers explored guanine-based triad (GEAn-ANI-NDI) in the G-quadruplex structure wherein the 8-(4′-aminophenylethynyl)guanine (GEAn) electron donor is covalently bonded to a 4-aminonaphthalene-1,8-imide (ANI) and a naphthalene-1,8:4,5-bis(dicarboximide) [NDI] electron acceptor (Fig. 6.13) [95]. The oligomeric synthetic G-quadruplexes (GQ-1) was formed analogously to the previous work and was evinced by

^1H and diffusion-ordered spectroscopy (DOSY) NMR, small-angle X-ray scattering (SAXS), and wide-angle X-ray scattering (WAXS) experiments. The monomeric triad and GQ-1 both undergo a two-step charge transfer, as elucidated by time-resolved transient absorption spectroscopy.

Step 1: GEAn—^{1*}ANI—NDI $\rightarrow$ GEAn$^{+\bullet}$—ANI$^{\bullet-}$—NDI
Step 2: GEAn$^{+\bullet}$—ANI$^{\bullet-}$—NDI $\rightarrow$ GEAn$^{+\bullet}$—ANI—NDI$^{\bullet-}$

The lifetime of the first step was nearly comparable in the triad (τ_{cs1} = 5 ps) and GQ-1 (τ_{cs1} = 10 ps). However, the second step, consisting of charge shift, revealed higher rates for GQ-1 (τ_{cs2} = 110 ps) as compared to the triad (τ_{cs2} = 300 ps) [95]. Moreover, the charge recombination is slowed down in GQ-1 ($\tau_{cr} \approx 1$ μs) versus triad (τ_{cr} = 300 ns) due to the radical–pair intersystem crossing that emerges as a result of several hyperfine interactions in the G-quadruplex assembly [95]. Alongside transient absorption spectroscopy, time-resolved electron paramagnetic resonance (TR-EPR) spectroscopy and fSRS support the proposed mechanism. The TR-EPR data further suggest greater dispersion of charge density in GQ-1 compared to the triad, indicating hole sharing between the guanines in the G-quadruplex core could occur concurrently with photoinduced hole injection [95]. Thus, the rapid long-range charge transport in the synthetic G-quadruplex is in agreement with the reported works of Porath and coworkers and Erbe and coworkers in natural G-quadruplexes [104, 105].

Furthermore, Wasielewski and coworkers extended the design of the synthetic G-quadruplexes to create a novel 2D hierarchical D-A motif [19]. The framework consisted of G-tetrad, which acts as an electron donor, and naphthalene-1,4:5,8-bis(dicarboximide) or 2,5,8,11-tetrahexylperylene-3,4:9,10-bis(dicarboximide) as the electron acceptor. The bis(guanine) derivatives, G2NDI and G2PDI, were proposed to form a 2D covalent framework based on the powder X-ray diffraction (PXRD) pattern. The 2D framework of G2NDI and G2PDI has a broad, red-shifted absorption in the visible region, unlike the monomers in solution. This shift is ascribed to exciton coupling and charge transfer interactions in the π-stacked chromophores. Upon photoexcitation, a rapid charge separation occurs that leads to the formation of long-lived G$^{+\bullet}$—NDI$^{\bullet-}$ and G$^{+\bullet}$—PDI$^{\bullet-}$ radical ion pairs in the 2D framework on a timescale of less than 150 fs. The kinetics of the radical ion pairs were fitted to two fast and a slow charge recombination process. The fast processes were attributed to geminate recombination of $G^{+\bullet}$—$NDI^{\bullet-}$ (τ_{cr1} = 8.6 ps [21%]; τ_{cr2} = 106 ps [44%]) and $G^{+\bullet}$—$PDI^{\bullet-}$ (τ_{cr1} = 1.4 ps [31%]; τ_{cr2} = 88 ps [39%]), whereas the slow process was attributed to the charge recombination of the dissociated charge carriers, on the microsecond timescale, in the 2D framework. Thus, the reduced PDI and NDI form a polaron wherein the electron is delocalized over several molecules in the segregated π-stacked array of the 2D framework [19].

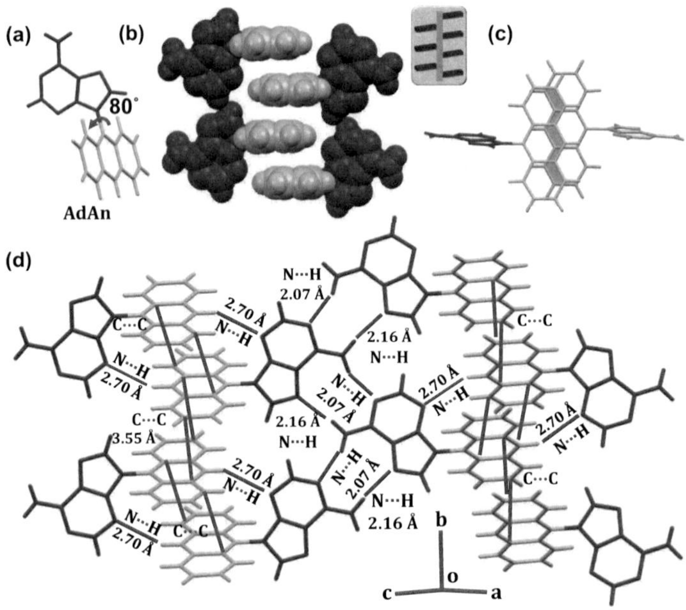

Figure 6.14 (a) Molecular structure; (b) close-packing analysis depicting alternate distichous stacks (inset: the model for alternate distichous arrangement); (c) the percentage of An-An overlap in π-stacked anthracenes; and (d) the close packing arrangement in crystalline AdAn. Reprinted with permission from Ref. [18]. Copyright (2017) American Chemical Society.

In a similar direction, taking inspiration from nature and the reported self-assembly approaches [14, 17], we have used the self-sorting characteristics of DNA to define a hierarchical D-A architecture in the adenine-anthracene (AdAn) dyad (Fig. 6.14)

[18]. By virtue of the strong base pairing and π-π interactions, AdAn forms an alternate distichous assembly, wherein the adenines nearly orthogonally (θ = 80°) flank the columnar anthracene-on-anthracene (An-An) stacks [18]. The close-packing analysis of the alternate distichous assembly reveals that the adenines self-sort via hydrogen bonding and form a motif for the anthracenes to stack via π-π interactions. The preliminary photophysical and redox analysis of AdAn in an organic solvent reveals minimal perturbation in the ground state properties. The frontier molecular orbital analysis suggests favorable charge transfer characteristic in the dyad, which is further corroborated by the negative value of the free energy for electron transfer from ^{1}Ad* to An (ΔG_{et} = –1.25 eV) and Ad to ^{1}An* (ΔG_{et} = –0.12 eV) obtained via Weller analysis.

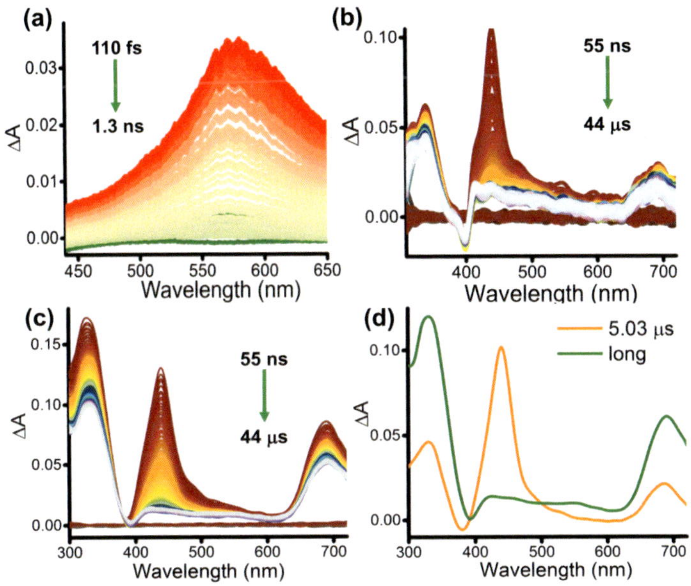

Figure 6.15 (a) fTA spectra (λ_{ex} = 400 nm); nTA spectra (55 ns–44 μs) by exciting at (b) 355 nm and (c) 266 nm. (d) DADS corresponding to the spectra shown in (c). Reprinted with permission from Ref. [18]. Copyright (2017) American Chemical Society.

The evaluation of the charge-separated state in the dyad was further explored using fTA and nanosecond transient absorption (nTA) spectroscopy (Fig. 6.15) [18]. Upon photoexcitation (λ_{ex} = 266 and 355 nm), self-assembled AdAn reveals the formation of radical ion pair intermediates (τ_{cr}^{A} = 120 ms), which displays a solvent polarity dependence as demonstrated via nTA spectroscopy [18]. The "emergence upon assembly" [106] approach offers segregated trajectories of the charge carriers, that is, delocalization of the electron through the columnar π-channels of anthracenes and the diffusion of the holes across hydrogen- bonded adenines, resulting in persistent radical ion pair intermediates.

Subsequently, the AdAn assembly undergoes photodamage to yield undesirable photoproducts at longer timescales (τ > 580 ms). Though the unsolicited photoexcited-state pathways of AdAn lend undesirable photoproducts, the long-lived charge-separated state from a favorable arrangement of D-A stack proposes an elegant D-A assembly design for emergent solution processable photofunctional devices.

6.5 Conclusions

Exploration of the photoinduced phenomena in chromophore-appended nucleobases and DNA oligonucleotides have given profound insights into the fundamental photoexcited-state processes in DNA. The riddle whether DNA can mediate long-range charge transfer had been intensively explored and debated for quite long. Several mechanisms were put forward based on the amalgamation of experimental and theoretical cognition for charge transfer in DNA. On the basis of the essence of the charge transport investigations acquired over a period of more than two decades, it has been understood that the primary charge separation in DNA can occur via superexchange (exergonic reactions), contact radical pair, exciplex formation (endergonic reactions), or deep hole transfer mechanisms, depending upon the energetics of the chromophore and the adjacent nucleobases. Further, long-range transport across the DNA stack can occur via a multistep coherent superexchange, incoherent hopping, or delocalized polaron-like states across the nucleobase stacks. Thus, the mechanism of the DNA charge transport depends on multiple parameters, like (i) energetics of the redox-active chromophore and the adjacent base, (ii) distance between the chromophore and the nucleobase, (iii) identity/sequence of the DNA assembly, and (iv) the relative conformation of the DNA nucleobase and the redox-active chromophore.

Additionally, the self-assembly and hole transport traits of the DNA nucleobases can be employed to create hierarchical assemblies aimed at producing long-lived charge-separated states for light-harvesting applications. This has provided a platform to further extend the self-driven directed assembly of DNA for various applications. As a result, the exploration of the photoexcited-state processes in chromophore-appended nucleic acids is still an open avenue and necessitates further understanding of photoexcited-state processes in DNA.

References

1. Barbatti, M., Borin, A. C. and Ullrich, S. (2016). *Photoinduced Phenomena in Nucleic Acids I: Nucleobases in the Gas Phase and in Solvents*, Springer International Publishing, Switzerland.
2. Lewis, F. D. (2008). Electron transfer and charge transport processes in DNA, in *Electron Transfer in Chemistry* (ed. Balzani, V.), Wiley-VCH Verlag GmbH, Weinheim, Germany, pp. 105–175.
3. Watson, J. D. and Crick, F. H. C. (1953). Molecular structure of nucleic acids: a structure for deoxyribose nucleic acid, *Nature*, **171**, 737–738.
4. Gustavsson, T., Improta, R. and Markovitsi, D. (2010). DNA/RNA: building blocks of life under UV irradiation, *J. Phys. Chem. Lett.*, **1**, 2025–2030.
5. Crespo-Hernández, C. E., Cohen, B., Hare, P. M. and Kohler, B. (2004). Ultrafast excited-state dynamics in nucleic acids, *Chem. Rev.*, **104**, 1977–2020.
6. Seeman, N. C. (2003). Biochemistry and structural DNA nanotechnology: an evolving symbiotic relationship, *Biochemistry*, **42**, 7259–7269.
7. Murphy, C., Arkin, M., Jenkins, Y., Ghatlia, N., Bossmann, S., Turro, N. and Barton, J. (1993). Long-range photoinduced electron transfer through a DNA helix, *Science*, **262**, 1025–1029.
8. Eley, D. D. and Spivey, D. I. (1962). Semiconductivity of organic substances. Part 9-Nucleic acid in the dry state, *Trans. Faraday Soc.*, **58**, 411–415.
9. Wagenknecht, H.-A. (2005). *Charge Transfer in DNA: From Mechanism to Application*, Wiley-VCH, Weinheim.

10. Núñez, M. E., Hall, D. B. and Barton, J. K. (1999). Long-range oxidative damage to DNA: effects of distance and sequence, *Chem. Biol.*, **6**, 85–97.
11. Middleton, C. T., de La Harpe, K., Su, C., Law, Y. K., Crespo-Hernández, C. E. and Kohler, B. (2009). DNA excited-state dynamics: from single bases to the double helix, *Annu. Rev. Phys. Chem.*, **60**, 217–239.
12. Ananthaswamy, H. N., Loughlin, S. M., Cox, P., Evans, R. L., Ullrich, S. E. and Kripke, M. L. (1997). Sunlight and skin cancer: inhibition of p53 mutations in UV-irradiated mouse skin by sunscreens, *Nat. Med.*, **3**, 510.
13. Teo, Y. N. and Kool, E. T. (2012). DNA-multichromophore systems, *Chem. Rev.*, **112**, 4221–4245.
14. Wasielewski, M. R. (2009). Self-assembly strategies for integrating light harvesting and charge separation in artificial photosynthetic systems, *Acc. Chem. Res.*, **42**, 1910–1921.
15. Mallia, A. R., Salini, P. S. and Hariharan, M. (2015). Nonparallel stacks of donor and acceptor chromophores evade geminate charge recombination, *J. Am. Chem. Soc.*, **137**, 15604–15607.
16. Aratani, N., Kim, D. and Osuka, A. (2009). Discrete cyclic porphyrin arrays as artificial light-harvesting antenna, *Acc. Chem. Res.*, **42**, 1922–1934.
17. Wu, Y.-L., Brown, K. E. and Wasielewski, M. R. (2013). Extending photoinduced charge separation lifetimes by using supramolecular design: guanine–perylenediimide G-quadruplex, *J. Am. Chem. Soc.*, **135**, 13322–13325.
18. Philip, A. M., Kuriakose, F. and Hariharan, M. (2017). Unsolicited photoexcited-state pathways relegate the long-lived charge separation in self-assembled nucleobase–arene conjugate, *J. Phys. Chem. C*, **121**, 23259–23267.
19. Wu, Y.-L., Horwitz, N. E., Chen, K.-S., Gomez-Gualdron, D. A., Luu, N. S., Ma, L., Wang, T. C., Hersam, M. C., Hupp, J. T., Farha, O. K., Snurr, R. Q. and Wasielewski, M. R. (2017). G-quadruplex organic frameworks, *Nat. Chem.*, **9**, 466–472.
20. Bucher, D. B., Pilles, B. M., Carell, T. and Zinth, W. (2014). Charge separation and charge delocalization identified in long-living states of photoexcited DNA, *Proc. Natl. Acad. Sci. U.S.A.*, **111**, 4369–4374.
21. Pecourt, J.-M. L., Peon, J. and Kohler, B. (2000). Ultrafast internal conversion of electronically excited RNA and DNA nucleosides in water, *J. Am. Chem. Soc.*, **122**, 9348–9349.

22. Voet, D., Gratzer, W. B., Cox, R. A. and Doty, P. (1963). Absorption spectra of nucleotides, polynucleotides, and nucleic acids in the far ultraviolet, *Biopolymers*, **1**, 193–208.

23. Ismail, N., Blancafort, L., Olivucci, M., Kohler, B. and Robb, M. A. (2002). Ultrafast decay of electronically excited singlet cytosine via a π,π^* to n_O,π^* state switch, *J. Am. Chem. Soc.*, **124**, 6818–6819.

24. Middleton, C. T., Cohen, B. and Kohler, B. (2007). Solvent and solvent isotope effects on the vibrational cooling dynamics of a DNA base derivative, *J. Phys. Chem. A*, **111**, 10460–10467.

25. Santoro, F., Barone, V., Gustavsson, T. and Improta, R. (2006). Solvent effect on the singlet excited-state lifetimes of nucleic acid bases: a computational study of 5-fluorouracil and uracil in acetonitrile and water, *J. Am. Chem. Soc.*, **128**, 16312–16322.

26. Hare, P. M., Crespo-Hernández, C. E. and Kohler, B. (2006). Solvent-dependent photophysics of 1-cyclohexyluracil: ultrafast branching in the initial bright state leads nonradiatively to the electronic ground state and a long-lived $1_n\pi^*$ state, *J. Phys. Chem. B*, **110**, 18641–18650.

27. Lewis, F. D., Letsinger, R. L. and Wasielewski, M. R. (2001). Dynamics of photoinduced charge transfer and hole transport in synthetic DNA hairpins, *Acc. Chem. Res.*, **34**, 159–170.

28. Seidel, C. A. M., Schulz, A. and Sauer, M. H. M. (1996). Nucleobase-specific quenching of fluorescent dyes. 1. Nucleobase one-electron redox potentials and their correlation with static and dynamic quenching efficiencies, *J. Phys. Chem.*, **100**, 5541–5553.

29. Saenger, W. (1984). *Principles of Nucleic Acid Structure*, Springer-Verlag, New York.

30. Kawai, K. and Majima, T. (2013). Hole transfer kinetics of DNA, *Acc. Chem. Res.*, **46**, 2616–2625.

31. Giese, B. (2002). Long-distance electron transfer through DNA, *Annu. Rev. Biochem.*, **71**, 51–70.

32. Kelley, S. O. and Barton, J. K. (1999). Electron transfer between bases in double helical DNA, *Science*, **283**, 375–381.

33. Lewis, F. D. (2013). Distance-dependent electronic interactions across DNA base pairs: charge transport, exciton coupling, and energy transfer, *Isr. J. Chem.*, **53**, 350–365.

34. Bixon, M., Giese, B., Wessely, S., Langenbacher, T., Michel-Beyerle, M. E. and Jortner, J. (1999). Long-range charge hopping in DNA, *Proc. Natl. Acad. Sci. U.S.A.*, **96**, 11713–11716.

35. Barbara, P. F., Meyer, T. J. and Ratner, M. A. (1996). Contemporary issues in electron transfer research, *J. Phys. Chem.*, **100**, 13148–13168.

36. Fromherz, P. and Rieger, B. (1986). Photoinduced electron transfer in DNA matrix from intercalated ethidium to condensed methylviologen, *J. Am. Chem. Soc.*, **108**, 5361–5362.

37. Barton, J. K., Kumar, C. V. and Turro, N. J. (1986). DNA-mediated photoelectron transfer reactions, *J. Am. Chem. Soc.*, **108**, 6391–6393.

38. Purugganan, M., Kumar, C., Turro, N. and Barton, J. (1988). Accelerated electron transfer between metal complexes mediated by DNA, *Science*, **241**, 1645–1649.

39. Murphy, C. J., Arkin, M. R., Ghatlia, N. D., Bossmann, S., Turro, N. J. and Barton, J. K. (1994). Fast photoinduced electron transfer through DNA intercalation, *Proc. Natl. Acad. Sci. U.S.A.*, **91**, 5315–5319.

40. Stemp, E. D. A., Arkin, M. R. and Barton, J. K. (1995). Electron transfer between metallointercalators bound to DNA: spectral identification of the transient intermediate, *J. Am. Chem. Soc.*, **117**, 2375–2376.

41. Arkin, M. R., Stemp, E. D. A., Turro, C., Turro, N. J. and Barton, J. K. (1996). Luminescence quenching in supramolecular systems: a comparison of DNA- and SDS micelle-mediated photoinduced electron transfer between metal complexes, *J. Am. Chem. Soc.*, **118**, 2267–2274.

42. Holmlin, R. E., Stemp, E. D. A. and Barton, J. K. (1996). $Os(phen)_2dppz^{2+}$ in photoinduced DNA-mediated electron transfer reactions, *J. Am. Chem. Soc.*, **118**, 5236–5244.

43. Meade, T. J. and Kayyem, J. F. (1995). Electron transfer through DNA: site-specific modification of duplex DNA with ruthenium donors and acceptors, *Angew. Chem. Int. Ed.*, **34**, 352–354.

44. Arkin, M. R., Stemp, E. D. A., Holmlin, R. E., Barton, J. K., Hörmann, A., Olson, E. J. C. and Barbara, P. F. (1996). Rates of DNA-mediated electron transfer between metallointercalators, *Science*, **273**, 475–480.

45. Lewis, F. D., Wu, T., Zhang, Y., Letsinger, R. L., Greenfield, S. R. and Wasielewski, M. R. (1997). Distance-dependent electron transfer in DNA hairpins, *Science*, **277**, 673–676.

46. Pyle, A. M., Long, E. C. and Barton, J. K. (1989). Shape-selective targeting of DNA by phenanthrenequinone diiminerhodium(III) photocleaving agents, *J. Am. Chem. Soc.*, **111**, 4520–4522.

47. Jenkins, Y., Friedman, A. E., Turro, N. J. and Barton, J. K. (1992). Characterization of dipyridophenazine complexes of ruthenium(II): the light switch effect as a function of nucleic acid sequence and conformation, *Biochemistry*, **31**, 10809–10816.

48. Amouyal, E., Homsi, A., Chambron, J.-C. and Sauvage, J.-P. (1990). Synthesis and study of a mixed-ligand ruthenium(II) complex in its ground and excited states: bis(2,2[prime or minute]-bipyridine)(dipyrido[3,2-a : 2[prime or minute],3[prime or minute]-c]phenazine-N4N5)ruthenium(II), *J. Chem. Soc. Dalton Trans.*, 1841–1845.

49. Arnold, A. R., Grodick, M. A. and Barton, J. K. (2016). DNA charge transport: from chemical principles to the cell, *Cell Chem. Biol.*, **23**, 183–197.

50. Steenken, S. and Jovanovic, S. V. (1997). How easily oxidizable is DNA? One-electron reduction potentials of adenosine and guanosine radicals in aqueous solution, *J. Am. Chem. Soc.*, **119**, 617–618.

51. Sugiyama, H. and Saito, I. (1996). Theoretical studies of GG-specific photocleavage of DNA via electron transfer: significant lowering of ionization potential and 5'-localization of HOMO of stacked GG bases in B-form DNA, *J. Am. Chem. Soc.*, **118**, 7063–7068.

52. Hall, D. B. and Barton, J. K. (1997). Sensitivity of DNA-mediated electron transfer to the intervening π-stack: a probe for the integrity of the DNA base stack, *J. Am. Chem. Soc.*, **119**, 5045–5046.

53. Kelley, S. O. and Barton, J. K. (1998). DNA-mediated electron transfer from a modified base to ethidium: π-stacking as a modulator of reactivity, *Chem. Biol.*, **5**, 413–425.

54. Kelley, S. O., Holmlin, R. E., Stemp, E. D. A. and Barton, J. K. (1997). Photoinduced electron transfer in ethidium-modified DNA duplexes: dependence on distance and base stacking, *J. Am. Chem. Soc.*, **119**, 9861–9870.

55. Waring, M. J. (1965). Complex formation between ethidium bromide and nucleic acids, *J. Mol. Biol.*, **13**, 269–282.

56. Amann, N., Huber, R. and Wagenknecht, H.-A. (2004). Phenanthridinium as an artificial base and charge donor in DNA, *Angew. Chem. Int. Ed.*, **43**, 1845–1847.

57. Wan, C., Fiebig, T., Kelley, S. O., Treadway, C. R., Barton, J. K. and Zewail, A. H. (1999). Femtosecond dynamics of DNA-mediated electron transfer, *Proc. Natl. Acad. Sci. U.S.A.*, **96**, 6014–6019.

58. Tanaka, M., Yukimoto, K., Ohkubo, K. and Fukuzumi, S. (2008). Intermolecular vs. intramolecular photoinduced electron transfer from nucleotides in DNA to acridinium ion derivatives in relation with DNA cleavage, *J. Photochem. Photobiol. A*, **197**, 206–212.

59. Hess, S., Götz, M., Davis, W. B. and Michel-Beyerle, M.-E. (2001). On the apparently anomalous distance dependence of charge-transfer rates

in 9-amino-6-chloro-2-methoxyacridine-modified DNA, *J. Am. Chem. Soc.*, **123**, 10046–10055.

60. Fukui, K. and Tanaka, K. (1998). Distance dependence of photoinduced electron transfer in DNA, *Angew. Chem. Int. Ed.*, **37**, 158–161.

61. Williams, T. T., Dohno, C., Stemp, E. D. A. and Barton, J. K. (2004). Effects of the photooxidant on DNA-mediated charge transport, *J. Am. Chem. Soc.*, **126**, 8148–8158.

62. Genereux, J. C., Augustyn, K. E., Davis, M. L., Shao, F. and Barton, J. K. (2008). Back-electron transfer suppresses the periodic length dependence of DNA-mediated charge transport across adenine tracts, *J. Am. Chem. Soc.*, **130**, 15150–15156.

63. Shao, F., O'Neill, M. A. and Barton, J. K. (2004). Long-range oxidative damage to cytosines in duplex DNA, *Proc. Natl. Acad. Sci. U.S.A.*, **101**, 17914–17919.

64. Bradley, D. F., Stellwagen, N. C., O'Konski, C. T. and Paulson, C. M. (1972). Electric birefringence and dichroism of acridine orange and methylene blue complexes with polynucleotides, *Biopolymers*, **11**, 645–652.

65. Nordén, B. and Tjerneld, F. (1982). Structure of methylene blue–DNA complexes studied by linear and circular dichroism spectroscopy, *Biopolymers*, **21**, 1713–1734.

66. Leng, F., Savkur, R., Fokt, I., Przewloka, T., Priebe, W. and Chaires, J. B. (1996). Base specific and regioselective chemical cross-linking of daunorubicin to DNA, *J. Am. Chem. Soc.*, **118**, 4731–4738.

67. Pascaly, M., Yoo, J. and Barton, J. K. (2002). DNA mediated charge transport: characterization of a DNA radical localized at an artificial nucleic acid base, *J. Am. Chem. Soc.*, **124**, 9083–9092.

68. Nakatani, K., Dohno, C. and Saito, I. (2000). Modulation of DNA-mediated hole-transport efficiency by changing superexchange electronic interaction, *J. Am. Chem. Soc.*, **122**, 5893–5894.

69. Wan, C., Fiebig, T., Schiemann, O., Barton, J. K. and Zewail, A. H. (2000). Femtosecond direct observation of charge transfer between bases in DNA, *Proc. Natl. Acad. Sci. U.S.A.*, **97**, 14052–14055.

70. Genereux, J. C., Wuerth, S. M. and Barton, J. K. (2011). Single-step charge transport through DNA over long distances, *J. Am. Chem. Soc.*, **133**, 3863–3868.

71. O'Neil, M. A. and Barton, J. K. (2004). DNA charge transport: conformationally gated hopping through stacked domains, *J. Am. Chem. Soc.*, **126**, 11471–11483.

72. Nakatani, K., Dohno, C. and Saito, I. (2001). Design of a hole-trapping nucleobase: termination of DNA-mediated hole transport at N2-cyclopropyldeoxyguanosine, *J. Am. Chem. Soc.*, **123**, 9681–9682.

73. Augustyn, K. E., Genereux, J. C. and Barton, J. K. (2007). Distance-independent DNA charge transport across an adenine tract, *Angew. Chem. Int. Ed.*, **46**, 5731–5733.

74. Elias, B., Genereux, J. C. and Barton, J. K. (2008). Ping-pong electron transfer through DNA, *Angew. Chem. Int. Ed.*, **47**, 9067–9070.

75. O'Neil, M. A. and Barton, J. K. (2002). 2-aminopurine: a probe of structural dynamics and charge transfer in DNA and DNA:RNA hybrids, *J. Am. Chem. Soc.*, **124**, 13053–13066.

76. Remington, J. M., Philip, A. M., Hariharan, M. and Kohler, B. (2016). On the origin of multiexponential fluorescence decays from 2-aminopurine-labeled dinucleotides, *J. Chem. Phys.*, **145**, 155101.

77. Wang, Q., Raytchev, M. and Fiebig, T. (2007). Ultrafast energy delocalization and electron transfer dynamics in 2-aminopurine-containing trinucleotides, *Photochem. Photobiol.*, **83**, 637–641.

78. O'Neill, M. A. and Barton, J. K. (2002). Effects of strand and directional asymmetry on base–base coupling and charge transfer in double-helical DNA, *Proc. Natl. Acad. Sci. U.S.A.*, **99**, 16543–16550.

79. Shafirovich, V., Dourandin, A., Huang, W., Luneva, N. P. and Geacintov, N. E. (2000). Electron transfer at a distance induced by site-selective photoionization of 2-aminopurine in oligonucleotides and investigated by transient absorption techniques, *Phys. Chem. Chem. Phys.*, **2**, 4399–4408.

80. Lewis, F. D., Liu, X., Wu, Y., Miller, S. E., Wasielewski, M. R., Letsinger, R. L., Sanishvili, R., Joachimiak, A., Tereshko, V. and Egli, M. (1999). Structure and photoinduced electron transfer in exceptionally stable synthetic DNA hairpins with stilbenediether linkers, *J. Am. Chem. Soc.*, **121**, 9905–9906.

81. Lewis, F. D., Wu, T., Liu, X., Letsinger, R. L., Greenfield, S. R., Miller, S. E. and Wasielewski, M. R. (2000). Dynamics of photoinduced charge separation and charge recombination in synthetic DNA hairpins with stilbenedicarboxamide linkers, *J. Am. Chem. Soc.*, **122**, 2889–2902.

82. Lewis, F. D., Wu, Y., Zhang, L., Zuo, X., Hayes, R. T. and Wasielewski, M. R. (2004). DNA-mediated exciton coupling and electron transfer between donor and acceptor stilbenes separated by a variable number of base pairs, *J. Am. Chem. Soc.*, **126**, 8206–8215.

83. Lewis, F. D., Zhu, H., Daublain, P., Fiebig, T., Raytchev, M., Wang, Q. and Shafirovich, V. (2006). Crossover from superexchange to hopping as the mechanism for photoinduced charge transfer in DNA hairpin conjugates, *J. Am. Chem. Soc.*, **128**, 791–800.

84. Takada, T., Kawai, K., Cai, X., Sugimoto, A., Fujitsuka, M. and Majima, T. (2004). Charge separation in DNA via consecutive adenine hopping, *J. Am. Chem. Soc.*, **126**, 1125–1129.

85. Blaustein, G. S., Lewis, F. D. and Burin, A. L. (2010). Kinetics of charge separation in poly(A)–poly(T) DNA hairpins, *J. Phys. Chem. B*, **114**, 6732–6739.

86. Renaud, N., Berlin, Y. A., Lewis, F. D. and Ratner, M. A. (2013). Between superexchange and hopping: an intermediate charge-transfer mechanism in poly(A)-poly(T) DNA hairpins, *J. Am. Chem. Soc.*, **135**, 3953–3963.

87. Genereux, J. C. and Barton, J. K. (2010). Mechanisms for DNA charge transport, *Chem. Rev.*, **110**, 1642–1662.

88. Harris, M. A., Mishra, A. K., Young, R. M., Brown, K. E., Wasielewski, M. R. and Lewis, F. D. (2016). Direct observation of the hole carriers in DNA photoinduced charge transport, *J. Am. Chem. Soc.*, **138**, 5491–5494.

89. Jortner, J., Bixon, M., Langenbacher, T. and Michel-Beyerle, M. E. (1998). Charge transfer and transport in DNA, *Proc. Natl. Acad. Sci. U.S.A.*, **95**, 12759–12765.

90. Schuster, G. B. (2000). Long-range charge transfer in DNA: transient structural distortions control the distance dependence, *Acc. Chem. Res.*, **33**, 253–260.

91. Zeidan, T. A., Carmieli, R., Kelley, R. F., Wilson, T. M., Lewis, F. D. and Wasielewski, M. R. (2008). Charge-transfer and spin dynamics in DNA hairpin conjugates with perylenediimide as a base-pair surrogate, *J. Am. Chem. Soc.*, **130**, 13945–13955.

92. Brown, K. E., Singh, A. P. N., Wu, Y.-L., Mishra, A. K., Zhou, J., Lewis, F. D., Young, R. M. and Wasielewski, M. R. (2017). Tracking hole transport in DNA hairpins using a phenylethynylguanine nucleobase, *J. Am. Chem. Soc.*, **139**, 12084–12092.

93. Renaud, N., Harris, M. A., Singh, A. P. N., Berlin, Y. A., Ratner, M. A., Wasielewski, M. R., Lewis, F. D. and Grozema, F. C. (2016). Deep-hole transfer leads to ultrafast charge migration in DNA hairpins, *Nat. Chem.*, **8**, 1015.

94. Sessler, J. L., Sathiosatham, M., Brown, C. T., Rhodes, T. A. and Wiederrecht, G. (2001). Hydrogen-bond-mediated photoinduced

electron-transfer: novel dimethylaniline–anthracene ensembles formed via watson–crick base-pairing, *J. Am. Chem. Soc.*, **123**, 3655–3660.

95. Wu, Y.-L., Brown, K. E., Gardner, D. M., Dyar, S. M. and Wasielewski, M. R. (2015). Photoinduced hole injection into a self-assembled π-extended G-quadruplex, *J. Am. Chem. Soc.*, **137**, 3981–3990.
96. Harriman, A., Magda, D. J. and Sessler, J. L. (1991). Photon antennae assembled by nucleic acid base pairing, *J. Phys. Chem.*, **95**, 1530–1532.
97. Harriman, A., Magda, D. J. and Sessler, J. L. (1991). Energy transfer across a hydrogen-bonded, cytosine-derived, zinc-free-base porphyrin conjugate, *J. Chem. Soc., Chem. Commun.*, 345–348.
98. Harriman, A., Kubo, Y. and Sessler, J. L. (1992). Molecular recognition via base pairing: photoinduced electron transfer in hydrogen-bonded zinc porphyrin-benzoquinone conjugates, *J. Am. Chem. Soc.*, **114**, 388–390.
99. Sessler, J. L., Wang, B. and Harriman, A. (1993). Long-range photoinduced electron transfer in an associated but non-covalently linked photosynthetic model system, *J. Am. Chem. Soc.*, **115**, 10418–10419.
100. Sessler, J. L. and Wang, R. (1996). Self-assembly of an "artificial dinucleotide duplex", *J. Am. Chem. Soc.*, **118**, 9808–9809.
101. Sessler, J. L. and Wang, R. (1998). Design, synthesis, and self-assembly of "artificial dinucleotide duplexes", *J. Org. Chem.*, **63**, 4079–4091.
102. Berg, A., Shuali, Z., Asano-Someda, M., Levanon, H., Fuhs, M., Möbius, K., Wang, R., Brown, C. and Sessler, J. L. (1999). A first high-field EPR study of photoinduced electron transfer in a base-paired porphyrin–dinitrobenzene supramolecular complex, *J. Am. Chem. Soc.*, **121**, 7433–7434.
103. Balasubramanian, S., Hurley, L. H. and Neidle, S. (2011). Targeting G-quadruplexes in gene promoters: a novel anticancer strategy?, *Nat. Rev. Drug Discov.*, **10**, 261.
104. Livshits, G. I., Stern, A., Rotem, D., Borovok, N., Eidelshtein, G., Migliore, A., Penzo, E., Wind, S. J., Di Felice, R., Skourtis, S. S., Cuevas, J. C., Gurevich, L., Kotlyar, A. B. and Porath, D. (2014). Long-range charge transport in single G-quadruplex DNA molecules, *Nat. Nanotechnol.*, **9**, 1040.
105. Liu, S.-P., Weisbrod, S. H., Tang, Z., Marx, A., Scheer, E. and Erbe, A. (2010). Direct measurement of electrical transport through G-quadruplex DNA with mechanically controllable break junction electrodes, *Angew. Chem. Int. Ed.*, **49**, 3313–3316.

106. Ley, D., Guzman, C. X., Adolfsson, K. H., Scott, A. M. and Braunschweig, A. B. (2014). Cooperatively assembling donor-acceptor superstructures direct energy into an emergent charge separated state, *J. Am. Chem. Soc.*, **136**, 7809–7812.

Chapter 7

Templated Arrays of Multichromophores and Oligonucleotides Supported by Metal Interactions and Their Functional Relevance

Mitsunobu Nakamura, Tadao Takada, and Kazushige Yamana
Department of Applied Chemistry, University of Hyogo, 2167 Shosha, Himeji, Hyogo 671-2280, Japan
mitunobu@eng.u-hyogo.ac.jp

7.1 Introduction

Because π-stacked multichromophore arrays exhibit unique photophysical and electroconductive properties, they have attracted considerable attention in relation to light harvesting [1–3] and organic field effect transistor materials and devices [4–6]. To seek new types of photo- and electronic functions, it is conceivable to evaluate the fundamental properties of the π-stacked arrays with a finite number of chromophores and then use the findings as feedback to design multichromophore arrays.

Templated DNA Nanotechnology: Functional DNA Nanoarchitectonics
Edited by Thimmaiah Govindaraju

ISBN 978-981-4800-21-1 (Hardcover), 978-0-429-42866-1 (eBook)
www.panstanford.com

Oligonucleotides are attractive molecules as construction materials for nanoarchitectures [7–10] as well as for multichromophore arrays. Using oligonucleotides, in principle, we can arrange a finite number of chromophores in a defined space and distance and also control the arrangement structures via the self-assembling properties. While the multichromophore arrays on oligonucleotides have been constructed via covalent [11, 12] and noncovalent [13] approaches, their photonic and electronic properties should await further exploration. Especially, the noncovalent approaches have achieved defined-length arrays and fiber-like arrays of great lengths using hydrogen bonding as the driving force for the chromophore assembly [13].

As a new strategy of noncovalent approaches, the coordinate bonding between metal complexes with nucleobases could be anticipated as the driving force. It has been well known that the Zn^{II}-cyclen complex selectively and strongly binds to the thymine base, disrupting the A-T base pair in double-stranded DNA via coordination of the Zn^{II} ion with the deprotonated imide nitrogen of thymine and via hydrogen bonding between the carbonyl groups of thymine and the NH groups of cyclen [14, 15]. This characteristic recognition of Zn^{II}-cyclen has been applied to probes for DNA assays [16–18], a cross linker [19], a cleavage reagent [20], and a supramolecule [21, 22]. Thus, we can utilize Zn^{II}-cyclen as the receptor unit of chromophore building blocks. Moreover, it has the advantage that the functionalization with hydrophilic Zn^{II}-cyclen can drastically enhance the solubility of hydrophobic chromophores in aqueous media.

Among various chromophores, diketopyrrolopyrrole (DPP) 1,4,5,8-naphthalenetetracarboxylic acid diimide (NDI) chromophores are widely used in optoelectronic devices, such as photovoltaic cells [23–25] and field-effect transistors [26, 27]. Especially DPP derivatives have several advantages in terms of ease of synthesis and modification, absorption of visible light, and fluoresce with a high quantum yield.

This chapter describes the synthesis of a bis(2-thienyl)-DPP and NDI derivative having two Zn^{II}-cyclens (Scheme 7.1), the construction of DNA templated multichromophore arrays, and the photoelectrochemical properties of the arrays in self-assembled monolayers on a gold (Au) substrate. Ultraviolet-visible (UV-Vis) absorption, circular dichroism (CD), and gel filtration chromatograph studies indicate that the bis(Zn^{II}-cyclen)-DPP and –NDI derivatives interact with oligodeoxythymidines (oligo-dT$_n$) to yield π-stacked multichromophore arrays. The assembly of DPP and NDI is found to be dependent on the length of the oligo-dT$_n$. The multichromophore arrays on a Au substrate exhibit photocurrent responses due to electron transfer through the π-stacked array.

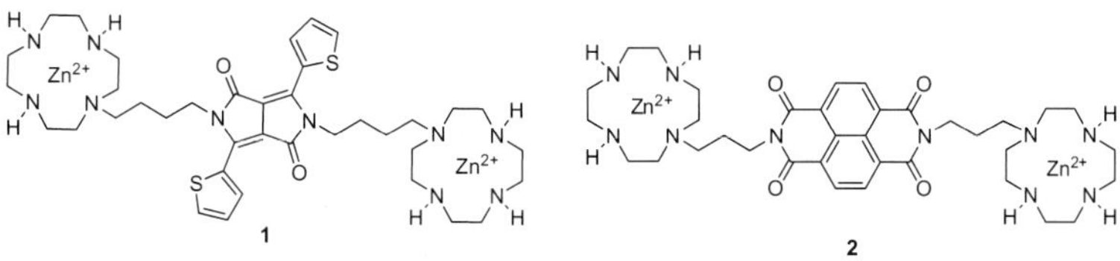

Scheme 7.1 Structures of **1** and **2**.

7.2 Synthesis of Building Blocks

7.2.1 Synthesis of Diketopyrrolopyrrole Building Block

The building block **1** can be prepared according to Scheme 7.2; bis(2-thienyl)-DPP **3**, which is readily obtained from the reaction of 2-thiophenecarbonitrile and diethyl succinate, is reacted with 1,4-dibromobutane to convert to dibromo-compound **4**. The compound **4** is then allowed to react with 1,4,7,10-tetraazacyclododecane to give **5**, and the subsequent complexation of **5** with $Zn^{II}(ClO_4)_2•6H_2O$ yields **1** as ClO_4^- salts. The structure of the final compound can be clarified by mass spectroscopic analysis.

Synthesis of 3: The mixture of 2-thiophenecarbonitrile (1.0 g, 9.2 mmol), diethyl succinate (0.80 g, 4.6 mmol), and potassium *tert*-butoxide (1.5 g, 13 mmol) in *tert*-amyl alcohol (30 mL) was stirred overnight at 100°C under an argon atmosphere. The mixture was then cooled to room temperature, and the precipitate **3** was filtered, washed with acetone, and dried under reduced pressure. Compound **3** was used without further purification. Yield, 53% (0.92 g).

Synthesis of 4: The mixture of compound **3** (0.92 g, 2.4 mmol) and 1,4-dibromobutane (1.3 g, 6.1 mmol) in *N,N*-dimethylformamide

(DMF) (20 mL) was stirred overnight at room temperature under an argon atmosphere. The solvent was removed under reduced pressure, and compound **4** was purified from the residue by silica gel column chromatography using dichloromethane containing methanol (20:1, CH_2Cl_2-MeOH, v/v). Yield, 12% (0.15 g); ^{1}H-nuclear magnetic resonance (NMR, 500 MHz, $CDCl_3$): δ (ppm) = 8.91 (d, J = 3.95 Hz, 2H), 7.67 (d, J = 5.00 Hz, 2H), 7.31 (dd, J = 3.95 and 5.00 Hz, 2H), 4.14 (t, J = 7.25 Hz, 4H), 3.46 (t, J = 6.43 Hz, 4H), 1.98 (m, 4H), and 1.93 (m, 4H); mass spectroscopy (MS) (matrix-assisted laser desorption ionization time-of-flight [MALDI-TOF]): calculated for $C_{22}H_{22}Br_2N_2O_2S_2$ [M^+] 567.92, found to be 567.95.

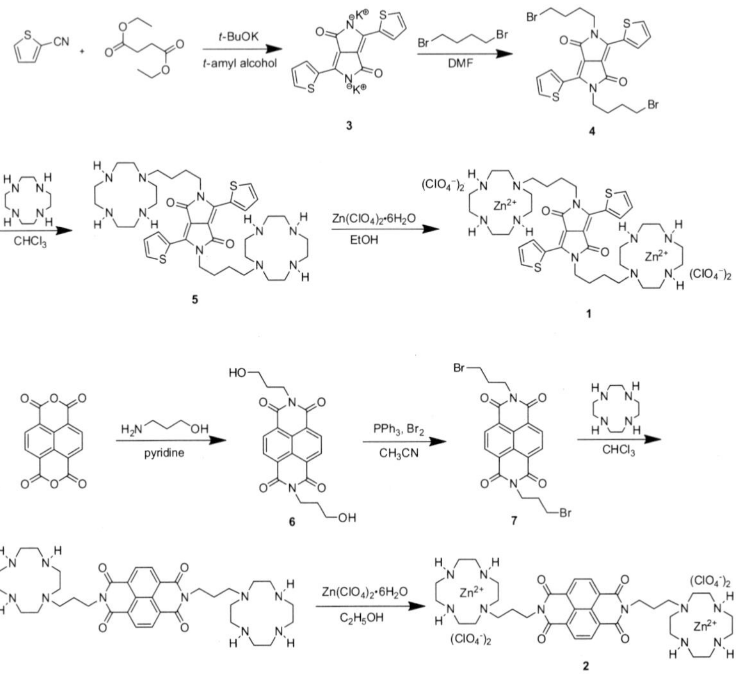

Scheme 7.2 Syntheses of **1** and **2**.

Synthesis of 5: The solution of compound **2** (0.12 g, 2.4 mmol) and 1,4,7,10-tetraazacyclododecane (0.15 g, 0.82 mmol), and triethylamine (0.14 g, 0.63 mmol) in $CHCl_3$ (40 mL), was refluxed overnight. After cooling, the crude mixture was washed with 0.1 M NaOH and several portions of water, dried over Na_2SO_4, and evaporated. Compound **3** was purified from the residue by alumina (neutral) column chromatography using CH_2Cl_2 as the eluant. Yield, 67% (0.11 g); MS (MALDI-TOF): calculated for $C_{38}H_{61}N_{10}O_2S_2^+$ [$(M+H)^+$] 753.44, found to be 753.19.

Synthesis of 1: The solution of compound **3** (0.1 g, 0.13 mmol) and Zn^{II} $(ClO_4)_2{\bullet}6H_2O$ (0.47 g, 0.78 mol) in ethanol was refluxed overnight under an argon atmosphere. The hot solution was filtered, and the filtrate was cooled to room temperature. The precipitated crystal was collected, washed with small portions of ethanol, and dried under reduced pressure. Yield, 54% (0.09 g); ^{1}H-NMR (500 MHz, D_2O): δ (ppm) = 8.22 (d, J = 3.70 Hz, 2H), 7.73 (d, J = 5.45 Hz, 2H) 7.17 (dd, J = 3.70 and 5.45 Hz, 2H), 2.83 (t, J = 6.67 Hz, 4H), 2.7-2.5 (broad, 36H), 1.45 (m, 4H), 1.41 (m, 4H); MS (MALDI-TOF): calculated for $C_{38}H_{57}N_{10}O_2S_2Zn_2^+$ [$(M-3H)^+$] 877.27, found to be 876.89.

7.2.2 Synthesis of Naphthalenetetracarboxylic Acid Diimide Building Block

For the synthesis of NDI-block **2**, *N,N′*-bis(3-hydroxypropyl)-1,4,5,8-naphthalenetetracarboxylic acid diimide **6**, which is obtained from the reaction of 1,4,5,8-naphthalenetetracarboxylic dianhydride and 3-amino-1-propanol, is converted to *N,N′*-bis(3-bromopropyl)-1,4,5,8-naphthalenetetracarboxdiimide **7** with triphenylphosphine and bromine. Compound **7** is allowed to react with 1,4,7,10-tetraazacyclododecane, and subsequent complexation with $Zn^{II}(ClO_4)_2{\bullet}6H_2O$ gives **2** (Scheme 7.2).

Synthesis of 6: To a suspension of 1,4,5,8-naphthalenetetracarboxylic dianhydride (1 g, 3.7 mmol) in pyridine, 3-amino-1-propanol (0.56 g, 7.4 mmol) and $Zn^{II}(OAc)_2{\cdot}2H_2O$ (1.6 g, 7.8 mmol) was added, and the mixture was refluxed overnight under an argon atmosphere. After refluxing, the pyridine was removed under reduced pressure. Compound **6** was purified by silica gel column chromatography using dichloromethane containing methanol (9:1, CH_2Cl_2-MeOH, v/v). Yield, 20% (0.28 g). Melting point (Mp): 279°C–281°C. Thin-layer chromatography (TLC) (9:1, CH_2Cl_2-MeOH, v/v): R_f = 0.67. ^{1}H-NMR (500 MHz, DMSO-d_6, 25°C): δ = 1.80 (tt, J = 6.1 and 7.4 Hz, 4H), 3.51 (dt, J = 5.1 and 6.1 Hz, 4H), 4.08 (t, J = 7.4 Hz, 4H), 4.53 (t, J

= 5.1 Hz, 2H), and 8.55 (s, 4H) ppm. 13C-NMR (125 MHz, DMSO-d6, 25°C): δ = 30.72, 38.09, 58.99, 125.75, 130.17, and 162.13 ppm.

Synthesis of 7: Bromine (0.27 g, 1.7 mmol) was added to the solution of triphenylphosphine (0.4 g, 1.5 mmol) in dry acetonitrile. To the solution was added the solution of **6** (0.28 g, 0.73 mmol) in $CHCl_3$ with vigorous stirring. The reaction mixture was refluxed for 2 h with stirring and then cooled to room temperature. The precipitated **7** was filtered and then used in the next step without further purification. Yield, 19% (0.36 g). TLC (9:1, CH_2Cl_2-MeOH, v/v): R_f = 0.90.

Synthesis of 2: The solution of **7** (0.36 g, 0.71 mmol) and 1,4,7,10-tetraazacyclododecane (0.85 g, 4.9 mmol), and triethylamine (0.14 g, 1.42 mmol) in $CHCl_3$, was refluxed overnight. After cooling, the crude mixture was washed with 0.1 M NaOH and several portions of water, dried over Na_2SO_4, and evaporated. The residue was dissolved in 10 mL of ethanol, and a solution of $Zn^{II}(ClO_4)_2$•$6H_2O$ (1g, 2.7 mmol) in ethanol was then added. The mixture was stirred for 2 h at room temperature. The precipitate was collected and recrystallized with H_2O/ethanol. Yield, 12% (0.52 g). Mp: 244°C (decomposed). ^{1}H-NMR (500 MHz, DMSO-d_6, 25°C): δ = 1.94 (broad, 4H), 2.69 (broad, 32H), 2.89 (broad, 4H), 4.09 (t, J = 6.2 Hz, 4H), 4.38 (broad, 2H), 4.49 (broad, 4H), 8.71 (s, aromatic, 4H) ppm. ^{13}C-NMR (125 MHz, DMSO-d_6, 25°C): δ = 42.29, 43.38, 44.43, 49.66, 50.85, 59.40, 126.64, 130.68, and 163.00 ppm. Elemental analysis (%), calculated for ($C_{36}H_{54}Cl_4N_{10}O_{20}Zn_2$): C, 35.46; H, 4.46; N, 11.49; found: C, 35.93; H, 4.34; N, 11.43.

7.3 Construction of Multichromophore Arrays

7.3.1 Complex Formation with Thymidine Monophosphate and Dithymidine Monophosphate

The formation of a complex between **1** and deprotonated 5′-deoxythymidine monophosphate (5′-dTMP) can be confirmed by using MALDI-TOF MS. Peaks corresponding to [**1**·2(5′-dTMP) – H^+]$^-$ (m/z = 1519.81) agree with the theoretical isotopic distributions for [**1**·2(5′-dTMP) – H^+]$^-$ (m/z for $[C_{58}H_{85}N_{14}O_{18}P_2S_2Zn_2]^-$ = 1519.37) (Fig. 7.1a). In addition, the formation of a complex with di-thymidine monophosphate (dT_2) is also confirmed. In the MALDI-TOF spectrum, a singly charged ion at m/z = 2846.20, whose isotope pattern corresponds to that of [2(**1**·dT_2) – H^+]$^+$ (m/z for $[C_{116}H_{167}N_{28}O_{28}P_2S_4Zn_4]^+$ = 2845.80) (Fig. 7.1b), is observed.

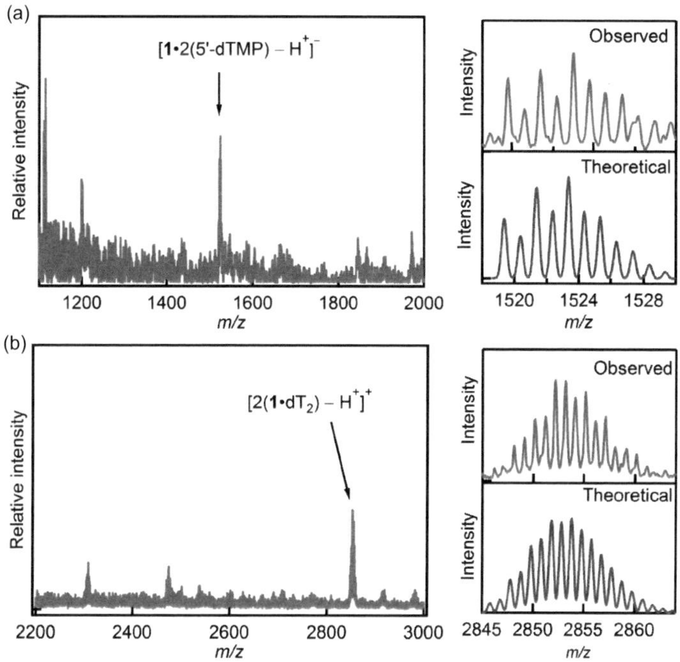

Figure 7.1 MALDI-TOF mass spectra of (a) **1**·2(5′-dTMP) and (b) 2(**1**·dT2).

7.3.2 UV-Vis Titrations with Oligodeoxythymidines

The formation of multichromophore arrays can be indicated by UV-vis titrations. Figure 7.2a shows UV-Vis titration spectra of **1** with dT_{50}. In the absence of the dT_{50}, **1** exhibits an absorption band due to the DPP moiety in the range of 420–600 nm. When dT_{50} is added, the intensity of the DPP absorption band gradually decreases and the peak red-shifts and sharpens. In addition, the peak shows clear vibrionic structures. These changes in the absorption band suggest the formation of chromophore arrays [28]. UV-Vis titration with other dT_ns (n = 10, 20, 30, and 40) shows similar spectral changes. On the other hand, no change in the DPP absorption is observed when dA_{20} is used as the titrant. These indicate that the Zn^{II}-cyclens of **1** are selectively bound by the thymine bases and that the DPP moieties are stacked with each other.

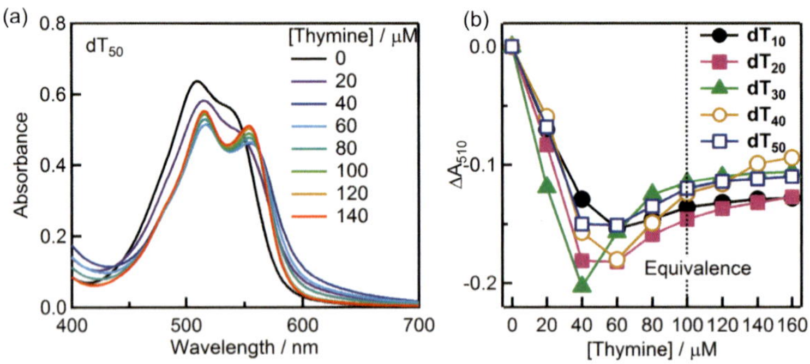

Figure 7.2 (a) UV-Vis titration spectra of **1** (50 μM) with dT_{50} in a buffer of pH 7.6 containing 0.1 M $NaNO_3$ at room temperature. (b) Plots of ΔA_{510} of **1** versus thymine concentration in dT_ns.

Figure 7.2b shows the changes in the absorbance of **1** at 510 nm (ΔA_{510}) during UV-Vis titration with the dT_ns. ΔA_{510} decreases as [Thymine] is increased to 50 μM (half equivalent to the Zn^{II}-cyclen of **1**) and then increases as [Thymine] is increased to 100 μM (equivalence point). Moreover, ΔA_{510} hardly changes at [Thymine] greater than the equivalence point, suggesting that the Zn^{II}-cyclens are stoichiometrically coordinated by the thymine bases. It is plausible that loosely ordered DPP arrays mainly form on single-stranded dT_ns in the concentration region 0 < [Thymine] < 50 μM, and the loosely ordered DPP arrays are gradually regulated by a second dT_n strand in the region 50 < [Thymine] < 100 μM to ultimately form J-type DPP arrays.

Figure 7.3a demonstrates the titration spectra of **2** with dT_{50}, the results of the UV-Vis titration. In the absence of dT_{50}, the characteristic absorption bands derived from the NDI moiety appear in the region from 300 nm to 400 nm. On increasing the concentration of dT_{50}, NDI absorption decreases with an isosbestic point at 295 nm. The peaks of the NDI absorption only slightly shift during the titration; the peak intensity ratio (A_{362}/A_{382}) at 362 and 382 nm, however, increases from 0.92 to 1.10. Similar spectral changes are observed in the UV-Vis titration of other dT_ns (n = 10, 20, 30, and 40). Figure 7.3b exhibits the change in the absorption coefficient (ε) values of **2** at 382 nm. The ε_{382} values decrease with an increase in the amount of the thymine base. These UV-Vis titration results also indicate that the Zn^{II}-cyclens of **2** selectively bind to the thymine bases to form the 1:1 complex and that the NDIs of **2** are stacked with each other. The number of stacked NDIs is consistent with the number of thymine bases or the length of the oligo-dTs [29].

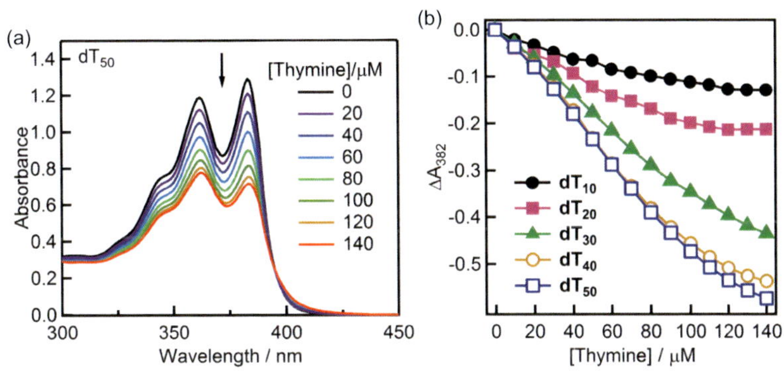

Figure 7.3 (a) UV-Vis titration spectra of **2** with dT_{50} in a pH 7.0 buffer containing 0.1 M NaCl at room temperature, and (b) plots of ΔA_{382} of **1** versus thymine concentration in dT_ns.

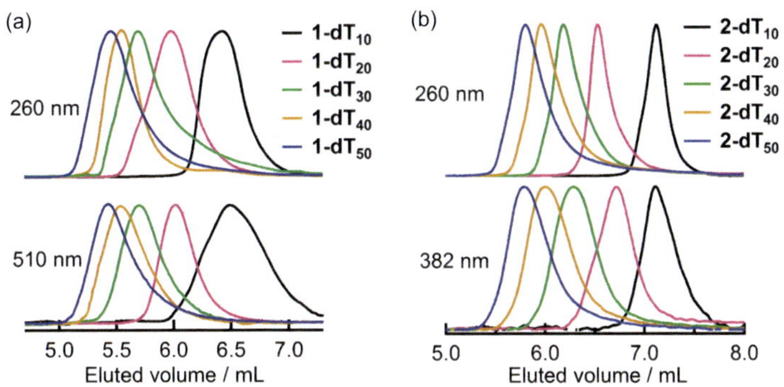

Figure 7.4 Gel filtration chromatograms of (a) **1-dT$_n$s** (n = 10, 20, 30, 40, and 50) monitored at 260 and 510 nm, and (b) **2-dT$_n$s** (n = 10, 20, 30, 40, and 50) monitored at 260 and 382 nm.

7.3.3 Gel Filtration Chromatography

The formation of the multichromophore arrays was characterized by using gel filtration chromatography (GFC) using a size exclusion column. Figure 7.4 shows GFC charts for 1:1 mixtures of chromophore blocks and dT_ns. At both wavelengths, each mixture displayed a unimodal peak with an eluted volume that reflects the size of the fused molecules. These GFC results indicate that uniform multichromophore arrays form when dT_n are added (Fig. 7.5).

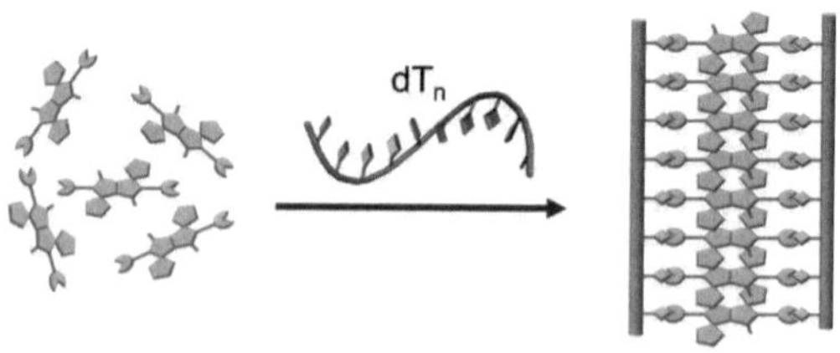

Figure 7.5 Schematic illustration of the construction of multichromophore arrays.

7.3.4 Circular Dichroism Spectra

CD spectroscopic analyses of **1**- and **2-dT**$_n$s were performed to clarify the structural preferences of multichromophore stacks. Figure 7.6a demonstrates CD spectra of **1-dT**$_n$s. In the DPP absorption region (400–600 nm), induced CD signals with a positive maximum at 570 nm and a negative maximum at 505 nm were observed. The positive and negative CD signals were attributed to exciton coupling among the DPP moieties because achiral **1** itself exhibits no CD signal. Moreover, the presence of exciton coupling supports the formation of multi-DPP arrays.

For **2-dT**$_n$s, exciton coupling CD signals with a positive maximum at 365 nm and a negative maximum at 398 nm were observed in the NDI absorption region (320–400 nm) (Fig. 7.6b).

7.3.5 Thermal Dissociation

Figure 7.7a indicates the thermal-dependent absorbance changes in **2-dT**$_{50}$ monitored at 260 and 382 nm. With an increase in the temperature, the absorbance intensity at 382 nm increased. In a similar experiment, a hyperchromic change was observed at 260 nm. With a decrease in the temperature, the absorbance intensity at 382 nm decreased and a hyperchromic change at 260 nm was observed. Such hyperchromic effect was not observed in the dissociation profile of **2** without dT_ns (Fig. 7.7b); thus the self-association of **2** is negligible.

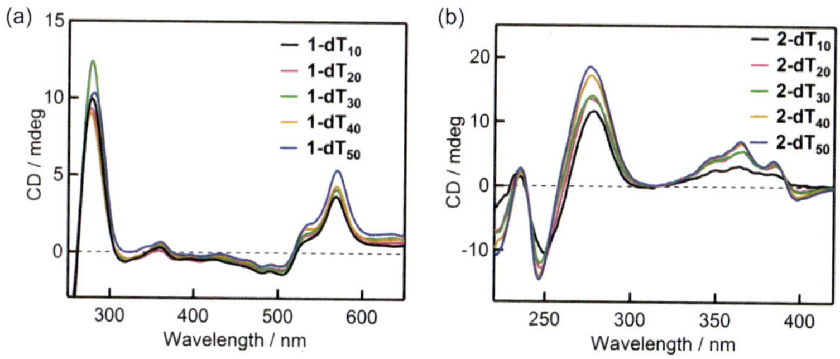

Figure 7.6 CD spectra of (a) **1**-dT_ns (n =10, 20, 30, 40, and 50) and (b) **2**-dT_ns (n =10, 20, 30, 40, and 50) in a buffer at room temperature ([**1**] and [**2**] = 50 μM and [dT_n] = 100 μM: nucleobase concentrations).

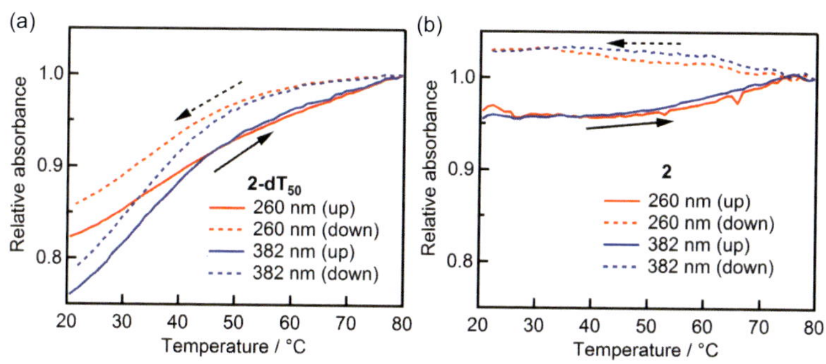

Figure 7.7 (a) Thermal dissociation and association profiles of **2**-dT_{50}, and (b) dissociation and association profiles of **2**.

The CD spectra of **2-dT$_{50}$** at various temperatures are depicted in Fig. 7.8a. The exciton-coupled CD signals in the region of 320–420 nm disappeared with an increase in the temperature. The temperature dependence of the CD at 365 nm is shown in Fig. 7.8b. The CD signal due to exciton coupling decreased with an increase in the temperature to 80°C and reversibly recovered with cooling to 25°C. These results imply that the complexation of Zn^{II}-cyclens of **2** and thymine bases occurs reversibly and exothermically. The formation of the multi-NDI arrays assisted by the thymidine oligomers is therefore controllable by temperature.

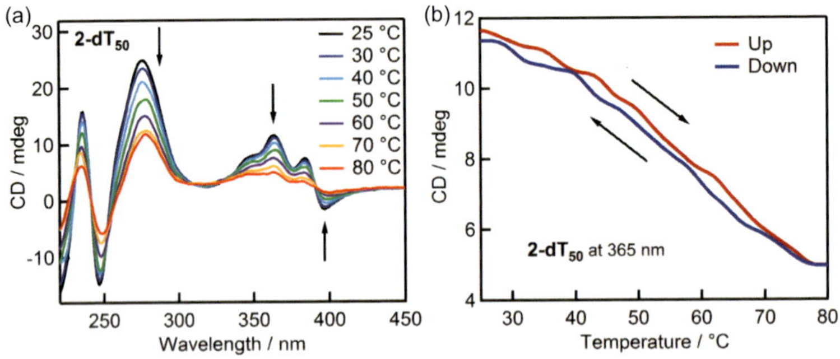

Figure 7.8 (a) CD spectra of **2-dT$_{50}$** mixture at various temperatures, and (b) thermal dissociation and association profiles in CD at 365 nm for **2-dT$_{50}$** mixture.

7.3.6 Stability of Multichromophore Arrays

DNA-templated multichromophore aggregation suppresses the charge transfer (CT) complexation among the chromophore building blocks, and the multichromophore arrays exist stably without disruption of their structures at room temperature. Figure 7.9 shows the absorption spectra of **1-dT$_{40}$** (red line), **2-dT$_{40}$** (blue line), and an equal mixture of **1-dT$_{40}$** and **2-dT$_{40}$** (black line, defined as **1-dT$_{40}$/2-dT$_{40}$**). The chromophore array solutions were adjusted to concentrations of 1.25 μM, corresponding to 50 μM of each chromophore. In the spectra, an absorption band due to dT_{40} appeared at around 260 nm. The mixing of **1-dT$_{40}$** and **2-dT$_{40}$** hardly changed the DPP and NDI absorption bands. The DPP and NDI absorption bands of **1-dT$_{40}$/2-dT$_{40}$** are nearly equal to the DPP absorption bands of **1-dT$_{40}$** and the NDI absorption bands of the spectral sum of **1-dT$_{40}$** and **2-dT$_{40}$** (green dot line), respectively. This indicates that **1-dT$_{40}$** and **2-dT$_{40}$** exist stably in the solution without disruption of their structures.

In contrast to multichromophore arrays, **1** formed the ground-state CT complex with **2** in solution. As shown in Fig. 7.10a, the

DPP absorption bands of **1** (red line) appeared at 510 and 536 nm, whereas the NDI absorption bands of **2** (blue line) appeared at 362 and 384 nm. By mixing **1** and **2** (black line, defined as **1-2**) at room temperature, the DPP absorption bands of **1** decreased and shifted to longer wavelengths (516 and 541 nm), and the shoulder at the longer wavelength side of the DPP absorption bands increased slightly. Judging from the spectral sum of **1** and **2** (green dot line), the NDI absorption bands of **2** also decreased. These results suggest that mixing **1** and **2** in a buffer solution results in the reversible formation of the ground-state CT complex.

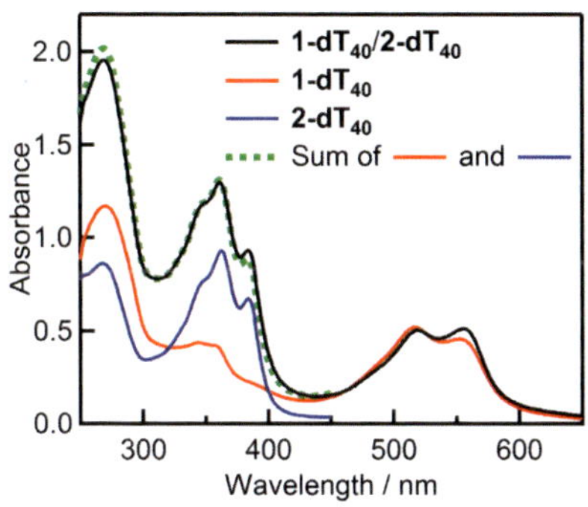

Figure 7.9 Absorption spectra of **1-dT$_{40}$** (1.25 μM), **2-dT$_{40}$** (1.25 μM), and **1-dT$_{40}$/2-dT$_{40}$** (the mixture of **1-dT$_{40}$** [1.25 μM] and **2-dT$_{40}$** [1.25 μM]) in a pH 7.6 solution containing 10 mM HEPES and 0.1 M $NaNO_3$ at room temperature.

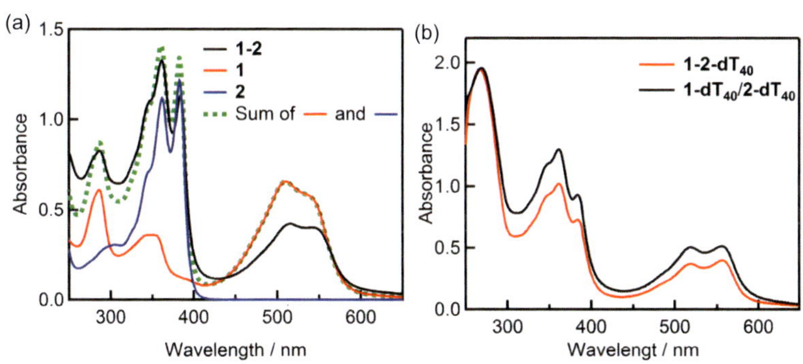

Figure 7.10 (a) Absorption spectra of **1** (50 μM), **2** (50 μM), and **1-2** (the mixture of **1** [50 μM] and **2** [50 μM]). (b) **1-2-dT$_{40}$** (1.25 μM), **1-2**, and **1-dT$_{40}$/2-dT$_{40}$**. The measurements were carried out in a pH 7.6 solution containing 10 mM HEPES and 0.1 M $NaNO_3$ at room temperature.

In the UV-Vis titration of **1-2** with dT_{40}, both DPP absorption and NDI absorption bands decreased gradually and the DPP absorption bands shifted to longer wavelengths (518 and 556 nm) with the addition of dT_{40}. However, the absorption spectrum of **1-2** with dT_{40} (**1-2-dT$_{40}$**) differed from the absorption spectra of **1-dT$_{40}$/2-dT$_{40}$** (Fig. 7.10b). The DPP and NDI absorption bands of **1-2-dT$_{40}$** (red line) were obviously lower than those of **1-dT$_{40}$/2-dT$_{40}$** (black line), although the absorption band of dT_{40} at around 260 nm was nearly equal to that of **1-dT$_{40}$/2-dT$_{40}$**. Therefore, in this system, random multichromophore arrays containing **1** and **2** and their CT complex are formed using dT_{40}.

7.4 Photoelectrochemical Properties of Multichromophore Arrays

Multichromophore arrays were constructed with 3′-thiol-modified dT_n-DNA and self-assembled monolayers of the arrays on a gold surface. Figure 7.11 illustrates the modified Au electrodes employed in this work; **1-dT$_n$E** is a **1-dT$_n$**-immobilized electrode and **2-dT$_n$E** is a **2-dT$_n$**-immobilized electrode. **1-dT$_n$/2-dT$_n$E** consists of **1-dT$_n$** and **2-dT$_n$**, such as a donor/acceptor heterojunction configuration prepared by immersing the electrode in a solution of **1-dT$_n$/2-dT$_n$**. **E2** is a **1-2-dT$_n$** functionalized electrode. Photoelectrochemical measurements were carried out in an argon-saturated aqueous solution containing methylvilogen dichloride (MV^{2+}), triethanolamine (TEOA), or ascorbic acid (AsA) as an electron sacrifice reagent. The multichromophore array–immobilized electrode was used as a working electrode, along with a platinum (Pt) counterelectrode and an Ag/AgCl reference electrode in a three-electrode system.

7.4.1 Photoelectrochemical Property of Diketopyrrolopyrrole Arrays

Photoirradiation of **1-dT$_{40}$E** using a 560 nm band pass filter caused cathodic photocurrents to occur immediately in the presence of 10 mmol MV^{2+} (Fig. 7.12a). The action spectra for **1-dT$_{40}$E** in the region of 460–600 nm (Fig. 7.12b) resembled the absorption spectrum of **1-dT$_{40}$** in the buffer solution, indicating that the DPP moiety of

1 is the photoactive species. On the other hand, photoirradiation of **1-dT$_{40}$E** in the presence of AsA causes anodic photocurrents (Fig. 7.13).

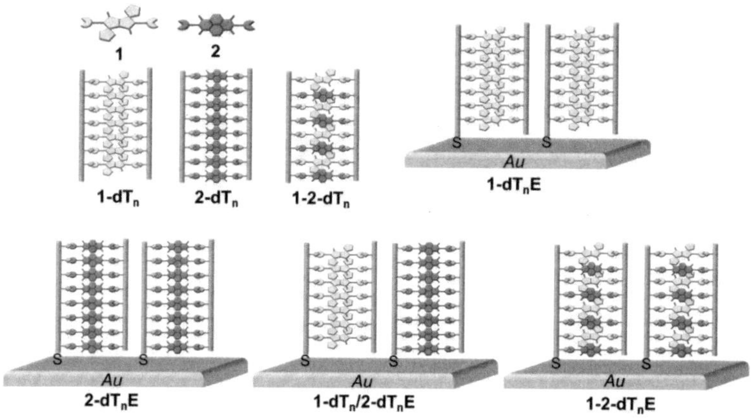

Figure 7.11 Schematic illustrations of multichromophore array–immobilized electrodes.

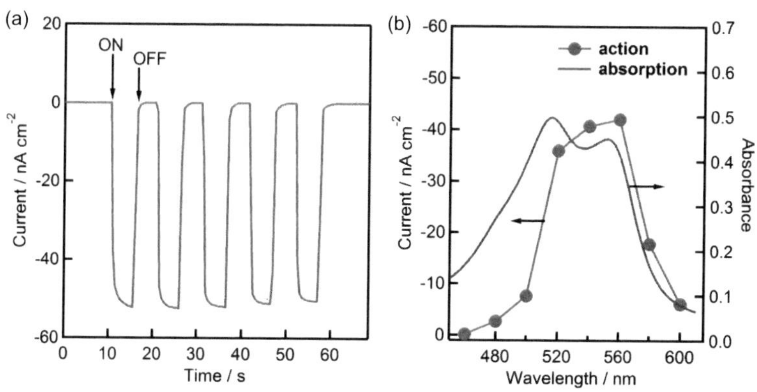

Figure 7.12 (a) Cathodic photocurrent of **1-dT$_{40}$E** at an applied voltage of 0 mV. (b) The action spectrum of **1-dT$_{40}$E** and the absorption spectrum of **1-dT$_{40}$**.

Nguyen and coworkers have reported that the lowest unoccupied molecular orbital (LUMO) level of dialkyl-substituted DPP is –2.51 eV [30]. This value is higher than the LUMO level (–3.99 eV) of MV^{2+} estimated from its reduction potential (–0.687 V versus SCE) [31, 32]. MV^{2+} probably accepts an electron from the photoexcited

1-dT$_{40}$ to form $MV^{+\bullet}$ and the electron flows to the Pt electrode from MV^{+} (Fig. 7.14a). On the other hand, the redox potential of AsA (AsA•+/AsA) is already known as −0.19 V [33, 34]. The photoexcited **1-dT$_{40}$** accepts an electron from AsA, and the electron moves to the electrode through **1-dT$_{40}$** (Fig. 7.14b). Therefore, **1-dT$_{40}$E** can switch the photocurrent direction by changing electron sacrifice reagents.

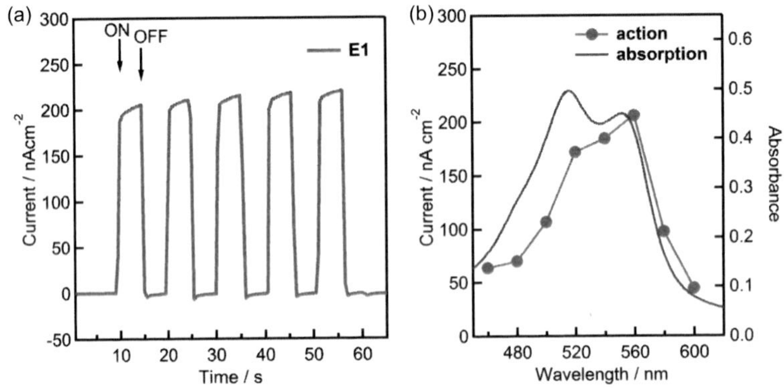

Figure 7.13 (a) Anodic photocurrent of **1-dT$_{40}$E** at an applied voltage of 0 mV. (b) The action spectrum of **1-dT$_{40}$E** and the absorption spectrum of **1-dT$_{40}$**.

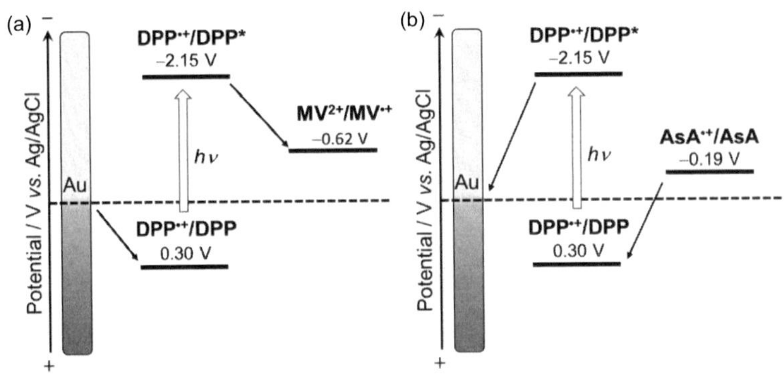

Figure 7.14 Schematic diagrams of photocurrent generation for (a) the **1-dT$_{40}$E**-MV^{2+} system and (b) the **1-dT$_{40}$E**-AsA system.

The anodic photocurrent generated from **1-dT$_{n}$E** electrodes changed depending on the length of dT$_{n}$-DNAs. Table 7.1 summarizes the photocurrent responses of **1-dT$_{n}$E** electrodes.

The anodic photocurrent increased with the increase in the length of dT_n-DNAs: from 97 ± 7.5 nA/cm^2 of dT_{10} to 210 ± 6.8 nA/cm^2 of dT_{40}. Since uniform DPP arrays corresponding to the length of dT_n-DNAs are generated and the total amounts of Au-S bonds are almost the same (2.1 ± 0.3 pmol/cm^2) for each **1-dT$_n$E** electrode [35], the photocurrent enhancement is probably due to the change of the total amounts of **1** on the electrode.

Table 7.1 DNA length dependence of anodic photocurrent of **1-dT$_n$E** electrodes[a]

1-dT$_n$E	**Photocurrent (nA/cm^2)**
1-dT$_{10}$E	97 ± 7.5
1-dT$_{20}$E	110 ± 5.6
1-dT$_{30}$E	140 ± 15
1-dT$_{40}$E	210 ± 6.8

[a]Photoirradiation was carried out using a 560 nm band pass filter in the presence of AsA

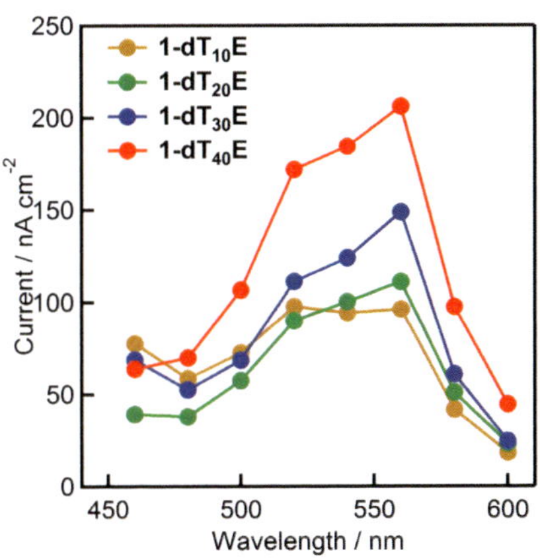

Figure 7.15 Action spectra of **1-dT$_n$E** (n = 10, 20, 30, and 40) electrodes in the region of 460–600 nm.

Figure 7.15 demonstrates the action spectra of **1-dT$_n$E** electrodes in the region of 460–600 nm. The action spectra of **1-dT$_n$E** electrodes are similar to the absorption spectra of **1-dT$_n$** in buffer solutions, indicating that the anodic photocurrent occurs from the photoexcited state of **1**. The anodic photocurrent responses depend on the length of dT$_n$-DNAs because of the formation of uniform DPP arrays corresponding to the DNA length. The results indicate that the photocurrent responses of the DPP arrays are controllable by DNA lengths and electron sacrifice reagents.

7.4.2 Photoelectrochemical Property of Naphthalenetetracarboxylic Acid Diimide Arrays

Figure 7.16a shows the photoelectronchemical responses of **2-dT$_n$E** electrodes. Photoelectrochemical measurements were carried out in an argon-saturated aqueous solution containing 10 mmol triethanolamine (TEOA) as an electron sacrifice reagent. The NDI array–immobilized electrode was used as a working electrode, along with a platinum (Pt) counterelectrode and an Ag/AgCl reference electrode in a three-electrode system. Photoirradiation of the **2-dT$_n$E** electrodes using a 360 nm band pass filter caused the immediate generation of anodic photocurrents, which depended on the DNA length.

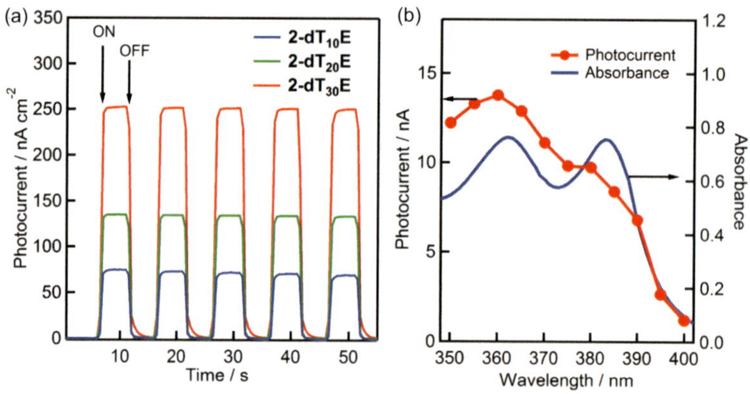

Figure 7.16 (a) Anodic photocurrents of **2-dT$_n$E** electrodes at a 0 mV applied voltage. (b) Action spectrum of **2-dT$_{30}$E** electrode and absorption spectrum of **2-dT$_{30}$** in a buffer solution.

The action spectra for the **2-dT$_{30}$E** electrode in the region of 350–400 nm (Fig. 7.16b) resembled the absorption spectrum of **2-dT$_{30}$** in solution, indicating that the NDI moiety of **2** is a photoactive species. Photocurrent generation under various applied bias voltages was also investigated to determine the direction of the current flow. The positive bias to the gold electrode enhanced the anodic photocurrent, and a semilinear correlation between the bias voltage and the anodic photocurrent was observed in the region of −100 ≈ +100 mV versus Ag/AgCl upon photoirradiation (Fig. 7.17). Therefore, the photoexcited **2-dT$_n$** accepted an electron from TEOA and the electron flowed to the Au electrode. The radical cation of TEOA accepted an electron from the Pt electrode. The NDI array generates anodic photocurrents whose apparent efficiency is dependent on the length of the arrays. This clearly indicates the utility of the NDI array as a charge transport domain. The DNA-templated assembly of π-aromatic molecules would be a useful strategy for the development of biomolecular electronic devices.

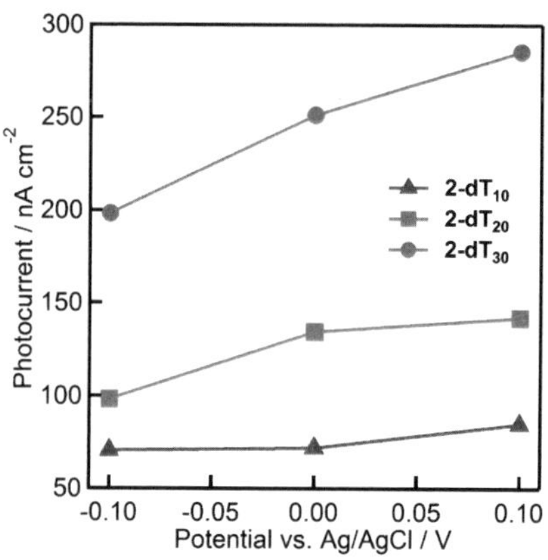

Figure 7.17 Photocurrent dependence of **2-dT$_n$E** electrodes on the applied bias voltage.

7.4.3 Donor/Acceptor Heterojunction Photocurrent Systems Based on Multichromophore Arrays

To realize practically useful photovoltaics, the donor/acceptor heterojunction configuration is believed to be key to successfully improving the power conversion efficiencies. Especially in bottom-up supramolecular approaches, continuous donor and acceptor arrays should be aligned coaxially and arranged vertically in a manner such that they face the electrode to transport photogenerated holes and electrons to the anode and cathode, respectively [36–39]. In addition, continuous donor and acceptor arrays can prevent donor/acceptor CT complexation that strongly interferes with photocurrent responses [40].

It is anticipated that DPP and NDI arrays can be applied to donor/acceptor heterojunction photosystems. The present system should get an advantage by the fact that the DNA-multichromophore arrays can be easily arranged vertically in a manner such that they face the electrode in the self-assembly process. The LUMO level of dialkyl-substituted DPP derivative (−2.51 eV) [30] is higher than that of the dialkyl-substituted NDI derivative (−3.40 eV) [41], indicating that the electron transfer from photoexcited DPP to NDI is exothermic. Photovoltaic cells using a DPP–NDI dyad system have recently been developed [42].

The strategy for chromophore assembly using DNA also suppresses the CT complexation between **1** and **2** and the photocurrent responses generated from the photoexcited **1-dT**$_n$ on a Au electrode were enhanced by coimmobilization with **2-dT**$_n$. The present results show that the approach of forming a donor/accepter heterojunction using DNA-multichromophore arrays is a useful method to efficiently generate photocurrent.

Figure 7.18a shows the photocurrent responses of electrodes **1-dT**$_{40}$**E**, **1-dT**$_{40}$**/2-dT**$_{40}$**E**, and **1-2-dT**$_{40}$**E** under photoirradiation with 560 nm light in the presence of L-ascorbic acid. The photoirradiation of **1-dT**$_{40}$**/2-dT**$_{40}$**E** caused fast generation of an anodic photocurrent, which declined quickly with the termination of the photoirradiation (414 ± 15 nAcm−2). In this case, **1-dT**$_{40}$ on the electrode was selectively photoirradiated because **2-dT**$_{40}$ has no absorption at 560 nm (Fig. 7.19). On the contrary, the 560 nm photoirradiation of **1-2-dT**$_{40}$**E** generated no photocurrent. The photoirradiation of **1-dT**$_{40}$**E** also generated an anodic photocurrent (203 ± 6.6 nAcm^{-2}). The action spectra of **1-dT**$_{40}$**/2-dT**$_{40}$**E** and **1-dT**$_{40}$**E** were similar to the absorption spectrum of **1-dT**$_{40}$ in a buffer solution (Fig. 7.18b), indicating that the photocurrent on **1-dT**$_{40}$**/2-dT**$_{40}$**E** is generated from photoexcited **1-dT**$_{40}$. A similar enhancement in photocurrent was observed in the presence of methyl viologen (MV^{2+}; Fig. 7.19a). The photoirradiation of **1-dT**$_{40}$**/2-dT**$_{40}$**E** with 560 nm light generated a cathodic photocurrent (−109 ± 9.3 nAcm^{-2}); the magnitude of this photocurrent was greater than those of the photocurrents generated by **1-2-dT$_{40}$E** and **1-dT$_{40}$E** (−55.6 ± 3.6 nAcm^{-2}). The spectral shapes of the action spectra of **1-dT$_{40}$/2-dT$_{40}$E** and **1-dT$_{40}$E** resembled that of the absorption spectrum of **1-dT$_{40}$** in solution (Fig. 7.19b).

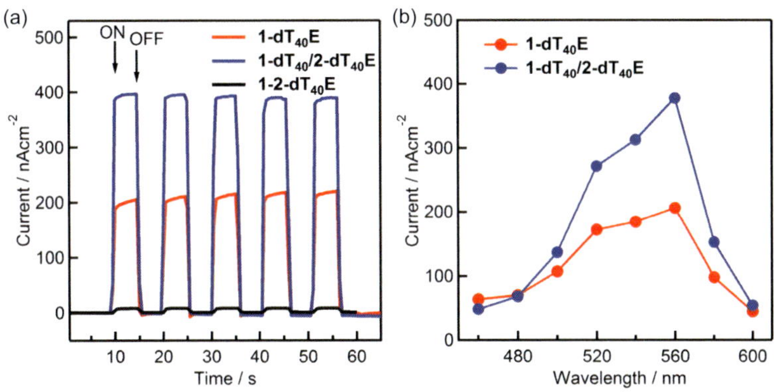

Figure 7.18 (a) Anodic photocurrent responses of electrodes **1-dT$_{40}$E**, **1-dT$_{40}$/2-dT$_{40}$E**, and **1-2-dT$_{40}$E** at a 0 mV applied voltage in an Ar-saturated pH 7.6 buffer solution containing 50 mM HEPES, 0.1 M NaNO3, and 20 mM AsA. (b) Action spectra of electrodes **1-dT$_{40}$E** and **1-dT$_{40}$/2-dT$_{40}$E**.

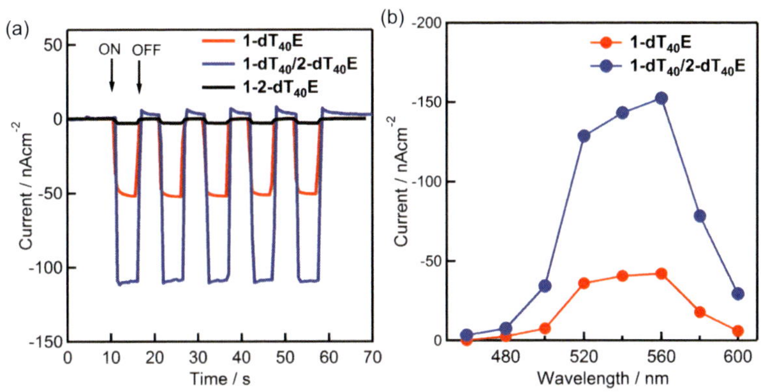

Figure 7.19 (a) Cathodic photocurrent responses of electrodes **1-dT$_{40}$E**, **1-dT$_{40}$/2-dT$_{40}$E**, and **1-2-dT$_{40}$E** at a 0 mV applied voltage in an Ar-saturated pH 7.6 buffer solution containing 10 mM HEPES, 0.1 M $NaNO_3$, and 10 mM MV^{2+}. (b) Action spectra of electrodes **1-dT$_{40}$E** and **1-dT$_{40}$/2-dT$_{40}$E**.

Importantly, the photocurrent intensity of **1-dT$_{40}$/2-dT$_{40}$E** was at least two times larger than that of **1-dT$_{40}$E**. The photocurrent enhancement in **1-dT$_{40}$/2-dT$_{40}$E** is attributed to efficient charge separation between the coaxially aligned **1-dT$_{40}$** and **2-dT$_{40}$**, which leads direction on the self-assembled monolayers. The CT from the photoexcited DPP array to the NDI array is evident from the experiments in which the fluorescence of **1-dT$_{20}$** was quenched in the presence of **2-dT$_{20}$** (Fig. 7.20). Assuming that the total amounts of multichromophore arrays on the Au surfaces in **1-dT$_{40}$/2-dT$_{40}$E** and **1-dT$_{40}$E** are the same, and that the amounts of **1-dT$_{40}$** and **2-dT$_{40}$** are equivalent in **E1**, the photocurrent quantum yield of **1-dT$_{40}$/2-dT$_{40}$E** would be four times greater than that of **1-dT$_{40}$E**. The photocurrent response of **1-2-dT$_{40}$E** is completely suppressed by the ground-state CT complex in **1-2-dT$_{40}$**.

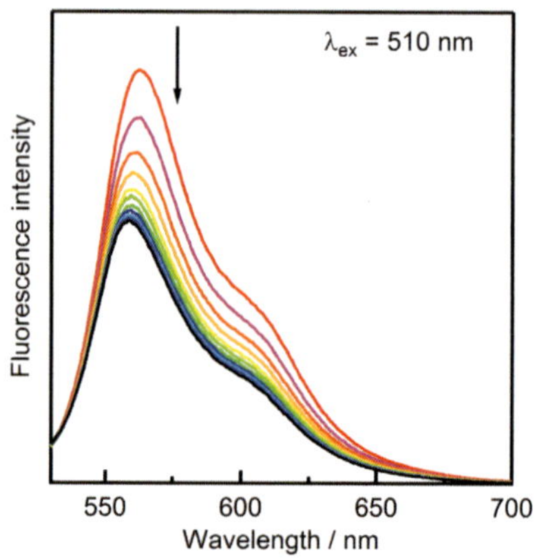

Figure 7.20 Fluorescence quenching of **1-dT$_{40}$** with **2-dT$_{40}$** in a pH 7.6 buffer solution containing 10 mM HEPES and 0.1 M $NaNO_3$. The excitation wavelength was 510 nm.

On basis of the energy diagrams of photocurrent generation in **1-dT$_{40}$/2-dT$_{40}$E** (Fig. 7.21), a plausible mechanism for anodic photocurrent generation is as follows. An electron is transferred from photoexcited **1-dT$_{40}$** ($DPP^{\bullet+}/DPP^*$ = −1.59 V) [43] to neighboring **2-dT$_{40}$** ($NDI/NDI^{\bullet+}$ = −0.89 V) [43] to form a charge-separated state. The electron on **2-dT$_{40}$** moves to the Au electrode by hopping through the NDI array, and the hole on **1-dT$_{40}$** is transferred to AsA ($AsA^{\bullet+}/AsA$ = −0.19 V) [44] (Fig. 7.21a). In the presence of MV^{2+}, the electron on **2-dT$_{40}$** is transferted to MV^{2+} ($MV^{2+}/MV^{\bullet+}$ = −0.63 V) [45] to generate cathodic photocurrent, while the hole on **1-dT$_{40}$** moves to the Au electrode (Fig. 7.21b).

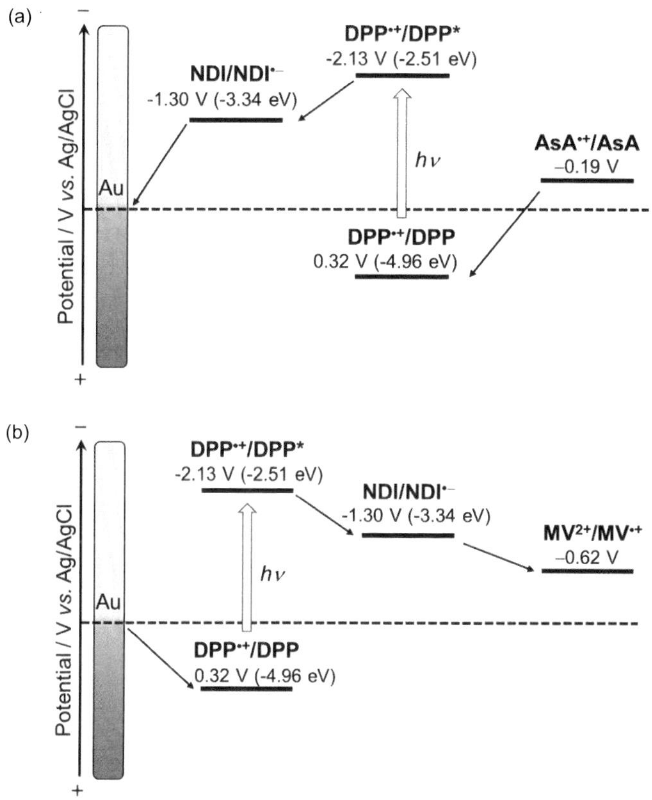

Figure 7.21 Energy diagrams of photocurrent generation in electrode **1-dT$_{40}$/2-dT$_{40}$E**: (a) anodic photocurrent in the presence of AsA and (b) cathodic photocurrent in the presence of MV^{2+}.

Whereas **1** and **2** form ground-state CT complexes that interfere with the photocurrent response, the multichromophore arrays **1-dT$_{40}$** and **2-dT$_{40}$** exist stably without disruption of their structures. The coimmobilized Au electrode of **1-dT$_{40}$** and **2-dT$_{40}$** can enhance both anodic and cathodic photocurrent responses compared to the **1-2-dT$_{40}$**- and **1-dT$_{40}$**-immobilized electrodes. Therefore, the self-organization of **1** or **2** using dT$_{40}$-DNAs plays a critical function in photocurrent enhancement. The present system provides a DNA-based heterojunction photosystem in which the donor and acceptor arrays are coaxially aligned and vertically arranged facing the electrode. This system has the potential to be applied in biosensors and photovoltaic cells.

7.5 Experimental

7.5.1 DNA Synthesis

dT_n and 3′-HS-$(CH_2)_3$-dT_n were synthesized on a 1.0 μmol scale using conventional phosphoramidite chemistry with the 4,4′-dimethoxytrityl (DMT)-on mode. All the reagents for the syntheses were purchased from Glen Research. After syntheses, the DNAs on CPG supports were placed in 1 mL of aqueous ammonia for 1 h at room temperature to remove the DNAs from the CPG. After evaporation of the ammonia, the DMT-protected DNAs were purified using a reversed-phase high-performance liquid chromatography (HPLC) system. The DMT was cleaved by dissolving the purified DNAs in an 80% acetic acid solution for 1 h at room temperature, and the acetic acid was then evaporated. The crude DNAs were purified using a reversed-phase HPLC system. 3′-HS-$(CH_2)_3$-dT_n was prepared by the reduction of 3′-HO-$(CH_2)_3$-SS-$(CH_2)_3$-dT_n, which was synthesized using DMT-O-$(CH_2)_3$-SS-$(CH_2)_3$-O-succinyl-CPG, with dithiothreitol [46]. The purified 3′-HO-$(CH_2)_3$-SS-$(CH_2)_3$-dT_n was dissolved in a pH 8.3 Tris-HCl buffer (0.1 M) solution and left to stand for 2 h at room temperature to cleave the disulfide bond. After evaporation, the 3′-HS-$(CH_2)_3$-dT_n was purified by a reversed-phase HPLC system.

7.5.2 Preparation of Sample Solutions for Spectroscopic Analysis

Sample solutions were prepared using a pH 7.6 buffer solution containing 10 HEPES and 0.1 M $NaNO_3$. The solutions of **1-dT*n*** and **2-dT*n*** were prepared by the addition of buffer solutions (0.5 mL) of dT_n (200 μM, base concentration) to buffer solutions (0.5 mL) of **1** (100 μM) and **2** (100 μM), respectively. The solution of **1-2-dT*n*** was prepared by the addition of a buffer solution (0.5 mL) of dT_n (200 μM, base concentration) to a buffer solution (0.5 mL) containing **1** (100 μM) and **2** (100 μM). The solution of **1-dT*n*/2-dT*n*** was prepared by mixing equal amounts of solutions (0.5 mL) of **1-dT*n*** (prepared from **1** [200 μM] and dT_{40} [400 μM, base concentration]) and **2-dT*n*** (prepared from **2** [200 μM] with dT_{40} [400 μM, base concentration]).

The concentration of dT_n was determined by absorption at 260 nm, and the single-stranded extinction coefficients were calculated on the basis of the nearest-neighbor model [47].

7.5.3 Preparation of the Chromophore Array Immobilized Electrode and Photoelectrochemical Measurements

All chromophore array solutions were prepared using a pH 7.6 solution containing 10 mmol HEPES and 0.1 M $NaNO_3$. The surface of a gold electrode (BAS, 3 mm inner diameter) was polished and polarized cyclically (scan rate 150 mVs^{-1}) in the −0.2 to +1.8 V potential range in 0.5 M H_2SO_4 for 10 cycles before use. A self-assembled monolayer of chromophore arrays was prepared by immersing the gold surface in the chromophore array solutions for 16 h at room temperature. It was then rinsed with a buffer before photoelectrochemical measurements.

Sample solution for electrode 1-dT_nE: The solution of **1-dT_n** was prepared by adding a solution (75 μL) containing 3′-HS-$(CH_2)_3$-dT_n (100 μM, base concentration) and dT_n (100 μM, base concentration) to a solution (75 μL) of **1** (100 μM).

Sample solution for electrode 2-dT_nE: The solution (75 μL) of **2-dT_n** was prepared in the same manner as described above.

Sample solution for electrode 1-dT_n/2-dT_nE: The solution of **1-dT_n/2-dT_n** was prepared by mixing the solution (75 μL) of **1-dT_n** and the solution (75 μL) of **2-dT_n**.

Sample solution for electrode 1-2-dT_nE: The solution of **1-2-dT_n** was prepared by adding a solution (75 μL) containing 3′-HS-$(CH_2)_3$-dT_n (100 μM, base concentration) and dT_n (100 μM, base concentration) to a solution (75 μL) containing **1** (50 μM) and **2** (50 μM).

Photoelectrochemical measurements at room temperature were carried out in an Ar-saturated pH 7.6 buffer solution containing 50 mmol HEPES, 0.1 M $NaNO_3$, and 20 mmol AsA or in an Ar-saturated pH 7.6 buffer solution containing 10 mmol HEPES, 0.1 M $NaNO_3$, and 10 mmol TEOA or MV^{2+}. The chromophore array–immobilized electrode was used as the working electrode along with a Pt counterelectrode and an Ag/AgCl (3 M NaCl) reference electrode in a three-electrode system. Photoirradiation was carried out using a 300-watt Xe lamp with a band pass filter.

References

1. Hernando, J., De Witte, P. A. J., van Dijk, E. M. H. P., Korterik, J., Nolte, R. J. M., Rowan, A. E., García-Parajó, M. F. and Van Hulst, N. F. (2004). Investigation of perylene photonic wires by combined single-molecule fluorescence and atomic force microscopy, *Angew. Chem. Int. Ed.*, **43**, 4045–4049.
2. Kim, O.-K., Melinger, J., Chung, S.-J. and Pepitone, M. (2008). Supramolecular device for artificial photosynthetic mimics as helix-mediated antenna/reaction center ensemble, *Org. Lett.*, **10**, 1625–1628.
3. Bhosale, S., Sisson, A. L., Talukdar, P., Fürstenberg, A., Banerji, N., Vauthey, E., Bollot, G., Mareda, J., Röger, C., Würthner, F., Sakai, N. and Matile, S. (2006). Photoproduction of proton gradients with π-stacked fluorophore scaffolds in lipid bilayers, *Science*, **313**, 84–86.
4. Finlayson, C. E., Friend, R. H., Otten, M. B. J., Schwartz, E., Cornelissen, J. J. L. M., Nolte, R. J. M., Rowan, A. E., Samorì, P., Palermo, V., Liscio, A., Peneva, K., Müllen, K., Trapani, S. and Beljonne, D. (2008). Electronic transport properties of ensembles of perylene-substituted poly-isocyanopeptide arrays, *Adv. Funct. Mater.*, **18**, 3947–3955.
5. Spada, G. P., Lena, S., Masiero, S., Pieraccini, S., Surin, M. and Samorì, P. (2008). Guanosine-based hydrogen-bonded scaffolds: controlling the assembly of oligothiophenes, *Adv. Mater.*, **20**, 2433–2438.
6. Shao, H., Seifert, J., Romano, N. C., Gao, M., Helmus, J. J., Jaroniec, C. P., Modarelli, D. A. and Parquette, J. R. (2010). Amphiphilic self-assembly of an n-type nanotube, *Angew. Chem. Int. Ed.*, **49**, 7688–7691.
7. Rothemund, P. W. K. (2006). Folding DNA to create nanoscale shapes and patterns, *Nature*, **440**, 297–302.
8. Lo, P. K., Karam, P., Aldaye, F. A., Mclaughlin, C. K., Hamblin, G. D., Cosa, G. and Sleiman, H. F. (2010). Loading and selective release of cargo in DNA nanotubes with longitudinal variation, *Nat. Chem.*, **2**, 319–328.
9. Gu, H., Chao, J., Xiao, S.-J. and Seeman, N. C. (2010). A proximity-based programmable DNA nanoscale assembly line, *Nature*, **465**, 202–205.
10. Wickham, S. F. J., Endo, M., Katsuda, Y., Hidaka, K., Bath, J., Sugiyama, H. and Turberfield, A. J. (2011). Direct observation of stepwise movement of a synthetic molecular transporter, *Nat. Nano*, **6**, 166–169.

11. Bandy, T. J., Brewer, A., Burns, J. R., Marth, G., Nguyen, T. and Stulz, E. (2011). DNA as supramolecular scaffold for functional molecules: progress in DNA nanotechnology, *Chem. Soc. Rev.*, **40**, 138–148.

12. Malinovskii, V. L., Wenger, D. and Häner, R. (2010). Nucleic acid-guided assembly of aromatic chromophores, *Chem. Soc. Rev.*, **39**, 410–422.

13. González-Rodríguez, D. and Schenning, A. P. H. J. (2011). Hydrogen-bonded supramolecular π-functional materials, *Chem. Mater.*, **23**, 310–325.

14. Shionoya, M., Kimura, E. and Shiro, M. (1993). A new ternary zinc(II) complex with [12]aneN4 (=1,4,7,10-tetraazacyclododecane) and AZT (=3'-azido-3'-deoxythymidine). Highly selective recognition of thymidine and its related nucleosides by a zinc(II) macrocyclic tetraamine complex with novel complementary associations, *J. Am. Chem. Soc.*, **115**, 6730–6737.

15. Shionoya, M., Ikeda, T., Kimura, E. and Shiro, M. (1994). Novel "multipoint" molecular recognition of nucleobases by a new zinc(II) complex of acridine-pendant cyclen (cyclen = 1,4,7,10-tetraazacyclododecane), *J. Am. Chem. Soc.*, **116**, 3848–3859.

16. Kinoshita-Kikuta, E., Kinoshita, E., Harada, N. and Koike, T. (2011). Zinc(II)-cyclen polyacrylamide gel electrophoresis for detection of mutations in short Ade/Thy-rich DNA fragments, *Anal. Biochem.*, **408**, 348–350.

17. Kinoshita-Kikuta, E., Kinoshita, E. and Koike, T. (2002). A novel procedure for simple and efficient genotyping of single nucleotide polymorphisms by using the Zn^{2+}–cyclen complex, *Nucleic Acids Res.*, **30**, e126.

18. Shiddiky, M. J., Torriero, A. A., Zeng, Z., Spiccia, L. and Bond, A. M. (2010). Highly Selective and sensitive DNA assay based on electrocatalytic oxidation of ferrocene bearing zinc(II)-cyclen complexes with diethylamine, *J. Am. Chem. Soc.*, **132**, 10053–10063.

19. Maie, K., Nakamura, M. and Yamana, K. (2006). Conformational changes of DNA by photoirradiation of DNA-Bis(Zn^{II}-cyclen)-azobenzene complex, *Nucleosides Nucleotides Nucleic Acids*, **25**, 453–462.

20. Tan, X.-Y., Zhang, J., Huang, Y., Zhang, Y., Zhou, L.-H., Jiang, N., Lin, H.-H., Wang, N., Xia, C.-Q. and Yu, X.-Q. (2007). Synthesis and DNA-cleavage properties of metal complexes of 1,4,7,10-tetraazacyclododecane (cyclen) functionalized with a pendant benzocrown ether, *Chem. Biodivers.*, **4**, 2190–2197.

21. Aoki, S., Zulkefeli, M., Shiro, M. and Kimura, E. (2002). New supramolecular trigonal prisms from zinc(II)–1,4,7,10-

tetraazacyclododecane (cyclen) complexes and trithiocyanurate in aqueous solution, *Proc. Natl. Acad. Sci. U.S.A.*, **99**, 4894–4899.

22. Zulkefeli, M., Sogon, T., Takeda, K., Kimura, E. and Aoki, S. (2009). Design and synthesis of a stable supramolecular trigonal prism formed by the self-assembly of a linear tetrakis(Zn^{2+}–cyclen) complex and trianionic trithiocyanuric acid in aqueous solution and its complexation with DNA (cyclen = 1,4,7,10-tetraazacyclododecane), *Inorg. Chem.*, **48**, 9567–9578.
23. Holcombe, T. W., Yum, J.-H., Kim, Y., Rakstys, K. and Gratzel, M. (2013). Diketopyrrolopyrrole-based sensitizers for dye-sensitized solar cell applications: anchor engineering, *J. Mater. Chem. A*, **1**, 13978–13983.
24. Dou, L., You, J., Yang, J., Chen, C.-C., He, Y., Murase, S., Moriarty, T., Emery, K., Li, G. and Yang, Y. (2012). Tandem polymer solar cells featuring a spectrally matched low-bandgap polymer, *Nat. Photonics*, **6**, 180–185.
25. Yiu, A. T., Beaujuge, P. M., Lee, O. P., Woo, C. H., Toney, M. F. and Fréchet, J. M. J. (2012). Side-chain tunability of furan-containing low-band-gap polymers provides control of structural order in efficient solar cells, *J. Am. Chem. Soc.*, **134**, 2180–2185.
26. Qiao, Y., Guo, Y., Yu, C., Zhang, F., Xu, W., Liu, Y. and Zhu, D. (2012). Diketopyrrolopyrrole-containing quinoidal small molecules for high-performance, air-stable, and solution-processable n-channel organic field-effect transistors, *J. Am. Chem. Soc.*, **134**, 4084-4087.
27. Kanimozhi, C., Yaacobi-Gross, N., Chou, K. W., Amassian, A., Anthopoulos, T. D. and Patil, S. (2012). Diketopyrrolopyrrole–diketopyrrolopyrrole-based conjugated copolymer for high-mobility organic field-effect transistors, *J. Am. Chem. Soc.*, **134**, 16532–16535.
28. Kirkus, M., Wang, L., Mothy, S., Beljonne, D., Cornil, J., Janssen, R. A. J. and Meskers, S. C. J. (2012). Optical properties of oligothiophene substituted diketopyrrolopyrrole derivatives in the solid phase: joint J- and H-type aggregation, *J. Phys. Chem. A*, **116**, 7927–7936.
29. Molla, M. R., Das, A. and Ghosh, S. (2011). Chiral induction by helical neighbour: spectroscopic visualization of macroscopic-interaction among self-sorted donor and acceptor π-stacks, *Chem. Commun.*, **47**, 8934–8936.
30. Tamayo, A. B., Tantiwiwat, M., Walker, B. and Nguyen, T.-Q. (2008). Design, synthesis, and self-assembly of oligothiophene derivatives with a diketopyrrolopyrrole core, *J. Phys. Chem. C*, **112**, 15543–15552.
31. Bird, C. L. and Kuhn, A. T. (1981). Electrochemistry of the viologens, *Chem. Soc. Rev.*, **10**, 49–82.

32. Trasatti, S. (1986). The absolute electrode potential: an explanatory note, *Pure Appl. Chem.*, **58**, 955–966.

33. Imahori, H., Azuma, T., Ajavakom, A., Norieda, H., Yamada, H. and Sakata, Y. (1999). An investigation of photocurrent generation by gold electrodes modified with self-assembled monolayers of C60, *J. Phys. Chem. B*, **103**, 7233–7237.

34. Liu, Y.-F., Chen, J.-X., Xu, M.-Q. and Zhao, G.-C. (2014). A novel photoelectrochemical platform for detection of protease, *Int. J. Electrochem. Sci.*, **9**, 4014–4023.

35. Widrig, C. A., Chung, C. and Porter, M. D. (1991). The electrochemical desorption of n-alkanethiol monolayers from polycrystalline Au and Ag electrodes, *J. Electroanal. Chem. Interfacial Electrochem.*, **310**, 335–359.

36. Prasanthkumar, S., Ghosh, S., Nair, V. C., Saeki, A., Seki, S. and Ajayaghosh, A. (2015). Organic donor–acceptor assemblies form coaxial p–n heterojunctions with high photoconductivity, *Angew. Chem. Int. Ed.*, **54**, 946–950.

37. Bolag, A., López-Andarias, J., Lascano, S., Soleimanpour, S., Atienza, C., Sakai, N., Martín, N. and Matile, S. (2014). A collection of fullerenes for synthetic access toward oriented charge-transfer cascades in triple-channel photosystems, *Angew. Chem. Int. Ed.*, **53**, 4890–4895.

38. Zhang, W., Jin, W., Fukushima, T., Saeki, A., Seki, S. and Aida, T. (2011). Supramolecular linear heterojunction composed of graphite-like semiconducting nanotubular segments, *Science*, **334**, 340–343.

39. Hayashi, H., Nihashi, W., Umeyama, T., Matano, Y., Seki, S., Shimizu, Y. and Imahori, H. (2011). Segregated donor–acceptor columns in liquid crystals that exhibit highly efficient ambipolar charge transport, *J. Am. Chem. Soc.*, **133**, 10736–10739.

40. Li, W.-S., Yamamoto, Y., Fukushima, T., Saeki, A., Seki, S., Tagawa, S., Masunaga, H., Sasaki, S., Takata, M. and Aida, T. (2008). Amphiphilic molecular design as a rational strategy for tailoring bicontinuous electron donor and acceptor arrays: photoconductive liquid crystalline oligothiophene–C_{60} dyads, *J. Am. Chem. Soc.*, **130**, 8886–8887.

41. Guo, S., Wu, W., Guo, H. and Zhao, J. (2012). Room-temperature long-lived triplet excited states of naphthalenediimides and their applications as organic triplet photosensitizers for photooxidation and triplet–triplet annihilation upconversions, *J. Org. Chem.*, **77**, 3933–3943.

42. Favereau, L., Warnan, J., Pellegrin, Y., Blart, E., Boujtita, M., Jacquemin, D. and Odobel, F. (2013). Diketopyrrolopyrrole derivatives for efficient NiO-based dye-sensitized solar cells, *Chem. Commun.*, **49**, 8018–8020.

43. Because clear cyclic voltammograms of 1-dT_{40} and 2-dT_{40} in a buffer solution or in SAMs on an Au electrode could not be obtained, the redox potentials of DPP and NDI shown in Figure 21 were estimated from the reported HOMO and LUMO energy levels of dialkyl-substituted DPP and NDI derivatives [41, 42] according to the method described in the literature.[32]

44. Matsui, T., Kitagawa, Y., Okumura, M. and Shigeta, Y. (2015). Accurate standard hydrogen electrode potential and applications to the redox potentials of vitamin C and NAD/NADH, *J. Phys. Chem. A*, **119**, 369–376.

45. Uosaki, K., Kondo, T., Zhang, X.-Q. and Yanagida, M. (1997). Very efficient visible-light-induced uphill electron transfer at a self-assembled monolayer with a porphyrin–ferrocene–thiol linked molecule, *J. Am. Chem. Soc.*, **119**, 8367–8368.

46. Okamoto, A., Tainaka, K. and Fujiwara, Y. (2006). Nile red nucleoside: design of a solvatofluorochromic nucleoside as an indicator of micropolarity around DNA, *J. Org. Chem.*, **71**, 3592–3598.

47. Borer, P. N. (1975). Optical properties of nucleic acids, absorption and circular dichroism spectra, in *Handbook of Biochemistry and Molecular Biology: Nucleic Acids* (ed. Fasman, G. D.), CRC Press, Cleveland, pp. 589–595.

Chapter 8

Carbon Nanomaterial–Nucleic Acid Complexes and Their Biological Applications

Ngoc Do Quyen Chau, Giacomo Reina, and Alberto Bianco
University of Strasbourg, CNRS, Immunopathology and therapeutic chemistry, UPR 3572, 67000 Strasbourg, France
a.bianco@ibmc-cnrs.unistra.fr

8.1 Introduction

Nucleic acids (NAs) are polymeric biomolecules made of monomers called nucleotides. Each nucleotide is composed of three components: a phosphate group, a ribose or a deoxyribose sugar, and a purine or a pyrimidine base. Depending on the sugar, we can have DNA or RNA. Research on the interaction between nanomaterials and NAs has been profoundly performed to yield interesting nanostructures with new properties and functions thanks to the combination of the two components [1–3]. In this context, carbon nanomaterials (CNMs), especially carbon

Templated DNA Nanotechnology: Functional DNA Nanoarchitectonics
Edited by Thimmaiah Govindaraju

ISBN 978-981-4800-21-1 (Hardcover), 978-0-429-42866-1 (eBook)
www.panstanford.com

nanotubes (CNTs) and graphene oxide (GO), have been extensively explored, opening to exploitation in the biomedical domain thanks to their unique properties. The study of the interactions of these CNMs with NAs through either noncovalent adsorption or covalent binding has substantially augmented over the years in the fields of biosensing, diagnostics, and biomedicine [4–10]. For their well-ordered, high-aspect-ratio, and flexible structure, CNTs have been studied as a novel platform for drug and gene delivery. On the other hand, GO, owing to its large surface area, good biocompatibility, tunable surface chemistry, high water dispersibility, and low-cost scalable production, became an alternative candidate over other nanocarriers in terms of interaction with biomolecules and, in particular, with DNA and RNA [11–14].

In this chapter, we will present the studies on the interactions between CNMs and NAs, especially DNA and RNA, followed by the applications of the complexes in gene and cancer therapy. Different noncovalent and covalent conjugations of single-stranded DNA (ssDNA) and double-stranded DNA (dsDNA) or RNA using CNTs and GO as carriers will be discussed in detail.

8.2 Interactions of Carbon Nanomaterials with Nucleic Acids

8.2.1 Complexes between Carbon Nanotubes and Nucleic Acids

CNTs are made of graphene sheets rolled into a cylindrical structure, entirely composed of C sp^2 atoms arranged in a series of condensed benzene rings. According to the structure, CNTs are classified into two main categories: single-walled carbon nanotubes (SWCNTs), with a single hollow cylinder and diameters in the range of 0.7 ≈ 2 nm and lengths in the range of 20–1000 nm; and multiwalled carbon nanotubes (MWCNTs), with several concentric cylinders of graphene sheets and diameters in the range of 1.4–100 nm and lengths from 1 to a few microns. The extraordinary physicochemical properties of CNTs, such as ultrahigh surface area, distinctive length-to-diameter ratios, and possibility of modification of their surface, make them prominent candidates for a wealth of applications. One of the drawbacks of CNTs in bioapplications is their low dispersibility in aqueous solutions because of the high hydrophobic surface character. Therefore, different types of functionalization have been performed on CNTs, including noncovalent and covalent conjugations, wrapping the tubes with various polymers and biomolecules.

One of the first studies on the interaction of CNTs with NAs concerned the capability of ssDNA to enhance the dispersion of CNTs in aqueous solutions [15, 16]. Molecular modeling allowed the measurement of the affinity of ssDNA and SWCNTs (Fig. 8.1) based on π-π stacking of the nucleobases, resulting in helical wrapping around the cylindrical surface, while the phosphate backbone remains exposed to water.

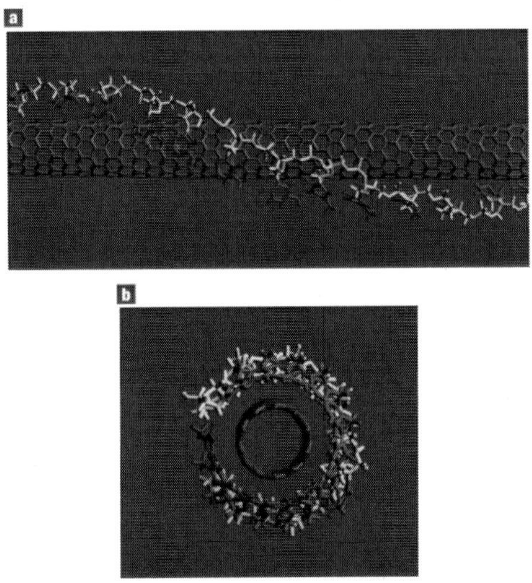

Figure 8.1 Binding model of a single-walled carbon nanotube wrapped by a poly(T) sequence shown from side (a) and top (b). Reprinted by permission from Springer Customer Service Centre GmbH: Springer Nature, *Nature Materials*, Ref. [15], Copyright (2003).

In another study, using thermodynamics and kinetics parameters and the stoichiometric composition in aqueous solution, Alidori et al. [17] investigated the binding affinity of ammonium-functionalized CNTs with a series of ssDNA, dsDNA, and RNA. The authors found that mainly electrostatic and hydrogen-bonding (H-bonding) interactions, and π-π stacking to a lesser extent, contributed to the assembly of these oligonucleotides with amino-functionalized CNTs.

Due to the wire-like shape, with the diameter matching the size of the NAs and their flexibility, CNTs can impact the conformational structure of DNA or RNA. Molecular simulation studies also showed that adsorption onto CNTs induces a transition of DNA from A to B conformation. Instead, by using molecule dynamic simulation, Johnson et al. [18] have revealed that SWCNTs induce a spontaneous conformational change of ssDNA that enables adsorption via π-π interactions between the two entities. Moreover, the authors found that among the four nucleobases, purines exhibit the strongest binding affinity. In a few nanoseconds, the ssDNA wraps around SWCNTs in compact right- or left-handed helices. The helix formation is derived from an electrostatic and torsion interaction within the sugar-phosphate backbone that results in ssDNA wrapping from 3′-end to 5′-end.

In another study, Wu et al. [19] have reported that when bound to SWCNTs, DNA strands are protected from enzymatic cleavage and interference from NA-binding proteins. The SWCNT/DNA conjugates can target a specific mRNA inside cells and improve self-delivery capability and intracellular biostability when compared to free DNA molecules.

Overall, NAs interact with CNTs via π-π forces to enhance the dispersibility of CNTs and with amino-functionalized CNTs mainly via electrostatic and H-bonding interactions. Dynamic simulations have shown that the conformation of NAs could be affected after these interactions. Besides, the ability of CNTs to protect NAs from enzymatic cleavage has opened the door to the use of CNTs as nanocarriers in gene therapy.

8.2.2 Complexes between Graphene Oxide and Nucleic Acids

GO is characterized by a basal plane composed of two domains consisting of a low-polar area that resembles the graphene surface and a polar defected surface characterized by the presence of oxygenated functions. Due to these features, besides the physisorption driven by the π-π interactions, additional interactions can contribute to the anchoring of external molecules onto the GO surface, including H-bonding with epoxides, alcohols, ethers, and carboxylic acids.

Therefore, noncovalent functionalization of GO with NAs involves π-π stacking, electrostatic interaction, and H-bonding.

One of the most common strategies in the development of GO-based biosensors depends on the high efficiency in fluorescence energy transfer from a dye to GO [20]. In this case, adsorption of a target fluorescent molecule can be monitored following its emission quenching due to GO proximity. For instance, via π-π interaction and H-bonding, fluorescently labeled ssDNA can strongly adsorb onto GO, causing dye emission quenching. The signal of the fluorescent dye can be recovered once the probe forms a duplex with a complementary DNA sequence, indicating the release of the molecule from GO. This technique has been examined by Lu et al. [21] to detect ssDNA labeled with FAM (a fluorescein-based dye) using a GO platform (Fig. 8.2).

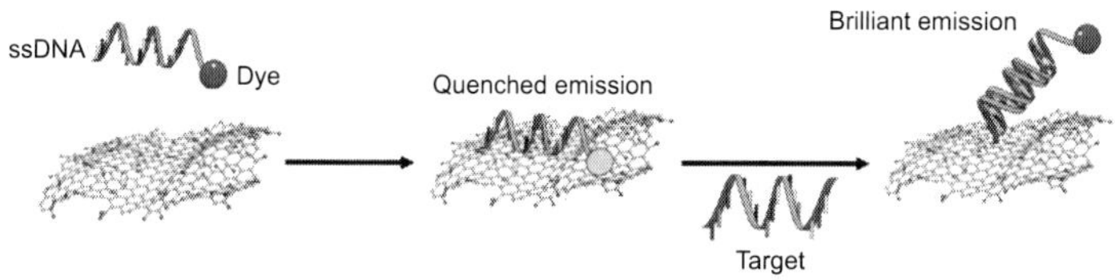

Figure 8.2 Schematic representation of target-induced fluorescence change of the ssDNA FAM/GO complex. FAM is a fluorescein-based fluorescent dye. In the first step, GO binds dye-labeled ssDNA and completely quenches the fluorescence of the dye. In the next step, in the presence of a complementary target strand that binds the dye-labeled ssDNA, dsDNA can be released from GO, leading to the restoration of fluorescence. Reprinted with permission from Ref. [21]. Copyright (2019) John Wiley and Sons.

In another example, Balapanuru et al. [22] developed a fluorescent-quenched charge transfer complex. The fluorescent positive-charged 4-(1-pyrenylvinyl)-*N*-butylpyridinium (PNP^+) was linked to GO to form PNP^+GO^-. As a result, the fluorescence of the PNP^+GO^- complex was quenched. In the presence of DNA, this fluorescence signal was recovered because the ionic attraction of DNA for PNP^+ is greater than the one of GO; hence, DNA-PNP^+ is released from the GO surface, turning on the fluorescence of the pyrene derivative.

Generally, the binding between GO and ssDNA or single-stranded RNA (ssRNA) is ascribed either to the π-π stacking involving both purine and pyrimidine bases or to H-bonding and van der Waals forces [23]. In terms of ssDNA, it was found that the adsorption was stabilized mainly by the H-bonding formed between the oxygen atoms on the surface of graphene and the polar groups of the oligonucleotide strand [24, 25], the π-π stacking between the hexagonal aromatic rings of GO, and the ring structure of the nucleobases [26, 27]. For example, Zhang et al. [28] designed a novel fluorescence-sensing system for the detection of biotin based on the interaction of GO and biotinylated ssDNA–binding streptavidin. In this platform, the streptavidin was conjugated with ssDNA, which was previously modified with biotin at the 3′-end and with 6-carboxyfluorescein (6-FAM) at the 5′-end, to protect the DNA from hydrolysis by exonuclease I. When streptavidin-DNA was adsorbed onto GO via π-π stacking, the fluorescence of 6-FAM was quenched. When free biotin was added, it competed with the DNA-labeled biotin for the binding sites of streptavidin and then exonuclease I could digest the liberated DNA sequence. Because of the weak affinity between the fluorophore and GO, the fluorescence was thus recovered. This study presented a unique platform for fluorescence sensing of biomolecules. However, in another work, Xue et al. [29] proved that the interaction between GO and ssDNA was mainly due to H-bonding formation rather than π-π stacking or hydrophobic contact. Indeed, surface plasmon resonance was used to directly investigate DNA/GO binding upon combination of GO and gold nanoparticles on the basis of indirect competitive inhibition assay to enhance sensitive DNA detection. The study not only demonstrates the ability of the system in ultrasensitive detection of ssDNA but also shows that the binding tendency of dsDNA is weaker than that of ssDNA, presumably due to competition between the interstrand binding in dsDNA and the binding with the GO. In fact, the double-strand structure of dsDNA infers a more rigid conformation to the molecule and this likely changes its interactions with GO surface.

Controversies still exist in terms of understanding the driving forces responsible for the interaction between GO and dsDNA or dsRNA. Some studies illustrated that the hydrophilic external surface of dsDNA can prevent the hydrophobic interactions of their bases with GO [30], while others confirmed that there are sufficient forces between the oxygenated groups of GO and DNA bases [31]. Moreover, the preferable binding of ssDNA to dsDNA was further studied. Wu et al. [32] reported that the hydrophobic forces were much higher than the electrostatic repulsions, especially in the presence of other ionic charges or low pH, where GO charge is reduced by protonation, thus diminishing the repulsion. In dsDNA, where the bases are inside the helical structure and the phosphate groups are exposed, there is a lower affinity on the GO surface. Lei et al. [33] have figured out that dsDNA can bind to GO to form a dsDNA/GO complex in the presence of a high concentration of salts via electrostatic interactions. Liu et al. [34] also reported the salt-controlled assembly of graphene sheets as a capture tool for dsDNA in a high salt concentration. In contrast, Tang et al. [23], by measuring the quenching effect between a fluorescently labeled dsDNA and GO, proved that there was no effect of salt on fluorescence changes, thus excluding the influence of the electrostatic interactions as major driving forces for dsDNA on the surface of GO. Moreover, the same study explored for the first time the partial denaturation of dsDNA incubated with GO. The deformation of dsDNA driven by hydrophobic π-π stacking was also confirmed in a previous study by molecular dynamic (MD) simulation [31]. On the other hand, Ren et al. [35] suggested that GO could intercalate efficiently with DNA molecules and enhance the DNA cleavage activity of copper ions via a mechanism based on an oxidative and hydrolytic effect. The DNA cleavage behavior of GO and reduced graphene oxide (rGO) in the presence of Cu^{2+} was trigged by π-π interaction of these materials with DNA [36].

In the case of ssRNA, there are very few studies on its interaction with GO because of its instability. Zhang et al. [37] explored the stronger π-π interaction of ssRNA with PEG-rGO than with PEG-GO, leading to superior ssRNA loading and delivery efficiency. Consistent with the experiments, MD simulation revealed that the attractive van der Waals interaction energy of ssRNA-rGO was more negative than that of ssRNA-GO. The computational method also evidenced a higher stacking capacity between the bases of ssRNA and the sp^2 domain, indicating the increasing of π-π interaction with the number of aromatic domains on graphene sheets. Overall, the interaction of GO and dsRNA, and in particular short interference RNA (siRNA), has not been completely evaluated yet. How the conformation of double-strand siRNA is affected by GO required more studies. Most of the reports on the main interactions of functionalized GO with siRNA were based on the electrostatic interaction between the negatively charged siRNA and the positively charged functionalized GO [38–40]. Very recently, our group has explored the siRNA complexation efficiency with GO and rGO by covalently functionalizing these materials with triethyleneglycoldiamine (TEG) and with low-molecular-weight polyethylenimine (PEI) via the epoxy ring opening reaction [41]. We figured out that the capacity of GO complexation is attributed to two main contributions corresponding to polar interactions by H-bonding and ionic interactions. H-bonding interactions seem to prevail in the case of nonfunctionalized GO, while ionic forces have a main role for PEI-functionalized GO. The comprehension of the mechanism of interaction between GO/functionalized GO and siRNA is crucial for their clinical future application in gene silencing.

8.3 Bioapplications of Carbon Nanomaterial–Nucleic Acid Conjugates in Gene and Cancer Therapy

NAs have considerable potential as therapeutic agents, either as gene expression systems for gene therapy (pDNA or messenger RNA [mRNA]), or for knocking down genes (siRNA). Gene delivery methods can be mechanical (microinjection, electroporation, biolistic, etc.), chemical (lipid carriers, nanoparticle carriers, etc.), or biological (viral vectors, bacterial vectors, etc.). Gene therapy has garnered significant interest as a promising method for treating gene-related disorders and other diseases, like cancer [42, 43].

Generally, transfection is the introduction of genetic materials into cells to induce gene expression and the consequent production of defected proteins. Depending on the nature of genomes, there are two main disparate ways of transfection: (i) stable transfection and (ii) transient transfection. In the case of the stable transfection, foreign DNA (normally pDNA), which is delivered to the nucleus after traveling through cellular and nuclear membranes, is integrated into the host genome and expresses continuously, even after mitosis of the host cells. In contrast, for the transient transfection, foreign DNA is delivered into the nucleus but does not interact with the genome, while mRNA is delivered into the cytosol, where it is translated and leads to protein expression. In this case, gene expression has a limited period of time, depending on the cell division and the environment. The stable transfection requires a complex and long process, while the transient process can afford extremely rapid responses.

Differently from gene transfection, gene silencing is a process that consists in switching off specific genes by the introduction of an antisense RNA sequence that can harm the translation of mRNA. Among different strategies, interfering RNA (RNAi) in gene silencing has been rapidly recognized as one of the most revolutionary approaches in the specific knockdown of the target gene because of its simple and rapid mechanism [44, 45]. Indeed, the induction of RNAi is mediated by siRNA, a class of dsRNA composed of 19–25 base pairs. Once siRNA enters into the cytoplasm, upon incorporation into the RNA-induced silencing complex, it can bind the complementary RNA, which triggers target mRNA degradation, subsequently inhibiting the specific protein expression [46, 47]. In fact, small RNA is preferred instead of long dsRNA because Dicer enzyme (one of the components of the complex involving mRNA translation) is not required for activation; so siRNA can start the gene down-regulation pathway as long as it is released into the cytoplasm [42].

Unfortunately, the "naked" siRNA faces rapid degradation in the extracellular environment via endogenous enzymes and it is not efficiently internalized into cells because it is too negatively charged to cross the cellular membranes [48]. Therefore, the challenge in the use of NAs is to overcome the struggle in delivering these anionic macromolecules to reach the targeted sites. Gene therapy requires a vector that protects the guest genetic material from nuclease degradation and facilitates its cellular uptake with high transfection efficiency, targeted tissue distribution, and intracellular trafficking [49]. The major issue is the lack of efficient and safe gene vectors [2].

The first investigations into gene delivery focused on the use of viral vectors due to their strong transgene expression [50]. However, the fundamental problems using capsids, related to toxicity [51], immunogenicity [52], carcinogenesis [53], and lack of scalability, oriented the research toward the generation of safer and more efficient delivery vectors [54]. With the flourishing of nanotechnology, nonviral systems based on nanoparticles and nanomaterials have been explored to address these concerns [55].

8.3.1 Carbon Nanotubes and Their Ability to Deliver DNA or RNA

In gene therapy, it is essential to create a platform that can first make a complex with NAs and then cross the plasma cell membrane to release them into the target cells [56]. Thanks to the interactions between CNTs and NAs, the nanotubes have been widely explored in gene therapy to overcome the poor pharmacokinetic profiles of the oligonucleotides and the low levels of gene expressions of other nonviral gene vectors [57]. Several strategies of driving NA-CNT conjugates into cell are summarized in Fig. 8.3, regardless of type of NA or gene therapy pathways [58].

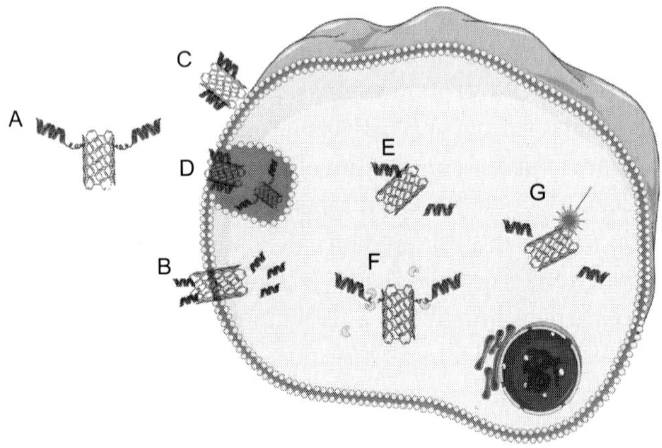

Figure 8.3 Suggested strategies for cellular delivery of NAs by CNTs: phagocytosis of NAs covalently linked to CNTs (A); injection of NAs through CNT nanochannels (B); penetration of NAs adsorbed onto the surface of CNTs (C_I) or that have made a complex with CNTs by electrostatic forces (C_{II}); and endocytosis of NAs electrostatically complexed with (D_I), covalently linked to (D_{II}), or adsorbed (D_{III}) onto CNTs. Suggested strategies for intracellular release of NAs from the CNTs: electrostatic dissociation (E), enzymatic cleavage (F), and NIR radiation (G). Reproduced from Ref. [58] with permission from the Royal Society of Chemistry.

8.3.1.1 Delivery of plasmid DNA

One of the first case studies of CNTs as components for engineering a gene delivery platform was reported by Bianco et al. in 2004 [59]. The complexation of positively charged SWCNT-NH_3^+, obtained by covalent amino-functionalization of SWCNTs using the 1,3-dipolar cycloaddition reaction, with negatively charged pDNA through electrostatic interaction facilitates higher DNA uptake and enhances the gene expression up to 10-fold in comparison to the control. This research opened a door for a novel class of nonviral vectors for the delivery of therapeutic agents by CNTs through noncovalent interactions. To optimize the efficiency of using CNTs as vehicles of DNA delivery, our group also investigated physicochemical interactions between various cationic CNTs (Fig. 8.4) and DNA [60].

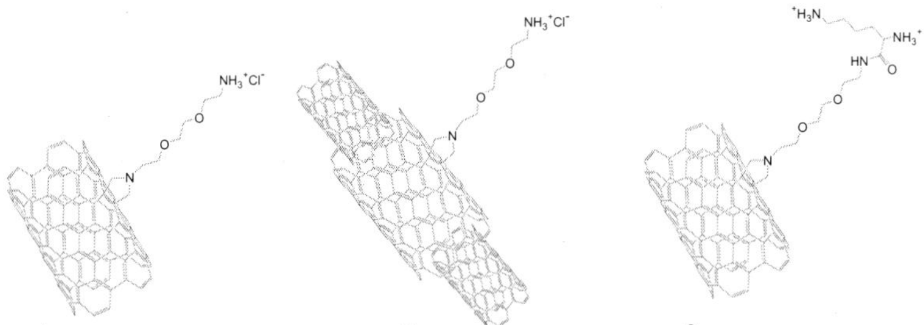

Figure 8.4 Molecular structures of ammonium-functionalized carbon nanotubes: (1) SWCNT-NH_3^+, (2) MWCNT-NH_3^+, and (3) lysine-functionalized SWCNTs. Adapted with permission from Ref. [60]. Copyright (2005) American Chemical Society.

Indeed, for a gene delivery system, it is required that pDNA be first adsorbed onto CNTs and then carried into the target cells, followed by the intracellular release from CNTs and entrance into the nucleus for expression. The study demonstrated that all three types of functionalized CNTs can adsorb pDNA-enhancing gene transfection over pDNA alone into a human cell line. The electrostatic interaction between the CNTs and DNA was mediated by the length and diameter of the CNTs [60]. Similar results were obtained by Gao et al. [61], where CNTs conjugated with four different chemical groups have been used to determine the delivery efficiency of pDNA. The results showed that among amino, carboxyl, hydroxyl, and alkyl groups, only the amine-functionalized CNTs were able to complex and deliver pDNA.

Another strategy proposed to enhance the positive charge onto the surface of CNTs was to functionalize the nanotubes with cationic polymers such as PEI or polyamidoamine (PAMAM). Liu et al. [62] showed a high transfection efficiency using PEI-MWCNTs to immobilize and protect DNA from degradation. Using confocal microscopy, the authors found that the uptake of the complex between DNA and PEI-MWCNTs was through endocytosis. Nunes et al. [63] have shown that MWCNTs grafted with polymers like PEI, polyallylamine (PAA), or PEI/PAA were able to efficiently make a complex with and transfer pDNA in vitro without appreciable cytotoxicity. The PEI-MWCNTs and PEI/PAA-MWCNTs exhibited a significant enhancement in gene transfection and gene expression in comparison to naked pDNA or pDNA-complexed polymers.

Recently, a novel CNT-based device has been developed to introduce pDNA into mammalian cells with high transfection efficiency and low cytotoxicity [64]. In this study, the authors reported an array of aligned hollow CNTs, schematically depicted in Fig. 8.5, that supports cell growth and enhances transfection through the lumens of CNTs into cells in a short time. A pDNA solution was injected on the back and allowed to diffuse through the CNTs into the cells. In addition, gene transfection using this CNT array did not show cell distress, suggesting limited side effects on cellular physiology.

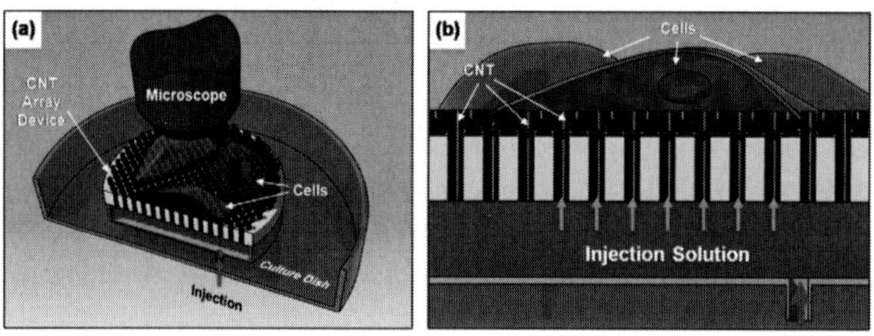

Figure 8.5 Schematic illustration of a carbon nanotube array–based device for parallel transfection of multiple cells. Reprinted with permission from Ref. [64]. Copyright (2016) John Wiley and Sons.

In conclusion, exploiting the CNT functionalization it is possible to increase the surface charge density of the nanotubes, enhancing the complexation efficiency with NAs, leading to significant gene therapy efficiency. However, most of the results have been obtained only in vitro. The case studies in vivo are few because of the requirements of overcoming different barriers [56]. Delivery of pDNA using the hybrid nanoparticles and CNTs (NP-CNTs), which was formed by electrostatic bond, has been reported in an in vivo canine model of restenosis [65].

8.3.1.2 Delivery of siRNA

The purpose of grafting polycationic molecules onto CNMs is to modulate the intrinsic toxicity of polycationic molecules [66–68] toward cell membranes, leading to safer functional materials. MWCNTs functionalized with PAMAM dendrons of first and second generation have shown good cellular delivery of siRNA and efficient gene silencing [69, 70]. In this pioneer study, MWCNTs exhibited a complexation capacity of siRNA proportional to the number of branches on the dendritic structure. The high efficiency was a consequence of an increase of the positive charges on the surface of conjugated CNTs [69, 70]. Our group also investigated the use of MWCNTs functionalized with ammonium and guanidinium dendrons for siRNA complexation [71]. It was found that the complexation ability for siRNA was independent of the core type of the dendritic structure but was influenced by the terminal groups. In the case of guanidinium dendrons, the highest generation showed the lowest siRNA complexation. The ammonium-functionalized dendrons exhibited instead similar siRNA complexation regardless of their size. Interestingly, ammonium dendron–modified CNTs showed higher siRNA complexation than the guanidinium counterparts.

Kam et al. [72] were the first to report the stimulated released of DNA or siRNA exploiting an enzyme disulfide bond cleavage. In this study, SWCNTs were coated with a lipophilic PEG covalently functionalized with DNA or siRNA via cleavable SS bonds (Fig. 8.6). After cell internalization, the release of DNA or siRNA is trigged by the disulfide reduction by glutathione. In the absence of disulfide bonds, only cellular internalization without nuclear translocation of DNA was observed. Only the DNA released from CNTs was biologically active. The comparison between siRNA-SWCNTs and siRNA complexed to lipofectamine proved that the CNT conjugates have an enhanced degree of gene silencing.

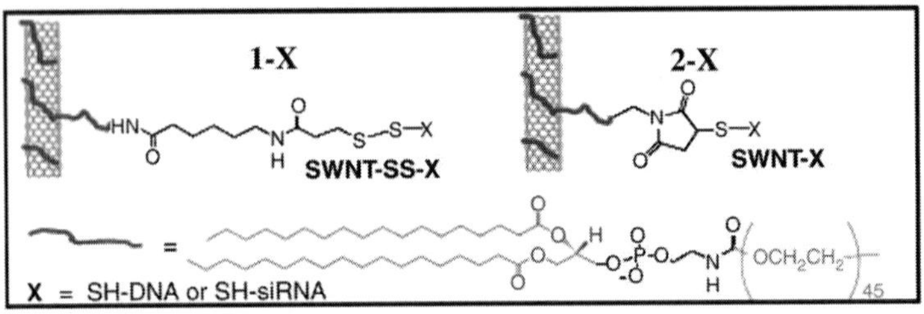

Figure 8.6 Molecular structures of SWCNT functionalized by thiolated molecules with (1-X) and without (2-X) disulfide bonds. Both DNA and siRNA molecules contain a thiol functional group at the 5′-end. Reprinted with permission from Ref. [72]. Copyright (2005) American Chemical Society.

Regarding in vivo therapy, Wang et al. [73] functionalized SWCNTs with PEI for complexation of hTERT-siRNA and with NGR peptide to target the PC3 tumor-bearing mice (hTERT stands for human telomerase reverse transcriptase, and the NGR peptide corresponds to the sequence Asn-Gly-Arg to selectively recognize tumor neovasculature). The siRNA was successfully translocated into PC3 cells, causing a significant tumor cell growth inhibition in vivo.

8.3.2 Graphene Oxide–Nucleic Acid Complexes for Gene and Cancer Therapy

Graphene and its derivatives have been exploited as a novel platform for gene delivery because of their ability to protect NAs from enzymatic degradation [74, 75]. For example, Tang et al. [76] reported that ssDNA adsorbed onto the GO surface can be effectively protected from enzymatic cleavage by deoxyribonuclease (DNAseI, a common enzyme that promotes DNA degradation). Other groups have also obtained similar results [77–79]. It has been evidenced that the protective effect of GO on ssDNA and dsDNA can be easily tuned by adjusting the salt concentration [33]. Furthermore, because the binding affinity of NAs to GO is affected by experimental conditions [80, 81], to create an ideal GO-protected NA system, it is necessary to optimize certain experimental conditions, such as the composition of the buffer solution, the incubation time, the concentration of the NA, and the presence of nucleases. Different approaches to enhance the genetic material protection mechanisms are still under exploration. First, the interaction between the NAs and graphene may cause a conformational change of the former [37], rendering it unrecognizable by the enzyme-binding pockets and thus protecting it from cleavage. Second, the change in local ion concentration induced by graphene can inhibit the enzyme activity. Third, the most interesting perspective is the sterical hindrance protection, which prevents nucleases from approaching NAs to initiate their enzymatic hydrolysis. Likewise, NA adsorption onto graphene can result in sterical hindrance, change in local ion concentration, or change in probe conformation, thereby protecting NAs from nuclease digestion [23]. Lu et al. [82] showed that graphene could protect oligonucleotides from enzymatic cleavage and was able to deliver ssDNA into the cells. The exact mechanism of how NA complexes pass through the cell membrane is still under investigation, but endocytosis and phagocytosis are believed to be the most involved in this process.

8.3.2.1 Delivery of plasmid DNA

Nowadays, among polycationic polymers used in gene delivery, PEI is considered the most suitable [83, 84]. Indeed, the ability of GO functionalized with high-molecular-weight PEI (HMW-PEI) in condensing and delivering pDNA into cells was demonstrated [85]. Tripathi et al. [86] have also reported the capability of linear PEI grafted to GO to transfer NAs into mammalian cells. A HMW-PEI was also used to doubly functionalize the hydrothermally treated rGO modified with PEG to deliver a functional plasmid, proving a remarkable enhancement of gene transfection efficiency under near-infrared (NIR) irradiation compared to nonhydrothermally rGO-PEG-PEI [87]. Interest has recently shifted toward the conjugation of GO with low-molecular-weight (LMW) PEI, which was proved to exhibit less cytotoxicity [88] and enhance cellular uptake and transfection. Feng et al. [13] have successfully performed the functionalization of GO with PEI of different molecular weights (1.2 kDa and 10 kDa) as nontoxic nanovehicles for efficient transfection of pDNA. The authors concluded that GOPEI-10 kDa had transfection efficiency similar to but toxicity lower than that of PEI-10 kDa. Kim et al. [39] also prepared LMW (1.8 kDa) branched PEI (BPEI) conjugated to GO (BPEI-GO) with high pDNA binding and observed a transfection capacity comparable to that of HMW-PEI (Fig. 8.7).

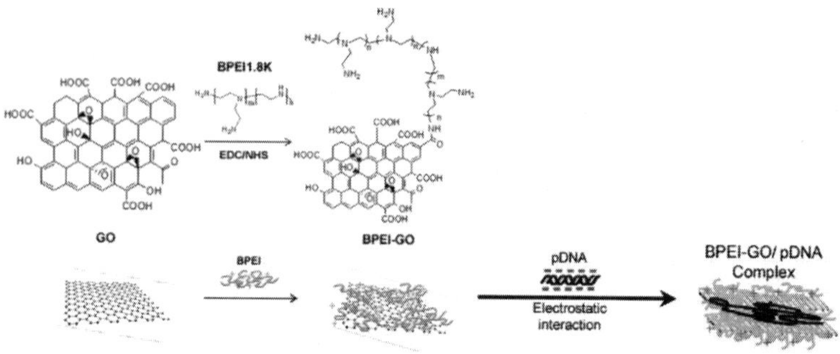

Figure 8.7 Synthetic scheme to achieve a BPEI-GO complex and its application in gene delivery and bioimaging. Reprinted with permission from Ref. [39]. Copyright (2011) American Chemical Society.

8.3.2.2 Delivery of siRNA

Besides the ability to deliver pDNA for gene transfection, GO has been applied as a nanocarrier of siRNA for gene silencing. For example, Imani et al. [40] reported a novel rGO functionalized with a phospholipid-based amphiphilic polymer (PL-PEG) and the cell penetrating peptide octa-arginine (R8). This rGO carrier is stable in a biological solution and does not affect cell viability. Moreover, with the high-positive-charge surface provided by R8, siRNA was easily complexed and taken up by the cells. The transfection efficiency on MCF-7 cells of a fluorescein isothiocyanate (FITC)-labeled siRNA loaded onto rGON-PL-PEG-R8 was comparable to the commercial reagent HiPerFect.

On the basis of the characteristics of GO, a combination of both covalent and noncovalent chemistry can be applied. GO also allows the development of multifunctionalization approaches, as has been widely demonstrated. In general, molecules like dispersants or targeting ligands are first covalently attached to GO, while the drugs, which are mainly lypophilic, are then adsorbed onto the GO surface via low-polar interaction, such as π-π stacking. This simple and versatile strategy helps to enhance the dispersibility of the platform in cell culture media, to reduce the cytotoxicity, and to increase the drug efficacy.

In the case of nontargeted drug and gene delivery, Zhang et al. [38] developed a PEI-conjugated GO system for the codelivery of siRNA and doxorubicin (DOX) in vitro. PEI(25K) was first covalently attached to GO, and then siRNA and DOX were simultaneously loaded by electrostatic interaction and π-π stacking, respectively. Interestingly, the sequential delivery of siRNA and DOX in HeLa cells displayed a strong synergistic anticancer effect. In fact, HeLa cells were first incubated with Bcl2-targeting siRNA, followed by PEI-GO/DOX. The results showed a decrease in the IC_{50} value (concentration of a drug that kills 50% of the cells) from 1.3 μg/mL for the control PEI-GO/scrambled siRNA/DOX to 0.52 μg/mL for PEI-GO/Bcl2-targeting siRNA/DOX, thus affording enhanced chemotherapy.

Although there are many carriers, such as inorganic nanoparticles, polymer-based systems, and lipid-based systems, low targeting accumulation and low conjugation efficiency remain constraints against therapeutic efficiency at tumor sites. Nevertheless, the side effects of nonspecific accumulation in normal tissues as well as the insufficiency of cellular uptake require a novel and intelligent design of carriers. Therefore, to enhance the cellular uptake, to selectively deliver genes at specific tumor sites, different tumor targeting moieties (antibodies [89], peptides [90], folic acid (FA) [91], aptamer [92], etc.) have been introduced onto the GO surface to recognize the corresponding "molecular signature" on the cell surface. These active targeting moieties on GO favor the accumulation of the platform at the desired tissues, thus enhancing the therapeutic effects while decreasing the side effects of the drug.

In the case of targeted delivery, a dual carrier for delivering DOX and siRNA to overcome drug-resistant MCF-7/ADR cancer cells was reported by Zeng et al. [93] (Fig. 8.8). First, PEG-NH_2 was covalently attached on GO via a ring-opening epoxide reaction; then PEI(10K) was linked to the carboxylic groups of PEG-GO (PEI-PEG-GO) to form polypropylene glycol (PPG). Then, FA, recognized by folate receptors overexpressed on cancer cells, was conjugated with PPG (PPG-FA) via amide bond formation between the carboxylic group of FA and the amino groups of PPG. Finally, PPG-FA was loaded with siRNA via electrostatic interaction and DOX via π-π stacking. The results showed that the PPG-FA/siRNA/DOX exhibits high cellular uptake, increased DOX penetration into the nucleus, and enhanced DOX cytotoxicity on the MCF-7/ADR cells in comparison to free DOX, or PPG-FA/DOX devoid of PEG. Interestingly, the authors also proved that the combination treatment of chemotherapy with photothermal treatment exerted a synergistic effect compared with chemotherapy alone.

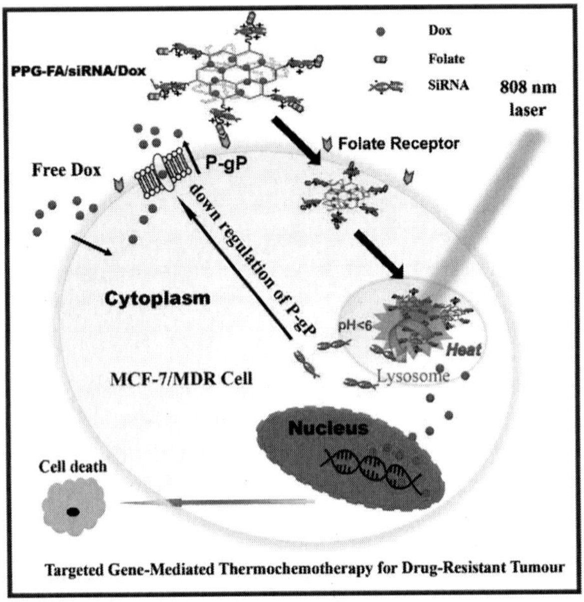

Figure 8.8 Targeted gene-mediated thermochemotherapy against drug-resistant tumors [93].

Recently, Gu et al. [94] conjugated GO with PAMAM for the codelivery of DOX and matrix metallopeptidase-9 short hairpin RNA (MMP-9 shRNA) plasmid for breast cancer therapy. The results evidenced that GO-PAMAM exhibits a pH-controlled drug release in vitro and could mediate a significant reduction of MMP-9 protein in MCF-7 cells. Moreover, the cell survival rate of codelivered GO-PAMAM/DOX/MMP-9 was lower than that of DOX or MMP-9 shRNA, separately. In addition, the codelivery of DOX and siRNA using GO-PEI has been reported to overcome multidrug resistance of breast cancer cells [95]. In this case, Zhi et al. [95] assembled an HMW-BPEI (25 kDa) and poly(sodium 4-styrenesulfonates) via a layer-by-layer assembly method onto GO, named PPSG. The drug adriamycin (ADR) was loaded onto PPSG via physical adsorption; then an antimicro-RNA (anti-miR-21) was complexed via electrostatic interaction. In ADR-resistant breast cancer cells (MCF-7/ADR), the complex significantly enhanced the accumulation of ADR and exhibited higher cytotoxicity in comparison to the free drug. Moreover, this enhanced therapeutic efficiency was correlated to the effective silencing of miR-21.

In another study, by coloading PEI and PEG onto GO, Feng et al. [96] were able to photothermally enhance gene delivery. The transfection efficiency of pDNA coding for green fluorescent protein in HeLa cell and siRNA-mediated knockdown of mRNA for the target polo-like kinase 1 in MDA-MB-435s cells was improved using NIR irradiation at 808 nm. This synergistic effect in siRNA delivery was explained by the increase in cell membrane permeability, which was caused by a photothermal effect.

In summary, when the siRNA is introduced into the cytoplasm of a cell, it can interact with the intracellular RNAses, which leads to its degradation. In addition, because of the extracellular degradation by enzymes located in serum and tissues, siRNA delivery is extremely challenging. Indeed siRNA has a lifetime in serum ranging from a few minutes to an hour [97], quite short to reach the target site and trigger the gene silencing. Another issue in gene silencing is the off-target silencing of siRNA, which is caused by the homology with six to seven nucleotides in the seed region of the siRNA sequence [98]. This off-target silencing leads to the suppression of noninterest genes, inducing dangerous mutation of other genes. Moreover, an immune response can be triggered by siRNA once the interferon responses are activated, leading to cell death [99]. Therefore, various nanocarriers have been widely and deeply explored, especially GO for its special physicochemical properties, to overcome these drawbacks in term of siRNA delivery.

8.4 Challenges and Future Perspectives of CNMs in Gene Therapy

Carbon-based nanotechnology research has become more and more interdisciplinary. In this regard, the considerable aqueous dispersibility and colloidal stability of GO and the flexible nanoneedle-like structure of CNTs, as well as their remarkable and tunable surface functionalization and biological properties, have intensified the studies of CNMs as promising candidates in biomedicine.

In this context, one of the first issues that need to be considered is the understanding and the control of noncovalent and covalent functionalization of CNMs, particularly for application in gene therapy. Moreover, the interaction of NAs with CNMs is one of the key points to design novel platforms for gene delivery. The role of CNMs as good carriers to deliver genetic materials into cellular and subcellular compartments in vitro and in vivo demands precise functionalization. The comprehension of pharmacokinetics and pharmacodynamics of CNMs, their long-term cytotoxicity, and their biodegradability are also crucial for clinical tests. The hope regarding clinical applications of CNMs in cancer therapy should be highlighted and strongly investigated for the enormous possible benefits.

In summary, CNMs have been studied as potential nanocarriers of NAs in gene therapy and cancer treatment. In this context, it will take a long time to carefully address the issues related to their applications in clinical trials, including the homogeneity of materials, the presence of trace impurities after the synthesis and purification process, the amount and nature of oxygenated species, and the size effect. Every journey starts with a step, and the science community is working toward progressing on the road to obtaining a "really suitable" platform of CNMs for the treatment of cancer in humans.

References

1. Mu, Q., Jiang, G., Chen, L., Zhou, H., Fourches, D., Tropsha, A. and Yan, B. (2014). Chemical basis of interactions between engineered nanoparticles and biological systems, *Chem. Rev.*, **114**, 7740–7781.
2. Mintzer, M. A. and Simanek, E. E. (2009). Nonviral vectors for gene delivery, *Chem. Rev.*, **109**, 259–302.
3. Stadler, A., Chi, C., van der Lelie, D. and Gang, O. DNA-incorporating nanomaterials in biotechnological applications, *Nanomedicine*, **5**, 319–334.
4. Krishna, K. V., Ménard-Moyon, C., Verma, S. and Bianco, A. Graphene-based nanomaterials for nanobiotechnology and biomedical applications, *Nanomedicine*, **8**, 1669–1688.
5. Lee, J., Yim, Y., Kim, S., Choi, M.-H. H., Choi, B.-S. S., Lee, Y. and Min, D.-H. H. (2016). In-depth investigation of the interaction between DNA and nano-sized graphene oxide, *Carbon*, **97**, 92–98.

6. Liu, Z., Robinson, J. T., Tabakman, S. M., Yang, K. and Dai, H. (2011). Carbon materials for drug delivery & cancer therapy, *Mater. Today*, **14**, 316–323.
7. Lu, F., Gu, L., Meziani, M. J., Wang, X., Luo, P. G., Veca, L. M., Cao, L. and Sun, Y.-P. (2009). Advances in bioapplications of carbon nanotubes, *Adv. Mater.*, **21**, 139–152.
8. Yang, K., Feng, L., Shi, X. and Liu, Z. (2013). Nano-graphene in biomedicine: theranostic applications, *Chem. Soc. Rev.*, **42**, 530–547.
9. Jana, B., Biswas, A., Mohapatra, S., Saha, A., Ghosh, S., Speranza, G., Manitto, P., Sironi, M., Chen, C., Theuer, C. P., et al. (2014). Single functionalized graphene oxide reconstitutes kinesin mediated intracellular cargo transport and delivers multiple cytoskeleton proteins and therapeutic molecules into the cell, *Chem. Commun.*, **50**, 11595–11598.
10. Jana, B., Mondal, G., Biswas, A., Chakraborty, I., Saha, A., Kurkute, P. and Ghosh, S. (2013). Dual functionalized graphene oxide serves as a carrier for delivering oligohistidine- and biotin-tagged biomolecules into cells, *Macromol. Biosci.*, **13**, 1478–1484.
11. Cui, L., Song, Y., Ke, G., Guan, Z., Zhang, H., Lin, Y., Huang, Y., Zhu, Z. and Yang, C. J. (2013). Graphene oxide protected nucleic acid probes for bioanalysis and biomedicine, *Chem. Eur. J.*, **19**, 10442–10451.
12. Hughes, Z. E. and Walsh, T. R. (2015). What makes a good graphene-binding peptide? Adsorption of amino acids and peptides at aqueous graphene interfaces, *J. Mater. Chem. B*, **3**, 3211–3221.
13. Feng, L., Zhang, S., Liu, Z., Ma, Y. F., Huang, Y., Chen, Y., Bai, Y. B., Li, T. J., Fang, H. P., Fan, C. H., et al. (2011). Graphene based gene transfection, *Nanoscale*, **3**, 1252–1257.
14. Theodosopoulos, G. V; Bilalis, P. and Sakellariou, G. (2015). Polymer functionalized graphene oxide: a versatile nanoplatform for drug/gene delivery, *Curr. Org. Chem.*, **19**, 1828–1837.
15. Zheng, M., Jagota, A., Semke, E. D., Diner, B. A., McLean, R. S., Lustig, S. R., Richardson, R. E. and Tassi, N. G. (2003). DNA-assisted dispersion and separation of carbon nanotubes, *Nat. Mater.*, **2**, 338–342.
16. Zheng, M. (2003). Structure-based carbon nanotube sorting by sequence-dependent DNA assembly, *Science*, **302**, 1545–1548.
17. Alidori, S., Asqiriba, K., Londero, P., Bergkvist, M., Leona, M., Scheinberg, D. A. and McDevitt, M. R. (2013). Deploying RNA and DNA with functionalized carbon nanotubes, *J. Phys. Chem. C*, **117**, 5982–5992.

18. Johnson, R. R., Johnson, A. T. C. and Klein, M. L. (2008). Probing the structure of DNA-carbon nanotube hybrids with molecular dynamics, *Nano Lett.*, **8**, 69–75.

19. Wu, Y., Phillips, J. A., Liu, H., Yang, R. and Tan, W. (2008). Carbon nanotubes protect DNA strands during cellular delivery, *ACS Nano*, **2**, 2023–2028.

20. Lee, J., Kim, J., Kim, S. and Min, D.-H. (2016). Biosensors based on graphene oxide and its biomedical application, *Adv. Drug Deliv. Rev.*, **105**, 275–287.

21. Lu, C. H., Yang, H. H., Zhu, C. L., Chen, X. and Chen, G. N. (2009). A graphene platform for sensing biomolecules, *Angew. Chem. Int. Ed.*, **48**, 4785–4787.

22. Balapanuru, J., Yang, J.-X. X., Xiao, S., Bao, Q., Jahan, M., Polavarapu, L., Wei, J., Xu, Q.-H. H. and Loh, K. P. P. (2010). A graphene oxide-organic dye ionic complex with DNA-sensing and optical-limiting properties, *Angew. Chem. Int. Ed.*, **49**, 6549–6553.

23. Tang, L., Chang, H., Liu, Y. and Li, J. (2012). Duplex DNA/graphene oxide biointerface: from fundamental understanding to specific enzymatic effects, *Adv. Funct. Mater.*, **22**, 3083–3088.

24. Yang, X., Wang, Y., Huang, X., Ma, Y., Huang, Y., Yang, R., Duan, H., Chen, Y. and Lubbe, A. S. (2011). Multi-functionalized graphene oxide based anticancer drug-carrier with dual-targeting function and pH-sensitivity, *J. Mater. Chem.*, **21**, 3448.

25. Wang, Z., Zhou, C., Xia, J., Via, B., Xia, Y., Zhang, F., Li, Y. and Xia, L. (2013). Fabrication and characterization of a triple functionalization of graphene oxide with Fe_3O_4, folic acid and doxorubicin as dual-targeted drug nanocarrier, *Colloids Surf., B*, **106**, 60–65.

26. Li, M., Pan, Y., Guo, X., Liang, Y., Wu, Y., Wen, Y., Yang, H., Xia, Y., Ruoff, R. S., Lin, Y. H., et al. (2015). Pt/single-stranded DNA/graphene nanocomposite with improved catalytic activity and CO tolerance, *J. Mater. Chem. A*, **3**, 10353–10359.

27. Banerjee, S., Wilson, J., Shim, J., Shankla, M., Corbin, E. A., Aksimentiev, A. and Bashir, R. (2015). Slowing DNA transport using graphene-DNA interactions, *Adv. Funct. Mater.*, **25**, 936–946.

28. Zhang, H., Li, Y. and Su, X. (2013). A small-molecule-linked DNA–graphene oxide-based fluorescence-sensing system for detection of biotin, *Anal. Biochem.*, **442**, 172–177.

29. Xue, T., Cui, X., Guan, W., Wang, Q., Liu, C., Wang, H., Qi, K., Singh, D. J. and Zheng, W. (2014). Surface plasmon resonance technique for directly

probing the interaction of dna and graphene oxide and ultra-sensitive biosensing, *Biosens. Bioelectron.*, **58**, 374–379.

30. He, S., Song, B., Li, D., Zhu, C., Qi, W., Wen, Y., Wang, L., Song, S., Fang, H. and Fan, C. (2010). A graphene nanoprobe for rapid, sensitive, and multicolor fluorescent DNA analysis, *Adv. Funct. Mater.*, **20**, 453–459.
31. Zhao, X. (2011). Self-assembly of DNA segments on graphene and carbon nanotube arrays in aqueous solution: a molecular simulation study, *J. Phys. Chem. C*, **115**, 6181–6189.
32. Wu, M., Kempaiah, R., Huang, P. J. J., Maheshwari, V. and Liu, J. (2011). Adsorption and desorption of DNA on graphene oxide studied by fluorescently labeled oligonucleotides, *Langmuir*, **27**, 2731–2738.
33. Lei, H., Mi, L., Zhou, X., Chen, J., Hu, J., Guo, S. and Zhang, Y. (2011). Adsorption of double-stranded DNA to graphene oxide preventing enzymatic digestion, *Nanoscale*, **3**, 3888–3892.
34. Liu, M., Zhao, H., Chen, S., Yu, H. and Quan, X. (2012). Capture of double-stranded DNA in stacked-graphene: giving new insight into the graphene/DNA interaction, *Chem. Commun.*, **48**, 564–566.
35. Ren, H., Wang, C., Zhang, J. J., Zhou, X., Xu, D., Zheng, J., Guo, S. and Zhang, J. J. (2010). DNA cleavage system of nanosized graphene oxide sheets and copper ions, *ACS Nano*, **4**, 7169–7174.
36. Zheng, B., Wang, C., Wu, C., Zhou, X., Lin, M., Wu, X., Xin, X., Chen, X., Xu, L., Liu, H., et al. (2012). Nuclease activity and cytotoxicity enhancement of the DNA intercalators via graphene oxide, *J. Phys. Chem. C*, **116**, 15839–15846.
37. Zhang, L., Wang, Z., Lu, Z., Shen, H., Huang, J., Zhao, Q., Liu, M., He, N. and Zhang, Z. (2013). PEGylated reduced graphene oxide as a superior ssRNA delivery system, *J. Mater. Chem. B*, **1**, 749–755.
38. Zhang, L., Lu, Z., Zhao, Q., Huang, J., Shen, H. and Zhang, Z. (2011). Enhanced chemotherapy efficacy by sequential delivery of siRNA and anticancer drugs using PEI-grafted graphene oxide, *Small*, **7**, 460–464.
39. Kim, H., Namgung, R., Singha, K., Oh, I.-K. and Kim, W. J. (2011). Graphene oxide–polyethylenimine nanoconstruct as a gene delivery vector and bioimaging tool, *Bioconjugate Chem.*, **22**, 2558–2567.
40. Imani, R., Shao, W., Taherkhani, S., Emami, S. H., Prakash, S. and Faghihi, S. (2016). Dual-functionalized graphene oxide for enhanced siRNA delivery to breast cancer cells, *Colloids Surf., B*, **147**, 315–325.
41. Chau, N. D. Q., Reina, G., Raya, J., Vacchi, I. A., Ménard-Moyon, C., Nishina, Y. and Bianco, A. (2017). Elucidation of siRNA complexation efficiency

by graphene oxide and reduced graphene oxide, *Carbon*, **122**, 643–652.

42. Whitehead, K. A., Langer, R. and Anderson, D. G. (2009). Knocking down barriers: advances in siRNA delivery, *Nat. Rev. Drug Discov.*, **8**, 129–138.

43. Yang, Z. R., Wang, H. F., Zhao, J., Peng, Y. Y., Wang, J., Guinn, B.-A. and Huang, L. Q. (2007). Recent developments in the use of adenoviruses and immunotoxins in cancer gene therapy, *Cancer Gene Ther.*, **14**, 599–615.

44. Hannon, G. J. (2002). RNA interference, *Nature*, **418**, 244–251.

45. Fire, A., Xu, S., Montgomery, M. K., Kostas, S. A., Driver, S. E. and Mello, C. C. (1998). Potent and specific genetic interference by double-stranded RNA in caenorhabditis elegans, *Nature*, **391**, 806–811.

46. Dykxhoorn, D. M., Palliser, D. and Lieberman, J. (2006). The silent treatment: siRNAs as small molecule drugs, *Gene Ther.*, **13**, 541–552.

47. Elbashir, S. M., Harborth, J., Lendeckel, W., Yalcin, A., Weber, K. and Tuschl, T. (2001). Duplexes of 21 ± nucleotide RNAs mediate RNA interference in cultured mammalian cells, *Nature*, **411**, 494–498.

48. Draz, M. S., Fang, B. A., Zhang, P., Hu, Z., Gu, S., Weng, K. C., Gray, J. W. and Chen, F. F. (2014). Nanoparticle-mediated systemic delivery of siRNA for treatment of cancers and viral infections, *Theranostics*, **4**, 872–892.

49. Naldini, L., Blömer, U., Gallay, P., Ory, D., Mulligan, R., Gage, F. H., Verma, I. M. and Trono, D. (1996). In vivo gene delivery and stable transduction of nondividing cells by a lentiviral vector, *Science*, **272**, 263–267.

50. Wong, H. H., Lemoine, N. R. and Wang, Y. (2010). Oncolytic viruses for cancer therapy: overcoming the obstacles. *Viruses*, **2**, 78–106.

51. Thomas, C. E., Ehrhardt, A. and Kay, M. A. (2003). Progress and problems with the use of viral vectors for gene therapy, *Nat. Rev. Genet.*, **4**, 346–358.

52. Bessis, N., Garciacozar, F. and Boissier, M.-C. (2004). Immune responses to gene therapy vectors: influence on vector function and effector mechanisms, *Gene Ther.*, **11**, S10–S17.

53. Baum, C., Kustikova, O., Modlich, U., Li, Z. and Fehse, B. (2006). Mutagenesis and oncogenesis by chromosomal insertion of gene transfer vectors, *Hum. Gene Ther.*, **17**, 253–263.

54. Takenobu, T., Tomizawa, K., Matsushita, M., Li, S.-T., Moriwaki, A., Lu, Y.-F. and Matsui, H. (2002). Development of p53 protein transduction therapy using membrane-permeable peptides and the application to oral cancer cells, *Mol. Cancer Ther.*, **1**, 1043–1049.

55. Yin, H., Kanasty, R. L., Eltoukhy, A. A., Vegas, A. J., Dorkin, J. R. and Anderson, D. G. (2014). Non-viral vectors for gene-based therapy, *Nat. Rev. Genet.*, **15**, 541–555.

56. Bates, K. and Kostarelos, K. (2013). Carbon nanotubes as vectors for gene therapy: past achievements, present challenges and future goals, *Adv. Drug Deliv. Rev.*, **65**, 2023–2033.

57. Niidome, T. and Huang, L. (2002). Gene therapy progress and prospects: nonviral vectors, *Gene Ther.*, **9**, 1647–1652.

58. Lacerda, L., Bianco, A., Prato, M. and Kostarelos, K. (2008). Carbon nanotube cell translocation and delivery of nucleic acids in vitro and in vivo, *J. Mater. Chem.*, **18**, 17–22.

59. Pantarotto, D., Singh, R., McCarthy, D., Erhardt, M., Briand, J.-P., Prato, M., Kostarelos, K. and Bianco, A. (2004). Functionalized carbon nanotubes for plasmid DNA gene delivery, *Angew. Chem. Int. Ed.*, **43**, 5242–5246.

60. Singh, R., Pantarotto, D., McCarthy, D., Chaloin, O., Hoebeke, J., Partidos, C. D., Briand, J. P., Prato, M., Bianco, A. and Kostarelos, K. (2005). Binding and condensation of plasmid DNA onto functionalized carbon nanotubes: toward the construction of nanotube-based gene delivery vectors, *J. Am. Chem. Soc.*, **127**, 4388–4396.

61. Gao, L., Nie, L., Wang, T., Qin, Y., Guo, Z., Yang, D. and Yan, X. (2006). Carbon nanotube delivery of the GFP gene into mammalian cells, *ChemBioChem*, **7**, 239–242.

62. Liu, Y., Wu, D. C., Zhang, W. De; Jiang, X., He, C. Bin; Chung, T. S., Goh, S. H. and Leong, K. W. (2005). Polyethylenimine-grafted multiwalled carbon nanotubes for secure noncovalent immobilization and efficient delivery of DNA, *Angew. Chem. Int. Ed.*, **44**, 4782–4785.

63. Nunes, A., Amsharov, N., Guo, C., Van Den Bossche, J., Santhosh, P., Karachalios, T. K., Nitodas, S. F., Burghard, M., Kostarelos, K. and Al-Jamal, K. T. (2010). Hybrid polymer-grafted multiwalled carbon nanotubes for in vitro gene delivery, *Small*, **6**, 2281–2291.

64. Golshadi, M., Wright, L. K., Dickerson, I. M. and Schrlau, M. G. (2016). High-efficiency gene transfection of cells through carbon nanotube arrays, *Small*, **12**, 3014–3020.

65. Paul, A., Shao, W., Shum-Tim, D. and Prakash, S. (2012). The attenuation of restenosis following arterial gene transfer using carbon nanotube coated stent incorporating TAT/DNAAng1+Vegf nanoparticles, *Biomaterials*, **33**, 7655–7664.

66. Goldman, C., Soroceanu, L., Smith, N., Gillespie, G., Shaw, W., Burgess, S., Bilbao, G. and Curiel, D. (1997). In vitro and in vivo gene delivery

mediated by a synthetic polycationic amino polymer, *Nat. Biotechnol.*, **15**, 462–466.

67. Putnam, D., Gentry, C. A., Pack, D. W. and Langer, R. (2001). Polymer-based gene delivery with low cytotoxicity by a unique balance of side-chain termini, *Proc. Natl. Acad. Sci. U.S.A.*, **98**, 1200–1205.

68. Xiang, Y.-Z., Feng, Z.-H., Zhang, J., Liao, Y.-L., Yu, C.-J., Yi, W.-J., Zhu, W. and Yu, X.-Q. (2009). Linear cyclen-based polyamine as a novel and efficient reagent in gene delivery, *Org. Biomol. Chem.*, **8**, 640–647.

69. Antonia Herrero, M., Toma, F. M., Al-Jamal, K. T., Kostarelos, K., Bianco, A., Da Ros, T., Bano, F., Casalis, L., Scoles, G. and Prato, M. (2009). Synthesis and characterization of a carbon nanotube-dendron series for efficient siRNA delivery, *J. Am. Chem. Soc.*, **131**, 9843–9848.

70. Al-Jamal, K. T., Toma, F. M., Yilmazer, A., Ali-Boucetta, H., Nunes, A., Herrero, M.-A., Tian, B., Eddaoudi, A., Eddaoui, A., Al-Jamal, W. T., et al. (2010). Enhanced cellular internalization and gene silencing with a series of cationic dendron-multiwalled carbon nanotube:siRNA complexes, *FASEB J.*, **24**, 4354–4365.

71. Battigelli, A., Wang, J. T. W., Russier, J., Da Ros, T., Kostarelos, K., Al-Jamal, K. T., Prato, M. and Bianco, A. (2013). Ammonium and guanidinium dendron-carbon nanotubes by amidation and click chemistry and their use for siRNA delivery, *Small*, **9**, 3610–3619.

72. Kam, W. N. S., Liu, Z. and Dai, H. (20050. Functionalization of carbon nanotubes via cleavable disulfide bonds for efficient intracellular delivery of siRNA and potent gene silencing, *J. Am. Chem. Soc.*, **127**, 12492–12493.

73. Wang, L., Shi, J., Zhang, H., Li, H., Gao, Y., Wang, Z., Wang, H., Li, L., Zhang, C., Chen, C., et al. (2013). Synergistic anticancer effect of RNAi and photothermal therapy mediated by functionalized single-walled carbon nanotubes, *Biomaterials*, **34**, 262–274.

74. Yang, T., Guan, Q., Guo, X., Meng, L., Du, M. and Jiao, K. (2013). Direct and freely switchable detection of target genes engineered by reduced graphene oxide-poly(*M* -aminobenzenesulfonic acid) nanocomposite via synchronous pulse electrosynthesis, *Anal. Chem.*, **85**, 1358–1366.

75. Dong, H., Zhang, J., Ju, H., Lu, H., Wang, S., Jin, S., Hao, K., Du, H. and Zhang, X. (2012). Highly sensitive multiple microRNA detection based on fluorescence quenching of graphene oxide and isothermal strand-displacement polymerase reaction, *Anal. Chem.*, **84**, 4587–4593.

76. Tang, Z., Wu, H., Cort, J. R., Buchko, G. W., Zhang, Y., Shao, Y., Aksay, I. A., Liu, J. and Lin, Y. (2010). Constraint of DNA on functionalized graphene improves its biostability and specificity, *Small*, **6**, 1205–1209.

77. Lu, C.-H., Li, J., Qi, X.-J., Song, X.-R., Yang, H.-H., Chen, X. and Chen, G.-N. (2011). Multiplex detection of nucleases by a graphene-based platform, *J. Mater. Chem.*, **21**, 10915–10919.

78. Wang, Y., Li, Z., Weber, T. J., Hu, D., Lin, C.-T., Li, J. and Lin, Y. (2013). In situ live cell sensing of multiple nucleotides exploiting DNA/RNA aptamers and graphene oxide nanosheets, *Anal. Chem.*, **85**, 6775–6782.

79. Cui, L., Lin, X., Lin, N., Song, Y., Zhu, Z., Chen, X., Yang, C. J., Kang, H., Lin, Y. H., Fan, C. H., et al. (2012). Graphene oxide-protected DNA probes for multiplex microRNA analysis in complex biological samples based on a cyclic enzymatic amplification method, *Chem. Commun.*, **48**, 194–196.

80. Park, J. S., Baek, A., Park, I.-S., Jun, B.-H., Kim, D.-E., Leslie, M. D., Schafer, E., Marchand, B., Adedeji, A., Michailidis, E., et al. (2013). A graphene oxide-based platform for the assay of RNA synthesis by RNA polymerase using a fluorescent peptide nucleic acid probe, *Chem. Commun.*, **49**, 9203.

81. Tang, L., Wang, Y., Liu, Y. and Li, J. (2011). DNA-directed self-assembly of graphene oxide with applications to ultrasensitive oligonucleotide assay, *ACS Nano*, **5**, 3817–3822.

82. Lu, C.-H., Zhu, C.-L., Li, J., Liu, J.-J., Chen, X., Yang, H.-H., Dai, H. J., Medley, C. D., Cao, Z. H., Li, J., et al. (2010). Using graphene to protect DNA from cleavage during cellular delivery, *Chem. Commun.*, **46**, 3116–3118.

83. Neu, M. and Kissel, T. (2005). Recent advances in rational gene transfer vector design based on poly (ethylene imine) and its derivatives, *J. Gene. Med.*, **7**, 992–1009.

84. Ren, T., Li, L., Cai, X., Dong, H., Liu, S. and Li, Y. (2012). Engineered polyethylenimine/graphene oxide nanocomposite for nuclear localized gene delivery, *Polym. Chem.*, **3**, 2561.

85. Chen, B., Liu, M., Zhang, L., Huang, J., Yao, J. and Zhang, Z. (2011). Polyethylenimine-functionalized graphene oxide as an efficient gene delivery vector, *J. Mater. Chem.*, **21**, 7736–7741.

86. Tripathi, S. K., Goyal, R., Gupta, K. C. and Kumar, P. (2013). Functionalized graphene oxide mediated nucleic acid delivery, *Carbon*, **51**, 224–235.

87. Li, T., Wu, L., Zhang, J., Xi, G., Pang, Y., Wang, X. and Chen, T. (2016). Hydrothermal reduction of polyethylenimine and polyethylene glycol dual-functionalized nanographene oxide for high-efficiency gene delivery, *ACS Appl. Mater. Interfaces*, **8**, 31311–31320.

88. Xiong, M. P., Forrest, M. L., Ton, G., Zhao, A., Davies, N. M. and Kwon, G. S. (2007). Poly(aspartate-G-PEI800), a polyethylenimine analogue of low toxicity and high transfection efficiency for gene delivery, *Biomaterials*, **28**, 4889–4900.

89. Dinauer, N., Balthasar, S., Weber, C., Kreuter, J., Langer, K. and von Briesen, H. (2005). Selective targeting of antibody-conjugated nanoparticles to leukemic cells and primary T-lymphocytes, *Biomaterials*, **26**, 5898–5906.

90. Guo, Y., Xu, H., Li, Y., Wu, F., Li, Y., Bao, Y., Yan, X., Huang, Z. and Xu, P. (2017). Hyaluronic acid and Arg-Gly-Asp peptide modified graphene oxide with dual receptor-targeting function for cancer therapy, *J. Biomater. Appl.*, **32**, 54–65.

91. Zhang, L., Xia, J., Zhao, Q., Liu, L. and Zhang, Z. (2010). Functional graphene oxide as a nanocarrier for controlled loading and targeted delivery of mixed anticancer drugs, *Small*, **6**, 537–544.

92. Alibolandi, M., Mohammadi, M., Taghdisi, S. M., Ramezani, M. and Abnous, K. (2017). Fabrication of aptamer decorated dextran coated nano-graphene oxide for targeted drug delivery, *Carbohydr. Polym.*, **155**, 218–229.

93. Zeng, Y., Yang, Z., Li, H., Hao, Y., Liu, C., Zhu, L., Liu, J., Lu, B. and Li, R. (2017). Multifunctional nanographene oxide for targeted gene-mediated thermochemotherapy of drug-resistant tumour, *Sci. Rep.*, **7**, 43506.

94. Gu, Y., Guo, Y., Wang, C., Xu, J., Wu, J., Kirk, T. B., Ma, D. and Xue, W. (2017). A polyamidoamne dendrimer functionalized graphene oxide for DOX and MMP-9 shRNA plasmid co-delivery, *Mater. Sci. Eng. C*, **70**, 572–585.

95. Zhi, F., Dong, H., Jia, X., Guo, W., Lu, H., Yang, Y., Ju, H., Zhang, X. and Hu, Y. (2013). Functionalized graphene oxide mediated adriamycin delivery and miR-21 gene silencing to overcome tumor multidrug resistance in vitro, *PLoS One*, **8**, e60034.

96. Feng, L., Yang, X., Shi, X., Tan, X., Peng, R., Wang, J. and Liu, Z. (2013). Polyethylene glycol and polyethylenimine dual-functionalized nano-graphene oxide for photothermally enhanced gene delivery, *Small*, **9**, 1989–1997.

97. Behlke, M. A., Kalinina, T. S., Dygalo, N. N., Agami, R., Mobbs, C. V. and Malik, A. B. (2006). Progress towards in vivo use of siRNAs, *Mol. Ther.*, **13**, 644–670.

98. Reynolds, A., Anderson, E. M., Vermeulen, A., Fedorov, Y., Robinson, K., Leake, D., Karpilow, J., Marshall, W. S. and Khvorova, A. (2006). Induction of the interferon response by siRNA is cell type- and duplex length-dependent, *RNA*, **12**, 988–993.

99. Gao, Y., Liu, X.-L. and Li, X.-R. (2011). IJN-17040-research-progress-on-sirna-delivery-with-non-viral-carriers, *Int. J. Nanomed.*, **6**, 1017–1025.

Chapter 9

Self-Assembled Functional Fullerenes and DNA Hybrid Nanomaterials for Various Applications

Sandeepa K. Vittala, Sajena K. Saraswathi, and Joshy Joseph

Photosciences and Photonics Section, Chemical Sciences and Technology Division, CSIR-National Institute for Interdisciplinary Science and Technology, Thiruvananthapuram 695019, Kerala, India
Academy of Scientific and Innovative Research (AcSIR), CSIR-NIIST Campus, Thiruvananthapuram 695019, Kerala, India
joshy@niist.res.in; joshyja@gmail.com

Recent developments in the area of functional supramolecular materials fabricated via the self-assembly of fullerene derivatives and DNA are discussed. Emphasis is given to different modes of DNA-fullerene interactions, fullerene-induced DNA condensation, and higher-order nanostructures, including fullerene cluster–assisted DNA assemblies. We also review various applications of these fullerene–DNA hybrid nanostructures reported in recent years, with a special focus on optoelectronic and biomedical applications.

Templated DNA Nanotechnology: Functional DNA Nanoarchitectonics
Edited by Thimmaiah Govindaraju

ISBN 978-981-4800-21-1 (Hardcover), 978-0-429-42866-1 (eBook)
www.panstanford.com

Finally, in the outlook, we describe the prospects of fullerene-assisted DNA nanostructures in DNA nanotechnology, material science, and molecular electronics.

9.1 Introduction

Fullerenes are a family of molecules composed entirely of carbon atoms that form 3D shapes, including spheres, ellipsoids, and tubes [1]. Although the term "fullerene" is most often associated with buckyballs,carbonnanotubesandnano-onionsarealsofundamentally included in the fullerene family. C_{60} and C_{70} are the two most stable buckyballs, and the structures of these fullerenes, along with the widely used [6, 6]-phenyl-C_{61}-butyricacidmethyl ester (PCBM) derivative, are shown in Fig. 9.1. The number in the subscript refers to the number of carbon atoms in the spherical molecule. Fullerene (C_{60}), a truncated icosahedron (I_h) having a diameter of 7.1 Å, has 20 hexagons and 12 pentagons consisting of sp^2-hybridized carbon atoms [2]. Experimentally, it was first discovered by Curl, Kroto, and Smalley in 1985, who shared the Nobel Prize for chemistry in 1996 [3]. But it was the synthesis of a macroscopic quantity of fullerenes by Kratschmer and Huffman in 1990 that fueled the interest of experimentalists in fullerenes [4]. Fullerenes are one of the most extensively studied classes of carbon nanomaterials in recent years, with a variety of interesting properties, such as the ability to accept and release electrons [5], photoactivity [6], physiological activity, and biocompatibility [7]. The pristine C_{60} carbon nanomaterials and their derivatives hold great promise in many applications, such as liquid crystals [8], organic light-emitting diodes (LEDs) [9, 10], photocurrent generation devices [11], and organic photovoltaics [12, 13]. By rational functionalization of fullerenes, one can fine-tune their solubility, energy levels, and molecular interactions and the morphology of the resultant nanostructures in the solution and film state [14]. These merits render functional fullerenes as one of the excellent electronic components in material science and organic photovoltaics [15, 16]. In recent years, the supramolecular chemistry of fullerenes has garnered a lot of attention where the development of various dimension-controlled functional nanostructures governed by noncovalent interactions is concerned [17]. Among the various reported supramolecular nanoarchitectures, 1D self-assembly of C_{60} and its functional derivatives, such as nanofibers/nanowires and nanonetworks, finds significant applications in optoelectronics [18, 19]. Considerable efforts are being focused on the construction of dimension-controlled nanostructures via various strategies, including encapsulation of fullerene in poly(methyl methacrylate) (PMMA) [20]; oligomer- or polymer-templated assembly [21, 22]; covalent attachment of fullerene on polymer backbones [23, 24]; and host–guest complexation with corranulene [25], calixarenes [26], etc.

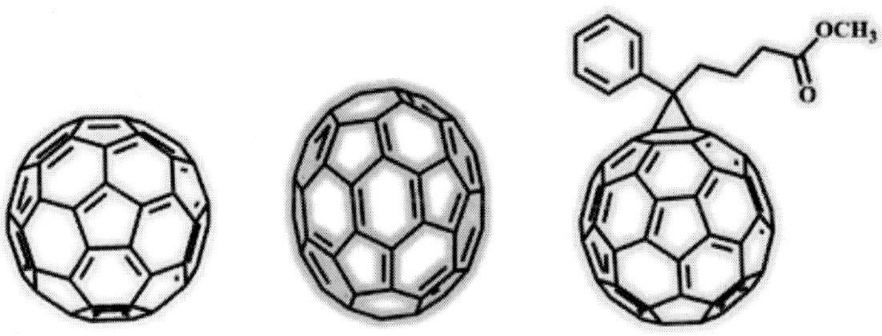

Figure 9.1 Chemical structures of fullerene; from left, C_{60}, C_{70}, and PCBM.

The double-helical DNA is particularly well-suited as a molecular scaffold or template for the self-assembly of chromophores due to its sequence programmability and self-assembly behavior with various molecules through noncovalent interactions [27]. The covalent functionalization of DNA nucleobases with organic chromophores has been utilized for the construction of ordered nanostructures of different sizes and shapes. For instance, chromophores such as pyrenes [28], Nile red [29], and porphyrins were successfully attached to DNA at specific nucleobases, which then was self-assembled into various nanostructures, such as vesicles, nanofibers, nanosheets, and nanonetworks, with tunable properties [30]. The yields of solid-phase synthesis and the solubility issues are the primary concerns in building covalent DNA-chromophore nanoconjugates. The organization of functional motifs with predetermined nanostructures was also achieved using noncovalent conjugation of DNA and chromophores without compromising the conformational flexibility. For example, single-stranded DNA (ssDNA)-templated self-assembly of up to 40 units of naphthalene

[31] and porphyrin [32] chromophores using hydrogen-bonding interaction with thymine nucleobases demonstrates the potential of ssDNA as a nanomaterial template. More recently, several reports envisage the supramolecular self-assembly of DNA and functional motifs in constructing hierarchical assemblies with horizontal charge transport properties [33], with applications in biomedical fields such as biosensing, gene therapy, and bionanotechnology. The biocompatibility and programmability of higher-order assemblies are the key advantages of using DNA in the fabrication of functional nanomaterials [34].

The electronic property and biological activity of functional fullerenes, along with the structure programmability of DNA, make fullerene–DNA conjugates an ideal candidate for material and medicinal applications. The poor aqueous solubility of fullerenes is the main hurdle against progress in this field, and several strategies, including encapsulation with water-soluble guests [35], chemical functionalization [36], and suspension with cosolvents [37], were demonstrated to overcome the solubility issues. The substitution on fullerene derivatives plays a vital role in determining the hydrophobic-hydrophilic balance and further interactions with DNA. The cationically charged moieties and polar functional end groups have shown to be excellent aqueous solubilizing appendages. The aggregates of functional fullerenes, such as clusters and vesicles, also have shown secondary interactions with DNA, forming subsequent hybrid nanostructures. In this chapter, we focus on the various functional fullerenes used in the construction of fullerene–DNA hybrid nanostructures, with a special emphasis on fullerene–DNA interactions, fullerene- or fullerene-aggregate-induced DNA condensation, and long-range ordered hybrid assemblies. The applications of these fullerene–DNA hybrids in various fields, such as nanodevices, charge transport, gene delivery, and photoinduced DNA cleavage, are also discussed.

9.2 Fullerene–DNA Interactions

The DNA (B-form) secondary structure possesses a variety of sites for the ligands to interact, including the sugar-phosphate backbone, the central base core, and the major/minor grooves (schematically represented in Fig. 9.2A). Ligands bind with DNA through various modes of interactions, such as electrostatic, groove-binding, and intercalative interactions. For example, charged ions or molecules are electrostatically attracted to the phosphate backbone of DNA and prefer electrostatic binding, whereas planar aromatic heterocyclic compounds prefer intercalative binding interactions with DNA due to the hydrophobic interior of the double helix. Some other unique types of molecules show major/minor groove-binding interactions guided through multiple interactions with the nucleobases. Utilizing these interactions, DNA has been successfully used as a template for organizing metal ions [38], clusters [39], nanoparticles [40, 41], and polymers [42] into controlled supramolecular nanostructures. Considering the diameter of fullerene derivative as ~1 nm, which is equal to the length span of three base pairs on DNA, the one helical pitch of DNA (3.4 nm) can accommodate almost three fullerene units or a cluster containing five to eight fullerenes, as shown in Fig. 9.2B. According to atomistic molecular dynamic simulations conducted by Zhao et al., C_{60} binding to DNA affects the DNA structure significantly [43]. C_{60} binds to double-stranded DNA (dsDNA) either at the hydrophobic ends or at the minor groove of the duplex. When C_{60} interacts with A-form dsDNA, C_{60} penetrates into the double helix from the end, forms stable hybrids, and frustrates the hydrogen bonds between end-group base pairs in the duplex. They also observed that the molecular association between C_{60} and DNA is stronger and

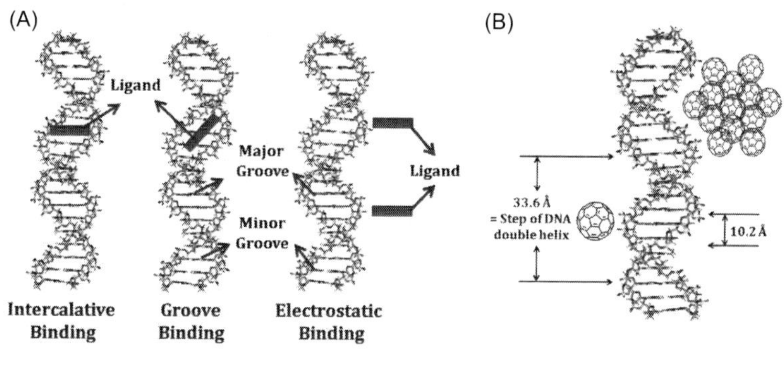

Figure 9.2 (A) Illustration of electrostatic, intercalative, and groove binding of small molecules to the DNA double helix and (B) schematic representation comparing the size of fullerene and a fullerene cluster with the height of a DNA double helix.

more favorable than that between two C_{60} molecules in water. These molecular interactions and associations can further lead to the aggregation of the initially formed DNA/C_{60} complexes. The unique 3D structures and hydrophobicity of fullerene derivatives allow them to interact with DNA through different modes of interactions.

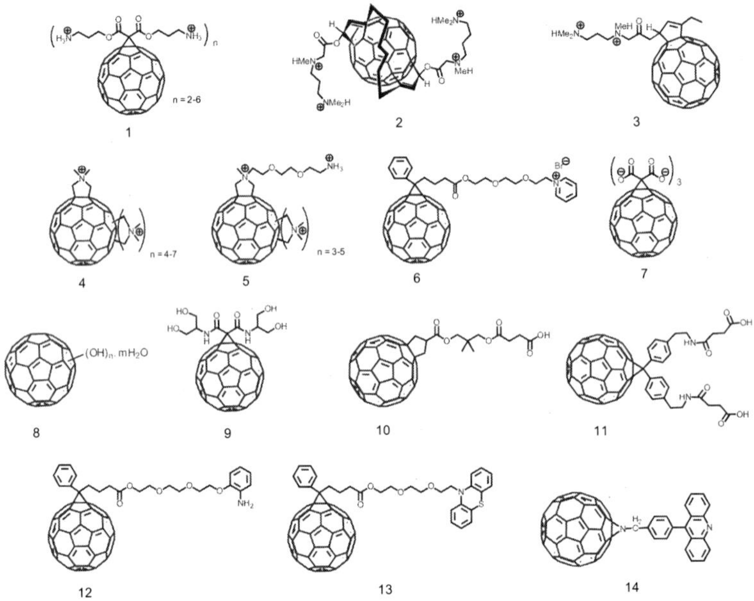

Figure 9.3 Some representative functional fullerenes used for DNA interaction studies.

On the basis of DNA interactions, different types of DNA-interacting fullerene adducts have been reported, as listed in Fig. 9.3. Most of the reported DNA-interacting fullerene derivatives have charged groups for aqueous solubility, which also promote the initial electrostatic binding with subsequent groove-binding and π–π interactions. The charged fullerenes prefer to be in the molecularly dissolved form in aqueous solutions, and their DNA-binding interactions can be easily explained on the basis of single fullerene molecule interactions. Several cationic charged fullerenes were constructed by appending mono-, di-, or multisubstituted quaternary salts, such as ammonium (**1–3**), pyrrolidinium (**4, 5**), and pyridinium (**6**), which exhibit differential solubility as well as DNA interactions, properties that depend on the hydrophilic-hydrophobic balance within the molecule. Anionic carboxylate (**7**)-substituted fullerenes also exhibit groove-binding interaction with DNA. On the other hand, in the case of charged fullerenes, which have the propensity to undergo self-assembly in an aqueous/buffer condition, the DNA interactions strongly depend on the initial fullerene nanostructure. For example, Nakamura and coworkers reported self-assembled vesicles of cationic amino fullerene derivatives (**2, 3**) that undergo effective interaction with plasmid pBR322 DNA [44]. The hydrophilicity of the fullerene derivative was enhanced in fullerenol (**8**) by substituting multiple hydroxyl functional groups [45]. In some cases, neutral fullerenes (**9–14**) were also used in binary solvent mixtures, such as dimethylsulfoxide (DMSO) buffer or dimethylformamide (DMF) buffer [46–48]. Moreover, Chu et al. investigated the fractal behavior of tetra-amino fullerene (**2**) aggregates in the presence of a tris-buffered saline-DMF mixture, which showed increased stability due to a favorable interaction with plasmid DNA [49].

We have reported the differential interaction of monosubstituted fullerene derivatives (**6, 12, 13**), which self-assemble into different sized nanoclusters in a dimethylsulfoxide–phosphate buffer solvent (DMSO-PBS) mixture depending on the end group functionalization [47]. The binding interactions of calf thymus DNA (CT-DNA) with these fullerene derivatives were probed via ultraviolet-visible (UV-Vis) titration [50]. The cationic pyridinium derivative (**6**) becomes molecularly dissolved and interacts with CT-DNA via electrostatic and groove-binding interactions, leading to the formation of micrometer-sized aggregates of CT-DNA/fullerene hybrid nanostructures. The aniline derivative (**12**) self-assembles to form 3–5 nm sized nanoclusters and interacts with DNA through groove-binding and π–π interactions. In contrast, the phenothiazine derivative (**13**), forming larger nanoclusters (70 nm) in DMSO-PBS, exhibited no significant interactions with CT-DNA under similar experimental conditions. Moreover, the conformational and chiral nature of the fullerene–DNA assemblies was analyzed by monitoring the circular dichroism spectra of DNA and the hybrid structures. For instance, the nanostructures of CT-DNA/**12** showed induced CD spectra around the 250–370 nm region, corresponding to the fullerene cluster absorption, which indicates the ordered, chiral assemblies of the unique-sized fullerene nanoclusters on the CT-

DNA template. Hence, the design of fullerene derivatives or fullerene aggregates with a unique hydrophobic-hydrophilic balance plays a significant role in DNA interactions and is crucial in applications such as DNA condensation, gene delivery, and dimension-controlled nanomaterial fabrication.

Fullerene substituted with known intercalating moieties often shows additional intercalative mode of interaction, along with groove binding of fullerenes. Despite the fact that the increase in hydrophobicity of such covalently linked conjugates limits their aqueous solubility, the intercalating groups such as aniline (**12**) and acridine (**14**) moieties were successfully attached to fullerene and utilized in self-assembly and various gene transfection studies. Miyata et al. described the DNA interaction of fullerene–acridine conjugate (**14**), prepared through aziridino addition between azide-functionalized acridine and C_{60} [48]. Here, the solubility of the adduct was achieved by using poly(vinylpyrrolidone) (PVP) as a detergent. The stronger DNA interaction of these derivatives compared to C_{60} was used to study the photoinduced DNA cleavage of pBR322 supercoiled plasmid DNA. A C_{60} derivative with a DNA intercalating acridine showed even stronger DNA-cleaving activity than C_{60} [48].

9.3 Fullerene-Induced DNA Condensation

DNA condensation can be explained as the compaction of highly charged, long genomic DNA into compact structures with the assistance of proteins, polycationic moieties, DNA-binding agents, etc., within the cell nuclei [51]. In nature, two types of DNA condensation can be observed. One is the compaction of long DNA chains into dense nanostructures with the aid of polycationic species, as realized in bacteriophage's heads, and another is the compaction of DNA by 3D nanoscale templates, such as histone proteins [52]. In a cellular context, nucleosomes are perfect examples of noncovalently assembled stable and compact DNA nanostructures, assisted by the octameric histone protein core template [53]. The schematic representation of various structural units involved in the chromosomal DNA compaction is shown in Fig. 9.4. Initially, the DNA double helix (~2 nm) undergoes packing into a chromosomal DNA (~10 nm), forming "beads on a string" structures through various types of histone-histone (purple cylinders) and histone-DNA interactions, which further organizes into a closed chromatin fiber (~30 nm). Here, the DNA-histone protein complex—chromatin—may well be packed into a much smaller volume than the DNA alone. DNA condensation into a nanosized object is essential for potential biomedical applications such as gene transfection and DNA nanotechnology [54]. Previously, several DNA condensing agents, such as small molecules, metal complexes, drugs, peptides, proteins, and carbon nanotubes, have been reported [55]. The fullerenes are particularly well suited for DNA condensation due to their multifunctional and nanoparticle properties [56]. The water-soluble DNA-interacting fullerene derivatives with cationic, amino, and hydroxyl functional groups and nano-C_{60} aggregates have shown to be excellent DNA condensing agents [57].

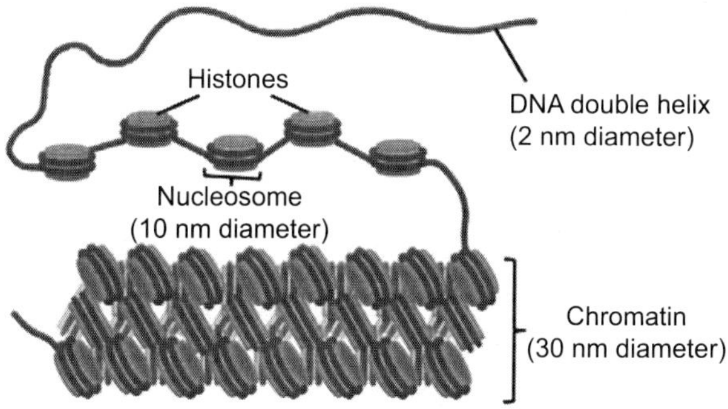

Figure 9.4 Schematic depiction of various structural units involved in chromosomal DNA condensation.

The supramolecular complexation between fullerene derivatives and DNA is another important approach introduced by Tour et al. in 1998, where DNA may be regarded as an anionic polyelectrolyte able to form polyelectrolyte-surfactant complexes (PSCs), in which the surfactant molecules are positively charged [58]. Unfortunately, once fullerenes were placed on the outer side of the DNA double helix, the hybrids became hydrophobic and strongly aggregated in water. Later in 2000, Nakamura and coworkers prepared a tetracationic C_{60} derivative, two-handed (TH) tetra-amino fullerene **(2)** with four anchoring points, which was shown to strongly interact with DNA due to the perfect complementarity of the diammonium fullerene side-chains with two parallel phosphate backbones along with the major groove of a DNA duplex, both separated from each other by 1.2 nm (Fig. 9.5A) [49]. The morphological studies revealed that this fullerene derivative can fold a supercoiled DNA molecule into a condensed structure through adhesion of DNA double strands (Fig. 9.5B–E). Prato et al. studied the DNA complexation and plasmid delivery ability of multi-*N*,*N*-dimethylfullereno-pyrrolidinium derivatives **4** and **5**, which bind plasmid DNA through strong electrostatic interactions. Bianco et al. reported DNA condensation using a multifunctionalized cationic fullerene adduct [59]. The transmission electron microscopy (TEM) analysis of poly-*N,N*-dimethylfulleropyrrolidinium, **4**, and pDNA hybrid (positive/negative charge ratio = 5:1) resulted in a globular, coiled network structure similar to a chain of pearls. Moreover, the absence of uncomplexed, free pDNA in the background confirms the complete complexation of pDNA due to the fullerene derivatives.

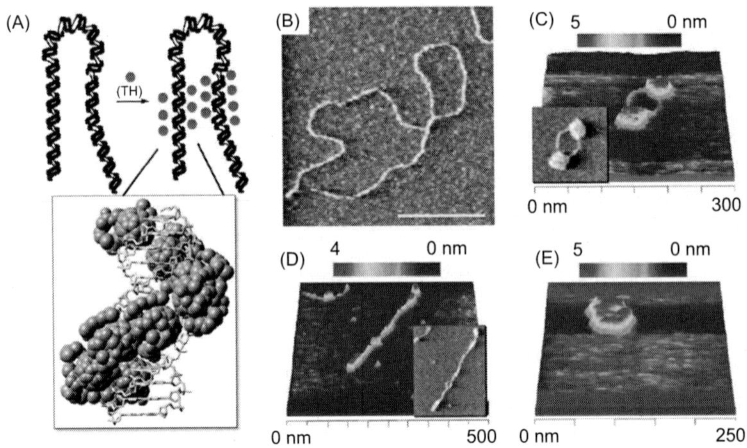

Figure 9.5 (A) Molecular model of a structure showing the possible interaction between DNA and two-handed fullerene **2**. Reprinted with permission from Ref. [49]. Copyright (2005) American Chemical Society. AFM images of pBR322 DNA in the (B) absence and (C–E) presence of **2** at various reagent/base pair ratio (*R*) values. (B) Intact DNA, $R = 0$. (C) Partially folded single DNA molecule, $R = 0.65$. (D) Partially condensed DNA with unfolded double strands, $R = 0.65$. and (E) fully condensed fullerene-DNA disk, $R = 0.65$. Scale bar = 200 nm. Reproduced with permission Ref. [44]. Copyright (2001) John Wiley and Sons.

Through atomic force microscopic (AFM) studies, we also observed the CT-DNA condensation with the aid of pyridinium derivative (**6**) into micrometer-sized aggregates [50]. On the other hand, the nanoclusters of the aniline-substituted derivative (**12**) could perturb the highly networked structure of CT-DNA, giving rise to a nanofibrous structure (Fig. 9.6A–C). These condensation processes resemble that of chromatin structure compaction from histone protein and occur as a consequence of inhibiting DNA-DNA interactions from bound fullerene–DNA macromolecular associations. The DNA binding shows a linear correlation with the number of cationic or amine group, as found previously for aliphatic polyamines [60]. A careful comparison of the structural characteristics of DNA-binding and transfecting compounds shows that successful application of these condensed aggregates in gene transfection requires DNA-release capabilities. Hence, the disassembly of the DNA-fullerene hybrid structure or neutralization of the basic amine under physiological conditions is also an important benchmark for practical applications.

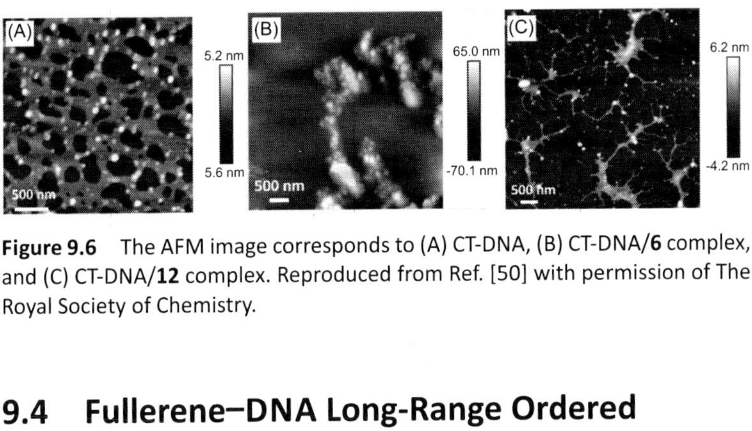

Figure 9.6 The AFM image corresponds to (A) CT-DNA, (B) CT-DNA/**6** complex, and (C) CT-DNA/**12** complex. Reproduced from Ref. [50] with permission of The Royal Society of Chemistry.

9.4 Fullerene–DNA Long-Range Ordered Assembly

Previously, DNA origami and tiled assembly showed the unique programmed assembly of functional organic components with precise hierarchical organization [61–63]. However, building organized nanostructures through noncovalently assembled DNA and organic chromophores remains a challenging task [64]. Cassell et al. have reported a facile route to construct a fullerene–DNA hybrid nanostructure that can be rapidly visualized using TEM without any staining agent [58]. Cationic C_{60}-*N,N*-dimethylpyrrolidinium iodide (monomer of **4**) was chosen to construct a DNA hybrid structure, and molecular modeling of iodide-free fullerene revealed the successive complexation of fullerene along the DNA backbone, with the thickness of the DNA-C_{60} hybrid around 5–6 nm (Fig. 9.7A). The TEM image of the hybrid indicated the DNA-fullerene long-range ordered assembly with excellent contrast (Fig. 9.7B).

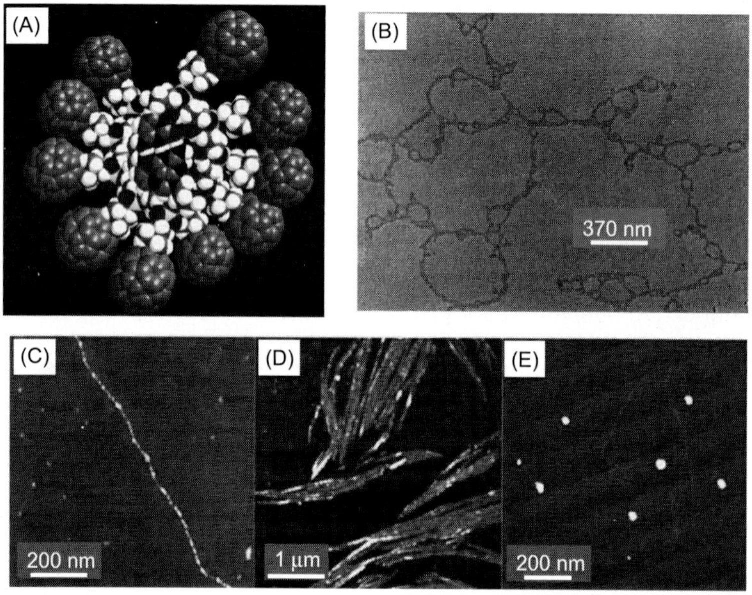

Figure 9.7 (A) Molecular model of iodide-free fullerene–DNA hybrid; (B) TEM image corresponding to plasmid DNA and C_{60}-*N,N*-dimethylpyrrolidinium iodide complex. Reproduced with permission from Ref [58]. Copyright (1998) John Wiley and Sons. AFM images of DNA incubated with (C) low concentration of **8**, (D) high concentration of **8**, and (E) with nano-C_{60}. Reprinted with permission from Ref. [57]. Copyright (2011) American Chemical Society.

In another attempt, An and Jin compared the self-assembly behavior of fullerenol (**8**) and nano-C_{60} with linear DNA [57]. The AFM results of DNA-incubated fullerenol and nano-C_{60} revealed that fullerenol binds more strongly than nano-C_{60}. DNA condensation into a nanofiber-like morphology was observed in the presence of a high dose of fullerenol (Figs. 9.7C and 9.7D). The differential interaction mainly depends on the initial fullerene agggregate size.

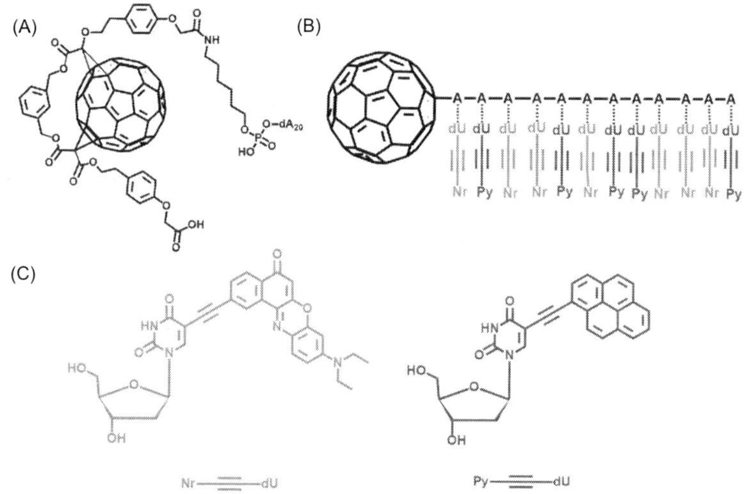

Figure 9.8 Chemical structure of (A) fullerene–$(dA)_{20}$ conjugate and (B) illustration of the Py-dU and Nr-dU assembly along the fullerene–$(dA)_{20}$ template. Chemical structures of (C) Py-dU and (D) Nr-dU used for the assembly. Adapted with permission from Ref [65]. Copyright (2015) John Wiley and Sons.

Fullerenol with a particle size of ~10 nm can undergo hydrogen-bonding interaction between hydroxyl groups and –NH2 groups of DNA base pairs or phosphate backbone. On the other hand, nano-C60 aggregates with a polyhydroxylated structure on the particle surface show weak interaction due to the large size, on the order of 100 nm (Fig. 9.7E). Another method of constructing long-range ordered assembly is by using covalently attached fullerene-short oligonucleotides as a template for the supramolecular organization of chromophore appended with complementary hydrogen-bonding units. For example, Wagenknecht and coworkers recently designed and constructed covalently attached fullerene–deoxyadenosine $(dA)_{20}$ conjugate through malonic ester linkage (Fig. 9.8A) [65]. This can effectively serve as a template for the self-assembly of an ordered array of ethynyl-pyrene and ethynyl–Nile red–nucleoside conjugates, as shown in Fig. 9.8B. Both Py-dU and Nr-dU (Fig. 9.8C) behave as derivatives of thymidine and hence selectively bind to oligo-2'-$(dA)_{20}$ via adenine-uridine complementary hydrogen bonding. Hence, this supramolecular polymerization or template-assisted assembly could become a rational bottom-up strategy to assemble functional chromophores with potential application in optoelectronics and biomedical fields.

9.5 Fullerene Cluster–Assisted DNA Assemblies

DNA has been successfully used as a template for the construction of self-assembled arrays of nanoparticles [66], with remarkable control over size and shape of resultant nanostructures. Song et al. demonstrated the templated assembly of C_{60}-*N,N*-dimethylpyrrolidinium iodide (a monomer of **4**) using DNA scaffolds as shown schematically in Fig. 9.9a [67]. It is known that this particular derivative forms vesicles in a DMSO-water mixture, which collapse and form completely extended multilayer domains on a solid surface. This technique was used for the self-organization of fullerenes on a DNA scaffold. For this purpose, the periodic distance (64 nm) of the parallel strips in DNA lattices was used for the selective docking of water-dispersed fullerene vesicles. The resultant nanostructures were compared using AFM images. By incubating the fullerene vesicle and a 2D DNA lattice bearing a phosphate-labeled protruded strand together (Fig. 9.9b), pearl-like arrays of fullerene domains were obtained on the DNA template (Fig. 9.9c). Therefore, a phosphated oligonucleotide plays a role in the alignment of fullerene vesicles. Similarly, the mutually assisted self-assembly of DNA by small molecules, nanoparticles, or nanoclusters into nanofibers/nanowires is very rare. In a recent report, we have presented a hierarchical assembly of unmodified short DNA duplex strands, using self-assembled fullerene nanoclusters as stapler motifs [47]. The AFM and TEM analysis of dsDNA/**12** revealed the formation of micrometer-long self-assembled nanowires (Fig. 9.10a–c). The zoomed portion of the TEM image clearly indicates the presence of fullerene clusters in the nanowires (Fig. 9.10d). Both fullerene clusters and dsDNA are incorporated in the nanostructures and are essential for continuous growth. These morphological studies support the mutually assisted interactions between fullerene nanoclusters and short oligonucleotide duplexes, such as dsDNA, leading to the formation of extended nanostructures, as shown in the Fig. 9.10e.

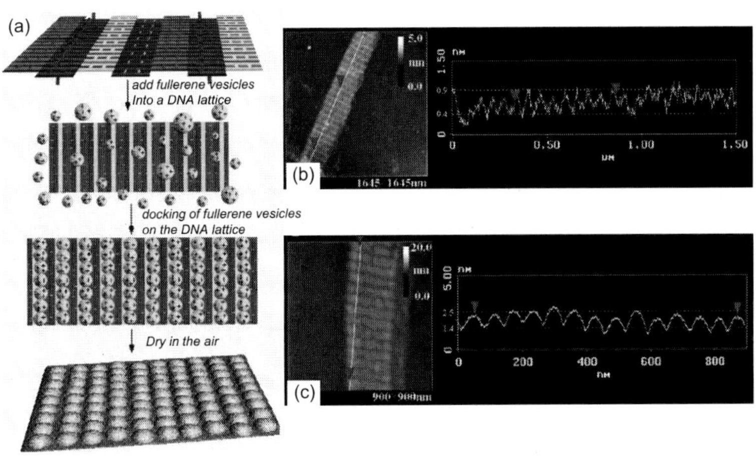

Figure 9.9 (a) Sketch of the self-assembly procedure of fullerene vesicles on the 2D DNA scaffolding. Topographical AFM images of the (b) DNA scaffolding and (c) fullerene arrays in the DNA scaffolding. Reprinted with permission from Ref. [67]. Copyright (2005) American Chemical Society.

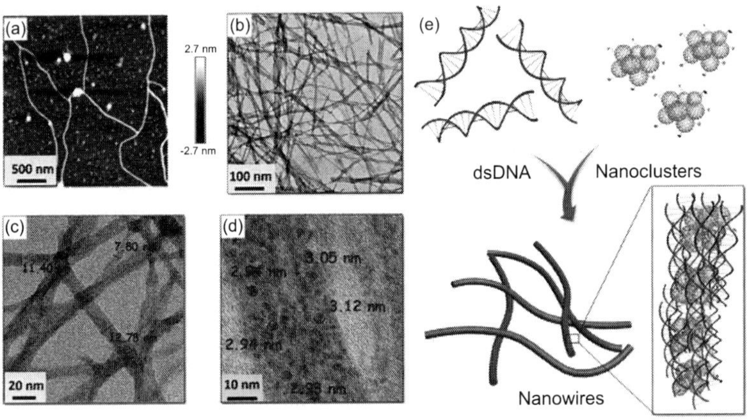

Figure 9.10 AFM height images and TEM images of the dsDNA/**12** complex at different magnifications, (a) 500 nm, (b) 200 nm, and (c) 20 nm, showing uniform nanowires. (d) Zoomed TEM image revealing the embedded nanoclusters. (e) Schematic illustration of the overall self-assembly process. Reproduced with permission from Ref [47]. Copyright (2017) John Wiley and Sons.

The unperturbed double-helical B-form structure of dsDNA in the resultant nanostructure further demonstrates the potential application of this strategy in building higher-order nanostructures.

This represents the unique example where a self-assembled fullerene cluster assists the long-range ordered assembly of short oligonucleotide duplexes into nanowires. These results show the possibility of constructing a 1D array of fullerene by using suitable DNA structures. Also, the nanostructure formed can be stabilized by using 2+2 cycloaddition reaction of fullerene [68]. Therefore, in the future, DNA might find a crucial role in the development of a stable morphology, required for various nanotechnology applications.

9.6 Applications of Fullerene–DNA Hybrids

9.6.1 Nanodevices and Electron Transport Properties

The construction of nanodevices that can actuate a specific response is the next logical step in future electronics. The programmable assembly of DNA building blocks and the electronic property of fullerene derivatives envisages the requirements for the functional nanomachine or nanorobot components [69, 70]. The first fullerene-attached DNA nanomachines were constructed by Ree et al., where the structural effects of fullerene on i-motif DNA were investigated [71]. In mildly acidic conditions (pH = 5), the 21-mer ssDNA forms a compact, folded i-motif conformation due to intramolecular noncanonical base pair interactions between a protonated and an unprotonated cytosine residue (i.e., a C^+:C base pair), as shown in Fig. 9.11A. Attaching C_{60}-fullerene at both 5′ and 3′ ends of the DNA strand improved the DNA nanomachine to change its conformation from "close" to "open" by varying the pH value from 5 to 8 (Fig. 9.11B). The fullerene shifted the pH-induced conformational transition between i-motif and the duplex structure, possibly due to the hydrophobic interactions between the terminal fullerenes and an internal TAA loop in the DNA strand with a dramatic reduction in the cyclic hysteresis. Here, fullerene molecules were chosen due to their nanosize, 3D spherical shapes, and known compatibility with DNA. Interestingly, this molecular machine showed a cycling efficiency of ~100% for 10 consecutive cycles, with a remarkable low response time of 20 s, similar to the nonfunctionalized DNA. DNA-based 1D nanomaterials are considered to be a smart strategy for an effective connection between the electrodes in nanodevices. The long-range charge transport properties of single G-quadruplex DNA molecules over several hundred nanometers shows the potential of DNA in electronics [33].

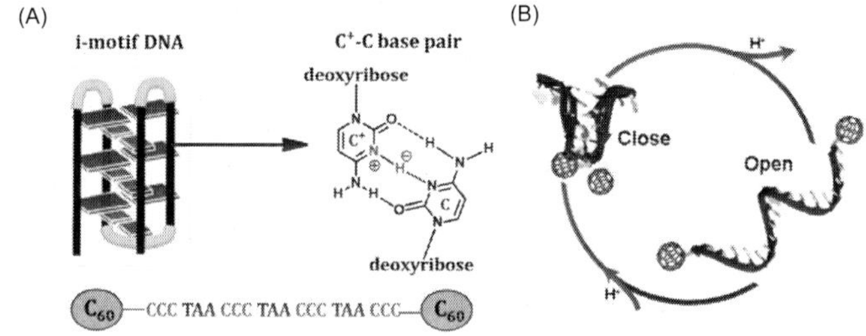

Figure 9.11 (A) Schematic diagram of i-motif DNA and sequence of the base with carboxyl-functionalized C_{60} on the two ends of i-motif DNA. (B) Representative models of a fullerene–DNA hybrid nanomachine and the switching cycle showing the conformational changes with a proton. Reproduced with permission from Ref. [71]. Copyright (2009) John Wiley and Sons.

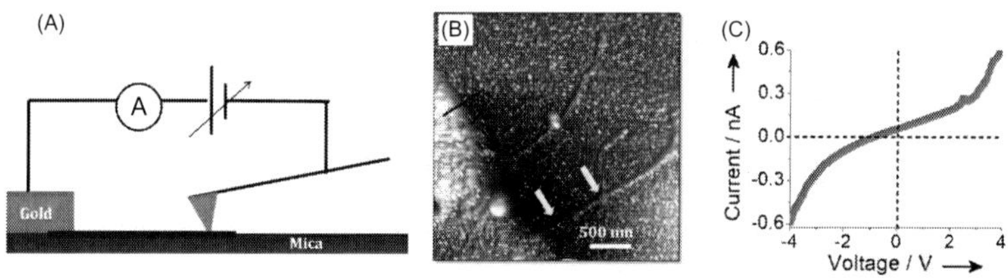

Figure 9.12 (A) Schematic representation of the c-AFM measurement setup, (B) AFM image showing the dsDNA/**12** nanowire protruding from the edge of the metal electrode, and (C) *I–V* measurements taken at the position indicated by the yellow arrow on the nanowire. Reproduced with permission from Ref [47]. Copyright (2017) John Wiley and Sons.

Moreover, the semiconducting properties of the fullerenes are an added advantage of the dimension-controlled DNA-C_{60} hybrid nanostructure, which might find unique application in molecular electronics. In this line, recently nanowires constructed from mutually assisted self-assembly of fullerene clusters and short DNA strands were tested for their horizontal conductivity based on the device structure, as shown in Fig. 9.12A [47]. Careful fabrication of dsDNA/**12** on the mica surface and gold deposition led to the formation of uniformly distributed nanowires protruding from the electrode surface (Fig. 9.12B). The horizontal current-voltage characteristics of the nanowires exhibit a conductivity of 8.12×10^{-5} Sm^{-1} for a nanowire height of 3 nm and an AFM tip contact radius of 35 nm (Fig. 9.12C). The similar conductivity values up to a horizontal distance of 700 nm, along with control measurements on the mica and gold surface, confirmed the long-range charge transport properties of these DNA-fullerene hybrids. The strategy used in assembling semiconducting fullerene–DNA nanowires could open up new avenues in DNA-based molecular electronics.

9.6.2 Biomedical Applications

9.6.2.1 Photoinduced DNA cleavage

Photoinduced DNA cleavage is an emerging field in the context of advance medical treatment such as photodynamic therapy [72]. DNA-C_{60} was extensively used in photoinduced DNA cleavage due to its ability to oxidize guanosine-containing sites with significant DNA cleavage [73]. Two possible mechanisms of photoinduced DNA cleavage have been proposed, via energy transfer mechanism and via electron transfer mechanism, both of which involve oxygen (Fig. 9.13A) [74, 75]. In the first mechanism, singlet oxygen is generated as a result of interaction of the photoexcited C_{60} moiety and molecular oxygen, which yields DNA-cleaved products via energy transfer. In the second mechanism, DNA cleavage occurs via electron transfer from DNA to the photoexcited C_{60}. The amount of photons absorbed by the functional molecular films and the yield of the photoinitiation process are critical parameters in the design of efficient photoresponsive DNA-cleaving systems. Nakamura et al. initially observed that out of monocarboxylic acid– and dicarboxylic acid–functionalized fullerene derivatives **10** and **11**, the compound **11** can induce effective DNA cleavage upon visible light irradiation [7, 76]. Further studies indicated that the cleavage is caused by active oxygen species generated by fullerene-sensitized activation of molecular oxygen in the medium and not because of the binding of carboxylic acid to DNA (Fig. 9.13B). Later they synthesized various fullerene/DNA conjugates in such a way that the fullerene molecule connected to a 14-mer oligonucleotide, which is complementary to single and double helices (Fig. 9.13C) [77]. This conjugate results in exclusive DNA cleavage at guanidine bases after irradiation with light. The fullerene molecules bearing an intercalating acridine group, **14**, can also trigger photoinduced DNA cleavage at very low concentrations due to enhanced binding to dsDNA. A series of novel cationic fullerene (C_{60}) derivatives, bearing a substituted s-triazine moiety (Fig. 9.14A) as a side arm, were synthesized by Shobhana et al. [78]. The systematic DNA photocleaving processes were also investigated through their interaction energies by computational approaches. Out of several cationic fullerene–s-triazine conjugates, the compounds **16a–f** cleaved DNA more efficiently than the corresponding substituted triazine analogs **15a–f.** Moreover, the molecules containing electron-donating groups (**16d–f**) scored better cleaving ability than the electron-withdrawing molecules (**16b–c**) due to the better interaction with DNA. The presence of the cationic fullerene and substituted s-triazine would result in a better DNA-cleaving ability compared to substituted s-triazine alone, due to the favorable electrostatic and ionic interactions between the positively charged cationic fullerene s-triazine derivatives and negatively charged phosphate DNA backbone along with the groove binding of fullerenes.

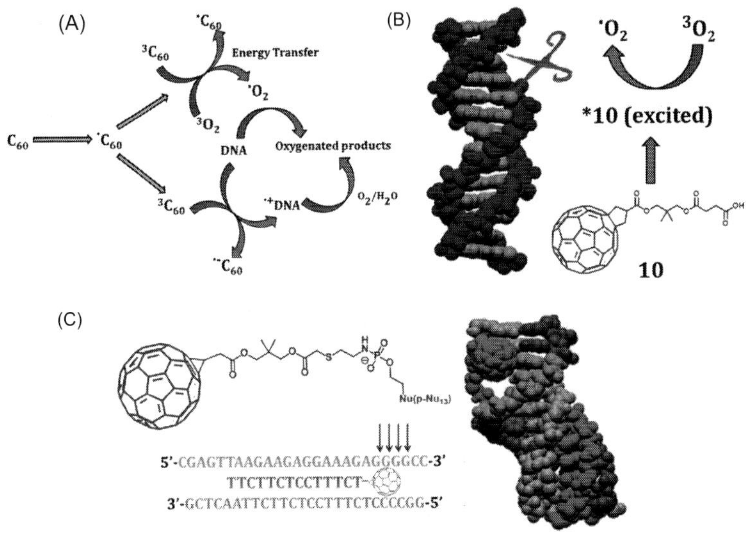

Figure 9.13 (A) Schematic representation of the proposed mechanisms of DNA cleavage, (B) representative model and mechanism of DNA cleavage by fullerene carboxylic acid **10**, and (C) chemical structure of triple-helix-forming fullerene–DNA conjugate and its 3D binding model. Reprinted with permission from Ref. [7]. Copyright (2003) American Chemical Society.

Higashi et al. studied the photoinduced DNA cleavage on solid supports with guanine specificity [79, 80]. Initially a mixed monolayer containing cysteamine and quaternary ammonium hydrocarbon species above the gold substrate was prepared, with subsequent immobilization of thiol-modified C_{60} and DNA (Fig. 9.14B). DNA cleavage into small fragments was observed upon visible light irradiation (λ > 500 nm), as analyzed using UV-vis absorption spectroscopy. The site-specific DNA cleavage was achieved by using several types of DNA having different base pair compositions and guanine contents such as salmon sperm DNA, calf thymus DNA, poly[$(dGDC)_2$], and poly[$(dAdT)_2$]. Along with monolayer assemblies, photosensitive multilayer films of DNA-C_{60} have also been fabricated. For example, Zhang et al. used a layer-by-layer (LBL) approach to fabricate polyelectrolyte-based multilayer films, using DNA as polyanion and C_{60}-ethylenediamine as polycation (Fig. 9.14C) [81]. In this case DNA cleavage occurred under visible light irradiation (λ > 410 nm) involving electron transfer mechanism rather than over singlet oxygen generation.

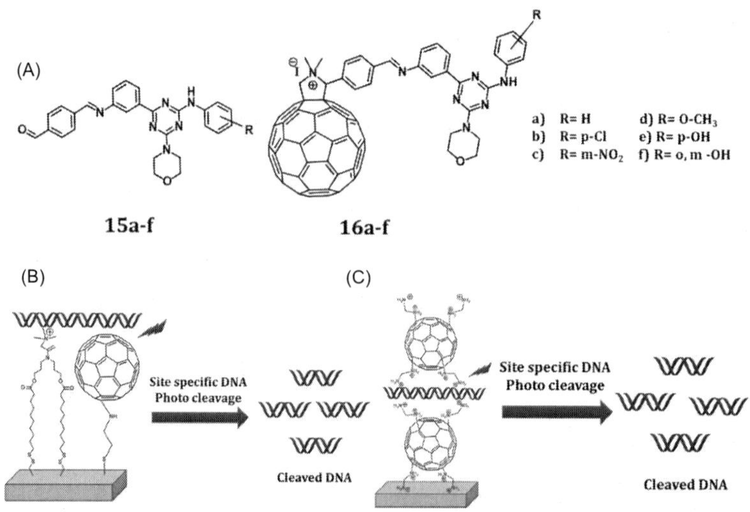

Figure 9.14 (A) Chemical structures of substituted s-triazine moieties and cationic fullerene-s-triazine conjugates, (B) immobilization and cleavage of DNA by incorporation of C_{60} into the monolayers of cysteamine onto a gold substrate upon irradiation of light, and (C) immobilization and cleavage of DNA by a multilayer assembly of alternate polycation and polyanion layers upon irradiation of light. Reproduced from Ref. [80] with permission of The Royal Society of Chemistry.

9.6.2.2 Gene delivery vectors

The fundamental machinery in modern molecular biology involves the delivery of a gene into living cells by selective encapsulation of the gene in a vector [82]. So, the transfection process, that is, transfer plus infection, allows protein synthesis in a cell subsequent to DNA delivery. Recent studies have indicated that suitably functionalized fullerene derivatives are effective for gene delivery [83]. Among a variety of other events occurring in gene transfection, the condensation of natural DNA into a small object is the necessary first step and the release of DNA into the cytoplasm is the second. Radiolabeling experiments have suggested that the water-soluble fullerene derivative **10** can pass through the blood-brain barrier and accumulates in the liver after a few hours [37]. This is quite a contrast to the parent fullerene, which is readily absorbed by serum protein in blood after injection. This clearly indicates the potential use of DNA-C_{60} hybrids in gene delivery applications [84]. The DNA binding of fullerenes is necessary but is not the only criterion to be used for gene delivery applications. For example, among the various amino fullerene derivatives synthesized, the amino fullerene **4** (a monomer of **4**) shows efficient binding with dsDNA without much change in the conformation of the DNA but is not effective in gene delivery [60].

In contrast, the tetracationic derivative **2** was found to be an effective transfection reagent for gene transfer, capable of binding a 4–40 kbp DNA vector and delivering it to mammalian cells after incubation [46]. The analysis of COS-1 cells under a digital interference contrast (DIC) micrograph indicated the formation of particles 1–10 μm in size upon incubation with a DNA/fullerene **2** hybrid (Fig. 9.15A). As this mixture was incubated for an extended period of two days, the dark-colored fullerene–DNA particles were taken into the cells (Fig. 9.15B). The fluorescence image of Fig. 9.15B and the superimposed images of Fig. 9.15B and 9.15C clearly indicate the effective distribution of black objects (fullerene/DNA complex) inside the cytoplasm of the cells. This is mainly due to the formation of condensed supercoiled DNA composing one to a few DNA double strands. Screening of various cell lines indicated that the efficiency of this first-generation fullerene vector is comparable to or better than the commercially available lipid-based vectors (e.g., lipofectin).

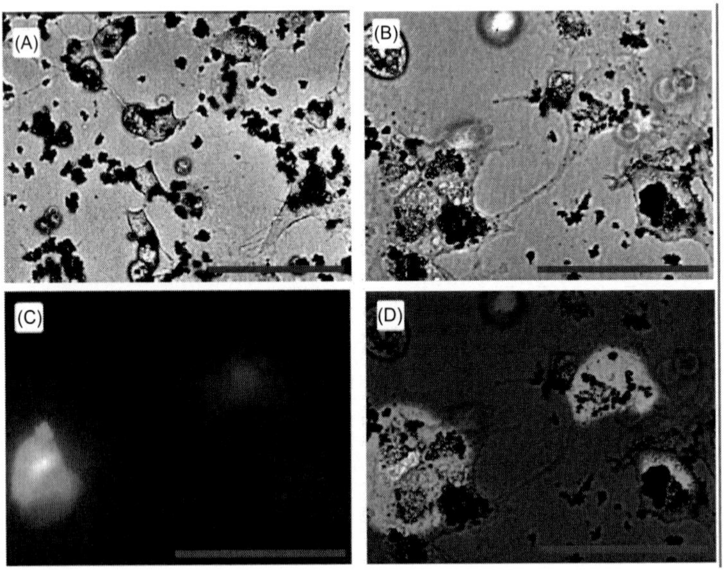

Figure 9.15 DIC micrograph of COS-1 cells incubated with a DNA/fullerene 2 hybrid after transfection times (A) 1 h and (B) 2 days. C) Fluorescence micrograph of (B) and (D) superimposed image of (B) and (C). Scale bar = 100 μm. Reprinted with permission from Ref. [7]. Copyright (2003) American Chemical Society.

Nakamura and coworkers successfully prepared and tested other gene delivering amino fullerenes, such as **5**, and they were able to give general rules for their future chemical design [60]. More recently, other multicationic fullerene derivatives have been tested as vectors for gene transfer. Prato et al. studied the DNA complexation and plasmid delivery ability of multi-*N*,*N*-dimethylfullereno-pyrrolidinium derivatives **4** (multiadducts) and **5**, capable of binding plasmid DNA through strong electrostatic interaction [59]. On the other hand, Engler and coworkers studied several cationic, neutral, and anionic Bingel-type C_{60}-multiadducts (**1**, **7**, and **9**, respectively) as nonviral gene delivery vectors [84, 85]. Among them, only octa- and dodeca-ammonium derivatives **1**, (n = 4) and (n = 6), displayed efficient in vitro transfection. The study of several fullerene derivatives toward gene transfection indicated that the presence of an ester group is crucial because it may be hydrolyzed in the cytoplasm to release the fullerene from the DNA.

9.7 Conclusions and Perspectives

We have summarized a series of ordered functional nanomaterials created from a combination of fullerene chemistry and DNA self-assembly with many strategic advantages. The functional fullerenes can interact with DNA through initial electrostatic, intercalative, and/or groove binding, with subsequent π–π interaction of adjacent fullerenes, which could find significantly improved potential applications in DNA nanotechnology and gene delivery. Such multiple interactions of the fullerene derivatives with unique functional moieties can effectively condense highly charged, long genomic DNA into compact structures. The condensed fullerene–DNA hybrids have been utilized in gene transfection, whereas the high singlet-oxygen-producing efficiency of fullerene was used in DNA cleavage, which may find application in photodynamic therapy. Moreover, fine-tuning the molecular design of fullerene derivatives and interaction with DNA would lead to the long-range ordered assembly of semiconductive nanowires.

The ordered self-assembly of fullerene and their analogs into regular DNA nanostructures may prove to be an effective DNA-based template for the selective localization of organic chromophores, inorganic ions, nanoparticles, nanoclusters, etc. Moreover, the programmed fullerene architectures constructed via intermolecular fullerene cross linking might be useful in optoelectronic applications where a routine self-assembly approach fails. Therefore, the fullerene–DNA interactions and structural design presented here could be applied to the development of mutually assisted supramolecular nanodevices and nanomachines, which can solve the next, more challenging problems of supramolecular chemistry and nanotechnology. The collaboration between these rapidly growing fields—fullerene chemistry and DNA nanotechnology—will open up new avenues in the field of supramolecular architectures and their applications.

Acknowledgments

The financial support from the Council of Scientific and Industrial Research (CSIR 12 FYP M2D-CSC-0134) and the Department of Science and Technology, Government of India, (Ramanujan

Fellowship Grant RJN-19/2012), are gratefully acknowledged. Sandeepa K. Vittala and Sajena K. Saraswathi acknowledge the University Grant Commission (UGC), Government of India, for their Research Fellowship.

References

1. Hirsch, A. (2010). The era of carbon allotropes, *Nat. Mater.,* **9**, 868–871.
2. Prato, M. (1997). [60]Fullerene chemistry for materials science applications, *J. Mater. Chem.*, **7**, 1097–1109.
3. Kroto, H. W., Heath, J. R., O'Brien, S. C., Curl, R. F. and Smalley, R. E. (1985). C60: Buckminsterfullerene, *Nature*, **318**, 162–163.
4. Kratschmer, W., Lamb, L. D., Fostiropoulos, K. and Huffman, D. R. (1990). Solid C-60 - a new form of carbon, *Nature*, **347**, 354–358.
5. Guldi, D. M. and Prato, M. (2000). Excited-state properties of C60 fullerene derivatives, *Acc. Chem. Res.*, **33**, 695–703.
6. Martín, N., Sánchez, L., Illescas, B. and Pérez, I. (1998). C60-based electroactive organofullerenes, *Chem. Rev.*, **98**, 2527–2548.
7. Nakamura, E. and Isobe, H. (2003). Functionalized fullerenes in water. The first 10 years of their chemistry, biology, and nanoscience, *Acc. Chem. Res.*, **36**, 807–815.
8. Li, W.-S. S., Yamamoto, Y., Fukushima, T., Saeki, A., Seki, S., Tagawa, S., Masunaga, H., Sasaki, S., Takata, M. and Aida, T. (2008). Amphiphilic molecular design as a rational strategy for tailoring bicontinuous electron donor and acceptor arrays: photoconductive liquid crystalline oligothiophene--C60 dyads, *J. Am. Chem. Soc.*, **130**, 8886–8887.
9. Matsuo, Y., Tahara, K., Sawamura, M. and Nakamura, E. (2004). Creation of hoop- and bowl-shaped benzenoid systems by selective detraction of [60]fullerene conjugation. [10]Cyclophenacene and fused corannulene derivatives, *J. Am. Chem. Soc.*, **126**, 8725–8734.
10. Matsuo, Y., Tahara K., Morita, K., Matsuo, K. and Nakamura, E. (2007). Regioselective eightfold and tenfold additions of a pyridine-modified organocopper reagent to [60]fullerene, *Angew. Chem. Int. Ed. Engl.*, **46**, 2844–2847.
11. Matsuo, Y., Kanaizuka, K., Matsuo, K., Zhong, Y. W., Nakae, T. and Nakamura, E. (2008). Photocurrent-generating properties of organometallic fullerene molecules on an electrode, *J. Am. Chem. Soc.*, **130**, 5016–5017.

12. Brabec, C. J., Gowrisanker, S., Halls, J. J. M., Laird, D., Jia, S. and Williams, S. P. (2009). Polymer-fullerene bulk-heterojunction solar cells, *Adv. Mater.*, **21**, 1323–1338.

13. Babu, S., Möhwald, H. and Nakanishi, T. (2010). Recent progress in morphology control of supramolecular fullerene assemblies and its applications, *Chem. Soc. Rev.*, **39**, 4021–4035.

14. Wantz, G., Derue, L., Dautel, O., Rivaton, A., Hudhomme, P. and Dagron-Lartigau, C. (2014). Stabilizing polymer-based bulk heterojunction solar cells via crosslinking, *Polym. Int.*, **63**, 1346–1361.

15. Lai, Y. Y., Cheng, Y. J. and Hsu, C. S. (2014). Applications of functional fullerene materials in polymer solar cells, *Energy Environ. Sci.*, **7**, 1866–1883.

16. Dirk, M. G., Beatriz, M. I., Carmen M, A., Mateusz, W. and Nazario, M. (2009). Fullerene for organic electronics, *Chem. Soc. Rev.*, **38**, 1587–1597.

17. Shrestha, L. K., Ji, Q., Mori, T., Miyazawa, K. i., Yamauchi, Y., Hill, J. P. and Ariga, K. (2013). Fullerene nanoarchitectonics: from zero to higher dimensions, *Chem. Asian J.*, **8**, 1662–1679.

18. Maeyoshi, Y., Saeki, A., Suwa, S., Omichi, M., Marui, H., Asano, A., Tsukuda, S., Sugimoto, M., Kishimura, A., Kataoka, K. and Seki, S. (2012). Fullerene nanowires as a versatile platform for organic electronics, *Sci. Rep.*, **2**, 600.

19. Li, C.-Z., Yip, H.-L. and Jen, A. K. Y. (2012). Functional fullerenes for organic photovoltaics, *J. Mater. Chem.*, **22**, 4161–4177.

20. Kawauchi, T., Kumaki, J., Kitaura, A., Okoshi, K., Kusanagi, H., Kobayashi, K., Sugai, T., Shinohara, H. and Yashima, E. (2008). Encapsulation of fullerenes in a helical PMMA cavity leading to a robust processable complex with a macromolecular helicity memory, *Angew. Chem. Int. Ed.*, **47**, 515–519.

21. Fujita, N., Yamashita, T., Asai, M. and Shinkai, S. (2005). Formation of [60]fullerene nanoclusters with controlled size and morphology through the aid of supramolecular rod-coil diblock copolymers, *Angew. Chem. Int. Ed.*, **44**, 1257–1261.

22. Hiroyuki, W., Takuya, M., Takeshi, Y. and Norio, T. (2009). A simple preparation of C60-poly(ethylene glycol) gel and its properties, *Polym. J.*, **41**, 40–45.

23. Giacalone, F., Martín, N. and Wudl, F. (2009). Fullerene-containing polymers: an overview, in *Fullerene Polymers: Synthesis, Properties*

and Applications (eds. Martin, N. and Giacalone, F.), Wiley-VCH Verlag GmbH & Co. KGaA, Weinheim, Germany.

24. Francesco, G. and Nazario, M. (2006). Fullerene polymers: synthesis and properties, *Chem. Rev.*, **106**, 5136–5190.
25. Stuparu, M. C. (2013). Rationally designed polymer hosts of fullerene, *Angew. Chem. Int. Ed.*, **52**, 7786–7790.
26. Chun, Y., Singh, N. J., Hwang, I.-C. C., Lee, J. W., Yu, S. U. and Kim, K. S. (2013). Calix[n]imidazolium as a new class of positively charged homo-calix compounds, *Nat. Commun.*, **4**, 1797.
27. Aldaye, F. A., Palmer, A. L. and Sleiman, H. F. (2008). Assembling materials with DNA as the guide, *Science*, **321**, 1795–1799.
28. Sezi, S. and Wagenknecht, H.-A. A. (2013). DNA-templated formation of fluorescent self-assembly of ethynyl pyrenes, *Chem. Commun.*, **49**, 9257–9259.
29. Varghese., R. and Wagenknecht, H.-A. (2010). Non-covalent versus covalent control of self-assembly and chirality of Nile red-modified nucleoside and DNA, *Chem. Eur. J.*, **16**, 9040–9046.
30. Malinovskii, V. L., Wenger, D. and Häner, R. (2010). Nucleic acid-guided assembly of aromatic chromophores, *Chem. Soc. Rev.*, **39**, 410–422.
31. Janssen, P., Vandenbergh, J., van Dongen, J., Meijer, E. and Schenning, A. (2007). ssDNA templated self-assembly of chromophores, *J. Am. Chem. Soc.*, **129**, 6078–6079.
32. Nguyen, T., Brewer, A. and Stulz, E. (2009). Duplex stabilization and energy transfer in zipper porphyrin-DNA, *Angew Chem. Int. Ed.*, **48**, 1974–1977.
33. Livshits, G. I., Stern, A., Rotem, D., Borovok, N., Eidelshtein, G., Migliore, A., Penzo, E., Wind, S. J., Di Felice, R., Skourtis, S. S., Cuevas, J. C., Gurevich, L., Kotlyar, A. B. and Porath, D. (2014). Long-range charge transport in single G-quadruplex DNA molecules, *Nat. Nanotechnol.*, **9**, 1040–1046.
34. Jones, M. R., Seeman, N. C. and Mirkin, C. A. (2015). Programmable materials and the nature of the DNA bond, *Science*, **347**, 1260901.
35. Samal, S., Choi, B. J. and Geckeler, K. E. (2000). The first water-soluble main-chain polyfullerene, *Chem. Commun.*, 1373–1374.
36. Nakamura, E. and Isobe, H. (2010). In vitro and in vivo gene delivery with tailor-designed aminofullerenes, *Chem. Rec.*, **10**, 260–270.
37. Scrivens, W. A., Tour, J. M., Creek, K. E. and Pirisi, L. (1994). Synthesis of 14C-labeled C60, its suspension in water, and its uptake by human keratinocytes, *J. Am. Chem. Soc.*, **116**, 4517–4518.

38. Ehrenschwender, T., Barth, A., Puchta, H. and Wagenknecht, H.-A. (2012). Metal-mediated DNA assembly using the ethynyl linked terpyridine ligand, *Org. Biomol. Chem.*, **10**, 46–48.

39. New, S. Y., Lee, S. T. and Su, X. D. (2016). DNA-templated silver nanoclusters: structural correlation and fluorescence modulation, *Nanoscale*, **8**, 17729–17746.

40. Sharma, J., Chhabra, R., Liu, Y., Ke, Y. and Yan, H. (2006). DNA-templated self-assembly of two-dimensional and periodical gold nanoparticle arrays, *Angew. Chem. Int. Ed.*, **118**, 744–749.

41. Song, Q., Shi, Y., He, D., Xu, S. and Ouyang, J. (2015). Sequence-dependent dsDNA-templated formation of fluorescent copper nanoparticles, *Chem. Eur. J.*, **21**, 2417–2422.

42. Hannant, J., Hedley, J., Pate, J., Walli, A., Farha Al-Said, S., Galindo, M., Connolly, B., Horrocks, B., Houlton, A. and Pike, A. (2010). Modification of DNA-templated conductive polymer nanowires via click chemistry, *Chem. Commun.*, **46**, 5870–5872.

43. Zhao, X., Striolo, A. and Cummings, P. T. (2005). C60 binds to and deforms nucleotides, *Biophys. J.*, **89**, 3856–3862.

44. Isobe, H., Sugiyama, S., Fukui, K.-I., Iwasawa, Y. and Nakamura, E. (2001). Atomic force microscope studies on condensation of plasmid DNA with functionalized fullerenes, *Angew. Chem. Int. Ed. Engl.*, **40**, 3364–3367.

45. Afreen, S., Kokubo, K., Muthoosamy, K. and Manickam, S. (2017). Hydration or hydroxylation: direct synthesis of fullerenol from pristine fullerene [C60] via acoustic cavitation in the presence of hydrogen peroxide, *Rsc. Adv.*, **7**, 31930–31939.

46. Nakamura, E., Isobe, H., Tomita, N., Sawamura, M., Jinno, S. and Okayama, H. (2000). Functionalized fullerene as an artificial vector for transfection, *Angew. Chem. Int. Ed. Engl.*, **39**, 4254–4257.

47. Vittala, S. K., Saraswathi, S. K. and Joseph, J. (2017). Fullerene cluster assisted self-assembly of short DNA strands into semiconducting nanowires, *Chem. Eur. J.*, **23**, 15759–15765.

48. Yamakoshi, Y., Yagami, T., Sueyoshi, S. and Miyata, N. (1996). Acridine adduct of [60] fullerene with enhanced DNA-cleaving activity, *J. Org. Chem.*, **61**, 7236–7237.

49. Ying, Q., Zhang, J., Liang, D., Nakanishi, W., Isobe, H., Nakamura, E. and Chu, B. (2005). Fractal behavior of functionalized fullerene aggregates. I. Aggregation of two-handed tetraaminofullerene with DNA, *Langmuir*, **21**, 9824–9831.

50. Vittala, S. K. and Joseph, J. (2018). Chiral self-assembly of fullerene clusters on CT-DNA templates, *Faraday Discuss.*, **207**, 459–469.

51. Estevez-Torres, A. and Baigl, D. (2011). DNA compaction: fundamentals and applications, *Soft Matter*, **7**, 6746–6756.

52. Bednar, J., Horowitz, R. A., Grigoryev, S. A., Carruthers, L. M., Hansen, J. C., Koster, A. J. and Woodcock, C. L. (1998). Nucleosomes, linker DNA, and linker histone form a unique structural motif that directs the higher-order folding and compaction of chromatin, *Proc. Natl. Acad. Sci. U.S.A.*, **95**, 14173.

53. Kornberg, R. D. (1974). Chromatin structure: a repeating unit of histones and DNA, *Science*, **184**, 868.

54. Abu-Salah, K. M., Ansari, A. A. and Alrokayan, S. A. (2010). DNA-based applications in nanobiotechnology, *J. Biomed. Biotechnol.*, **2010**, 715295.

55. Zhou, T., Llizo, A., Wang, C., Xu, G. and Yang, Y. (2013). Nanostructure-induced DNA condensation, *Nanoscale*, **5**, 8288–8306.

56. Nakanishi, W., Minami, K., Shrestha, L. K., Ji, Q., Hill, J. P. and Ariga, K. (2014). Bioactive nanocarbon assemblies: nanoarchitectonics and applications, *Nano Today*, **9**, 378–394.

57. An, H. and Jin, B. (2011). DNA exposure to buckminsterfullerene (C-60): toward DNA stability, reactivity, and replication, *Environ. Sci. Technol.*, **45**, 6608–6616.

58. Cassell, A. M., Scrivens, W. A. and Tour, J. M. (1998). Assembly of DNA/fullerene hybrid materials, *Angew. Chem. Int. Ed. Engl.*, **37**, 1528–1531.

59. Klumpp, C., Lacerda, L., Chaloin, O., Ros, T. D., Kostarelos, K., Prato, M. and Bianco, A. (2007). Multifunctionalised cationic fullerene adducts for gene transfer: design, synthesis and DNA complexation, *Chem. Commun.*, 3762–3764.

60. Isobe, H., Nakanishi, W., Tomita, N., Jinno, S., Okayama, H. and Nakamura, E. (2006). Gene delivery by aminofullerenes: structural requirements for efficient transfection, *Chem. Asian J.*, **1**, 167–175.

61. Yang, D. Y., Campolongo, M. J., Tran, T. N. N., Ruiz, R. C. H., Kahn, J. S. and Luo, D. (2010). Novel DNA materials and their applications, *WIREs Nanomed. Nanobiotechnol.*, **2**, 648–669.

62. Torring, T., Voigt, N. V., Nangreave, J., Yan, H. and Gothelf, K. V. (2011). DNA origami: a quantum leap for self-assembly of complex structures, *Chem. Soc. Rev.*, **40**, 5636–5646.

63. Lau, K. L. and Sleiman, H. F. (2016). Minimalist approach to complexity: templating the assembly of DNA tile structures with sequentially grown input strands, *ACS Nano*, **10**, 6542–6551.

64. Pinheiro, A. V., Han, D., Shih, W. M. and Yan, H. (2011). Challenges and opportunities for structural DNA nanotechnology, *Nat. Nanotechnol.*, **6**, 763–772.

65. Ensslen, P., Gärtner, S., Glaser, K., Colsmann, A. and Wagenknecht, H.-A. (2016). A DNA–fullerene conjugate as a template for supramolecular chromophore assemblies: towards DNA-based solar cells, *Angew. Chem. Int. Ed. Engl.*, **55**, 1904–1908.

66. Zheng, J., Constantinou, P. E., Micheel, C., Alivisatos, A. P., Kiehl, R. A. and Seeman, N. C. (2006). Two-dimensional nanoparticle arrays show the organizational power of robust DNA motifs, *Nano Lett.*, **6**, 1502–1504.

67. Cheng, S., Ya-Qing, C., Shou-Jun, X., Long, B., Zhong-Ze, G., Yi, P. and Xiao-Zeng, Y. (2005). Assembly of fullerene arrays templated by DNA scaffolds, *Chem. Mater.*, **17**, 6521–6524.

68. Sun, Y.-P., Ma, B., Bunker, C. E. and Liu, B. (1995). All-carbon polymers (polyfullerenes) from photochemical reactions of fullerene clusters in room-temperature solvent mixtures, *J. Am. Chem. Soc.*, **117**, 12705–12711.

69. Bath, J. and Turberfield, A. J. (2007). DNA nanomachines, *Nat. Nanotechnol.*, **2**, 275.

70. Vives, G. and Tour, J. M. (2009). Synthesis of single-molecule nanocars, *Acc. Chem. Res.*, **42**, 473–487.

71. Shin, S. R., Jin, K. S., Lee, C. K., Kim, S. I., Spinks, G. M., So, I., Jeon, J.-H., Kang, T. M., Mun, J. Y., Han, S.-S., Ree, M. and Kim, S. J. (2009). Fullerene attachment enhances performance of a DNA nanomachine, *Adv. Mater.*, **21**, 1907–1910.

72. Mroz, P., Tegos, G. P., Gali, H., Wharton, T., Sarna, T. and Hamblin, M. R. (2007). Photodynamic therapy with fullerenes, *Photochem. Photobiol. Sci.*, **6**, 1139–1149.

73. Yang, X., Ebrahimi, A., Li, J. and Cui, Q. (2014). Fullerene–biomolecule conjugates and their biomedicinal applications, *Int. J. Nanomed.*, **9**, 77–92.

74. Saito, I., Takayama, M., Sugiyama, H., Nakatani, K., Tsuchida, A. and Yamamoto, M. (1995). Photoinduced DNA cleavage via electron transfer: demonstration that guanine residues located 5′ to guanine are the most electron-donating sites, *J. Am. Chem. Soc.*, **117**, 6406–6407.

75. An, Y.-Z., Chen, C.-H. B., Anderson, J. L., Sigman, D. S., Foote, C. S. and Rubin, Y. (1996). Sequence-specific modification of guanosine in DNA by a C60-linked deoxyoligonucleotide: evidence for a non-singlet oxygen mechanism, *Tetrahedron*, **52**, 5179–5189.

76. Yamakoshi, Y., Sueyoshi, S., Fukuhara, K., Miyata, N., Masumizu, T. and Kohno, M. (1998). •OH and O2•- generation in aqueous C60 and C70 solutions by photoirradiation: an EPR study, *J. Am. Chem. Soc.*, **120**, 12363–12364.

77. Boutorine, A. S., Takasugi, M., Hélène, C., Tokuyama, H., Isobe, H. and Nakamura, E. (1995). Fullerene–oligonucleotide conjugates: photoinduced sequence-specific DNA cleavage, *Angew. Chem. Int. Ed.*, **33**, 2462–2465.

78. Manishkumar, B. P., Uma, H., Nikunj, N. V., Deval, S. M., Kuldeep, V. J., Sivakumar Prasanth, K., Kishor, H. C., Linz Buoy, G., Yogesh, T. J. and Shobhana, K. M. (2013). Novel cationic fullerene derivatized s-triazine scaffolds as photoinduced DNA cleavage agents: design, synthesis, biological evaluation and computational investigation, *Rsc. Adv.*, **3**, 8734–8746.

79. Higashi, N., Inoue, T. and Niwa, M. (1997). Immobilization and cleavage of DNA at cationic, self-assembled monolayers containing C60 on gold, *Chem. Commun.*, 1507–1508.

80. Singh, V., Zharnikov, M., Gulino, A. and Gupta, T. (2011). DNA immobilization, delivery and cleavage on solid supports, *J. Mater. Chem.*, **21**, 10602–10618.

81. Zhang, Y., Yang, S., Liu, C., Dai, X., Cao, W., Xu, J. and Li, Y. (2002). Photo-induced DNA cleavage in self-assembly multilayer films, *New J. Chem.*, **26**, 617–620.

82. Nakamura, E. and Isobe, H. (2010). In vitro and in vivo gene delivery with tailor-designed aminofullerenes, *Chem. Rec.*, **10**, 260–270.

83. Minami, K., Okamoto, K., Doi, K., Harano, K., Noiri, E. and Nakamura, E. (2014). siRNA delivery targeting to the lung via agglutination-induced accumulation and clearance of cationic tetraamino fullerene, *Sci. Rep.*, **4**, 4916.

84. Sitharaman, B., Zakharian, T. Y., Saraf, A., Misra, P., Ashcroft, J., Pan, S., Pham, Q. P., Mikos, A. G., Wilson, L. J. and Engler, D. A. (2008). Water-soluble fullerene (C(60)) derivatives as nonviral gene-delivery vectors, *Mol. Pharm.*, **5**, 567–578.

85. Isobe, H., Nakanishi, W., Tomita, N., Jinno, S., Okayama, H. and Nakamura, E. (2006). Nonviral gene delivery by tetraamino fullerene, *Mol. Pharm.*, **3**, 124–134.

Chapter 10

Nucleic Acid–Based Biosensors and Molecular Devices

Deepti Sharma, Prasanna Kumar Athyala, and Ashwani Sharma
Department of Chemistry, Indian Institute of Science Education and Research, Tirupati 517507, India
a.sharma@iisertirupati.ac.in

10.1 Introduction

Nucleic acids (DNA and RNA), one among the three major biopolymers, are essential molecules of life where DNA acts as a molecular archive of genetic information and RNA acts as an informational intermediate between a DNA sequence and its encoded protein through mRNA. It is well known that the entire genome does not contain protein-encoding DNA sequences and the noncoding regions were long believed to have no functional role, often referred to as junk DNA. However, it is evident now that junk DNA has certain important biological roles; for example, it acts as a genetic switch to determine where and when genes will get expressed and also

Templated DNA Nanotechnology: Functional DNA Nanoarchitectonics
Edited by Thimmaiah Govindaraju

ISBN 978-981-4800-21-1 (Hardcover), 978-0-429-42866-1 (eBook)
www.panstanford.com

controls transcription by determining the location for attachment of transcription factors. Some of the noncoding DNA sequences of mammals and other complex organisms are also transcribed into noncoding RNA (ncRNA) molecules that do not code for proteins, for example, transfer RNA (tRNA), ribosomal RNA, and regulatory RNA, such as siRNA (small interfering RNA), miRNA (microRNA), and snRNA (small nuclear RNA) [1]. In addition to these, there are riboswitches and catalytic ribozymes. All of these ncRNAs have important roles in regulating various levels of gene expression, translation, RNA splicing, catalysis, etc. Due to their regulatory functions in many diseases, including cancer, ncRNAs have become valuable tools in disease diagnostics and therapeutics [2].

Apart from their biological functions, nucleic acids (both DNA and RNA) can serve as robust constructional material to knit a variety of ordered structures, such as arrays of tiles, scaffolds, linked rings, and polyhedra [3–6]. The underlying principle relies on molecular recognition, which is the specific interaction between two or more species through noncovalent bonding and is universally observed in biological systems, for example, receptor-ligand, antigen-antibody, and enzyme-substrate. The molecular recognition property of nucleic acids is mainly due to highly specific Watson–Crick (WC) base pairing interactions between complementary base pairs. The predictable base complementarity of nucleic acids imparts functional and structural programmability to nucleic acids, which enables them to self-assemble into well-defined 2D and complex 3D arrangements and scaffolds on the nanoscale. These scaffolds have been used to make mechanical and molecular devices ranging from nanomotors, logic gates, and shape-shifting structures to gears and walkers [7]. The molecular recognition property of nucleic acids has also been exploited to develop various biosensors that have broad applications, including the detection of several biologically relevant analytes, such as small molecules, peptides, and proteins, as well as nucleic acids. These biosensors are capable of responding to external stimuli, like temperature, pH, light, small ions/molecules, and electricity, which makes nucleic acids useful for nonbiological applications as well, apart from their biological applications. Biosensors are recently emerging as significant tools in molecular diagnostics, detection of pathogens, environmental monitoring, and food safety control, which is evident from a number of publications in reputed journals in this area.

In this chapter, we will focus on nucleic acid–based molecular devices and biosensors and their design principles and application in detection as well as diagnostics. In addition, a major portion of this chapter will deal with aptamers and their application in designing biosensors for both detection and diagnostics. Later, we will focus on application of nucleic acids in imaging. We will end with a discussion of the application of nucleic acids in therapeutics for different diseases, including cancer. Before going into more detail, let us briefly discuss nucleic acid structure and stability and chemical modifications to improve their stability.

10.1.1 Nucleic Acid Structure and Hydrogen Bonding

Nucleic acids are biopolymers composed of monomers called nucleotides, consisting of three components: a pentose sugar, a nitrogenous heterocyclic base, and a phosphate group. DNA and RNA are distinguished on the basis of the pentose sugar, where DNA possesses D-2-deoxyribose, while RNA contains D-ribose. The nitrogenous bases in both DNA and RNA are purines and pyrimidines. The pyrimidines found in DNA are thymine (T) and cytosine (C), where thymine is replaced with uracil (U) in the case of RNA; the purines found in both DNA and RNA are adenine (A) and guanine (G). These bases are inherently programmable, as adenine always pairs with thymine in DNA (or with uracil in RNA) and guanine pairs with cytosine. A-T and C-G base pairs are held together through WC base pairing, which leads to the formation of the double-helix structure of DNA stabilized through hydrogen bonding between the complementary base pairs and stacking interaction between flat aromatic surfaces of the bases. The N-H groups of the bases are potent hydrogen bond donors, while the sp^2-hybridized electron pairs on the oxygen in the carbonyl group of the bases and those on the ring nitrogen are hydrogen bond acceptors, thus forming two hydrogen bonds in an A-T base pair and three in a C-G base pair, as shown in Fig. 10.1.

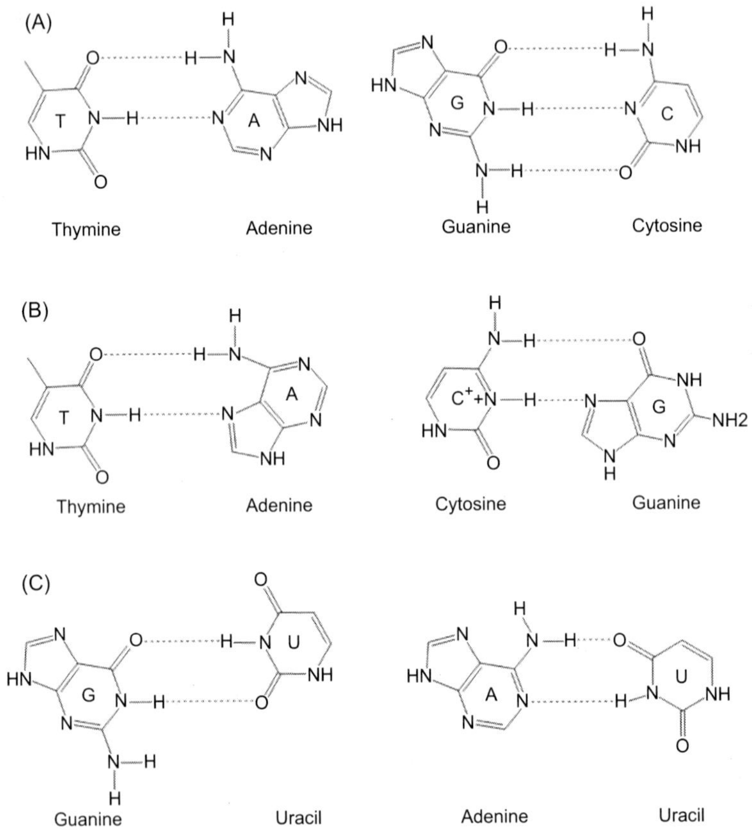

Figure 10.1 (A) Watson and Crick hydrogen-bonding scheme for A-T and G-C base pairs. (B) Hoogsteen hydrogen-bonding scheme for A-T and G-C base pairs. (C) GU Wobble base pair.

Though WC base pairing is the primary recognition mode between the bases, there are unusual modes of recognition, referred to as noncanonical base pairing. Karst Hoogsteen recognized that bases can also pair up in an alternate way. Under physiological and certain nonphysiological conditions: at acidic pH values in the absence of Mg^{2+} or in a high degree of negative supercoiling, the purine bases, adenine and guanine, can flip their normal conformations and form a new set of hydrogen bonds with their partners, leading to Hoogsteen base pairing (Fig. 10.1). Hoogsteen base pairing has been recognized as the basis for triple helices. In this case, pyrimidine bases can form hydrogen bonds with purines already engaged in WC base pairing, for example, cytosine in protonated form (in acidic condition) can make two hydrogen bonds with guanine of the WC C-G base pair in the parallel orientation to form a triple-helical structure. The triplexes are found in group I intron, group II intron, ribosomes, and telomerase RNA [8–10]. On other hand, guanine-rich sequences of nucleic acids form self-assembled structures called G-quadruplexes in the presence of monovalent ions (K^+, Na^+, etc.) in physiological conditions. These structures are stabilized by Hoogsteen hydrogen bonding arising from association of four guanines observed in telomeres. The inventory of sequences forming telomeric structures correspond to d(TTTTGGGG), d(TTGGGG), and d(TTAGGG), present in sequences from *Oxytricha*, *Tetrahymena*, and humans, respectively. Apart from WC and Hoogsteen base pairs, other noncanonical base pairs also exist in the RNA world, such as wobble base pairs. Among the wobble base pairs, the G U wobbles received much attention due to their abundance, importance in the interaction of messenger RNA (mRNA) with tRNA on the ribosome during protein synthesis (codon-anticodon interactions), and their biological implications [11]. G U wobble pair is held together by two hydrogen bonds, as revealed by structural investigations of yeast $tRNA^{Phe}$ [12, 13]. The thermodynamic stability is comparable to WC base pairs. Therefore, it often substitutes for G–C or A-U base pairs, giving rise to various secondary structures of RNA, like helix, stem loops, and pseudoknots.

10.1.2 Stability of Nucleic Acids

Thermal as well as chemical stability of nucleic acids plays a crucial role in their application. The WC base pairing of the nucleotide bases enables the pairing between two individual strands of DNA to form a double-stranded structure. The thermal stability of these double-stranded structures is important for DNA replication, for DNA repair, and for the transcription of DNA. Double-stranded regions are also found in RNA molecules, and their intrinsic base pairing abilities result in the formation of various secondary structures, required for their function. Also, the stability of nucleic acid duplexes (DNA-DNA, RNA-RNA) and hybrid duplexes (DNA-RNA) plays an important role in the designing of various nucleic acid–based molecular devices and biosensors and their biomedical diagnostic applications.

In terms of the physical stability of duplexes, RNA-RNA duplexes are thermally more stable than DNA-DNA duplexes of the same sequence, with DNA-RNA hybrid duplexes being the weakest of the three [14]. The tertiary structure of RNA is also stabilized due to the presence of special motifs, such as bends, stacks, junctions, and loops. The important structural differences between DNA and RNA are the 2′-hydroxyl (-OH) on the ribose sugar in RNA and the thymine C5-methyl group in DNA. The 2′-hydroxyl on the ribose greatly influences the structure and function of RNA and is responsible for the conformational and thermodynamic differences between RNA and DNA in an aqueous solution. The presence of the C2′-OH in RNA stabilizes the C3′-endo sugar conformation, leading to an A-form helix whereas the 2′-deoxy sugars in DNA tend to be in the C2′-endo conformation, leading to a B-form helix. Thus, the 2′-hydroxyl group increases the melting temperature (T_m) and stability of RNA by locking RNA duplex into a compact A-form helix that is more stable than the DNA's B-form helix. In contrast, the thymine C5-methyl group primarily affects the thermodynamic stability of DNA from increased base-stacking efficiency as a result of increased polarizability [15]. The differences in the geometry and chemistry of DNA and RNA also affect the hydration properties of the major and minor grooves [16, 17]. These differences in chemistry and conformation are presumably responsible for the big difference in the thermodynamic stability of DNA and RNA duplexes.

Though RNA is more stable than DNA in terms of thermal stability, from the chemical stability standpoint, RNA, with its additional 2′-hydroxyl group, is hydrolytically much less stable than DNA. RNA is more susceptible to alkaline hydrolysis, which cleaves the backbone at the phosphodiester bond. In the base-catalyzed hydrolysis, the 2′-hydroxyl of the ribose in RNA attacks the adjacent phosphorus in the phosphodiester bond of the sugar-phosphate backbone, leading to cleavage of the RNA strand. This alkaline hydrolysis makes RNA chemically less stable compared to DNA, which does not have a 2′-hydroxyl group and thus is not susceptible to hydrolysis. However, both DNA and RNA are extremely sensitive to nuclease degradation, especially in blood plasma, which drastically limits their in vivo applications. In the last few decades, a lot of progress has been made to improve the stability of nucleic acids for in vivo application and to make them more nuclease resistant by chemical modification of DNA/RNA.

10.1.3 Chemical Modifications of Nucleic Acids

The instability of nucleic acids in vivo is mainly the result of enzymatic degradation by nucleases. To make DNA and RNA more nuclease resistant, various chemical modifications have been reported in the recent past to improve their chemical stability. This includes modification of the ribose sugar, for example, 2′-fluoro (2′-F), 2′-methoxy (2′-OMe), or 2′-amino (2′-NH_2); modification on the bases, for example, 5-bromouracil (5-BrU) and 5-iodouracil (5-IU) [18]; and modification of the phosphate linkage, for example, phosphothioate and boranophosphate [19], locked nucleic acids (LNAs) [20], peptide nucleic acids (PNAs) [21], and polycarbamate nucleic acids (PCNAs) [22]. Hong et al. have reported the 5′UA/3′UA and 5′CA/3′UG pairings to be the two most vulnerable sites to serum ribonuclease [23]. With 5′UA/3′UA being the most sensitive, modification at this particular site improves the stability of dsRNA in serum. All these chemical modifications are very efficient in rendering RNA resistant to RNAse in vitro and in vivo. However, the challenge still remains that these modifications might often change the folding properties and biological functioning of the nucleic acids. Therefore, development of a suitable method conferring RNAse resistance without affecting its folding and biological functioning is critical. It has been reported that among the above-mentioned methods, 2′-F modifications have the least detrimental effects on the RNA folding and function [24]. All the aforementioned methods of chemical modifications have rendered nucleic acids stable for in vivo applications [25, 26].

10.2 Nucleic Acid–Based Molecular Devices

Molecular devices are based on the predictable interactions between the complementary sequences of DNA and RNA. Both single-stranded and double-stranded DNA (ssDNA and dsDNA) and single-stranded and double-stranded RNA contribute to the molecular segments of these devices. The single strands form the flexible components of the devices and also act as molecular tags on which complementary strands can bind. The double strands form the rigid molecular segments of devices and may contribute to their chemical function by chemical modifications or through incorporation of binding sites. Specific combinations of these flexible and rigid elements impart distinct mechanical and chemical properties to the molecular devices. Along with the mechanical and thermodynamic stability of nucleic acids, their secondary structures and the unconventional conformations, such as G-quadruplexes and i-motifs, which are responsive to a specific stimulus often termed as "fuel," lay the foundation for designing these molecular devices. The stimuli or the fuel can be ions, small molecules, pH, light, temperature, or even the small DNA/RNA strands.

10.2.1 G-quadruplex-Based Devices

A G-quadruplex can be found mostly in vivo in the telomeric ends of the chromosome and in some oncogene regulatory regions of the genome [27]. G-quadruplex structures are formed in the nucleic acid sequences due to the high guanine content. Four guanine bases can come closer through Hoogsteen hydrogen bonding to fetch a square planar guanine tetrad (Fig. 10.2A). The assembly of two or more guanine tetrads, one on top of the other, can form a G-quadruplex. G-quadruplexes are stabilized mainly by monovalent cations like Na^+, K^+, and NH_4^+, which are located at the center of each pair of tetrads. These structures are known to be formed through an intramolecular or intermolecular arrangement of guanine-rich strands, giving rise to diverse structural topologies, for example, parallel, antiparallel, and hybrid G-quadruplexes (Fig. 10.2B) [28, 29]. In intermolecular G-quadruplexes, more than one nucleic acid strand is involved, of which one guanine from each strand will contribute to the G-quartets, whereas intramolecular G-quadruplexes are involved in single nucleic acid strands [30]. The π-stacked guanine tetrads are connected by intervening sequences of varied lengths that form loops and are situated on the exterior of the core.

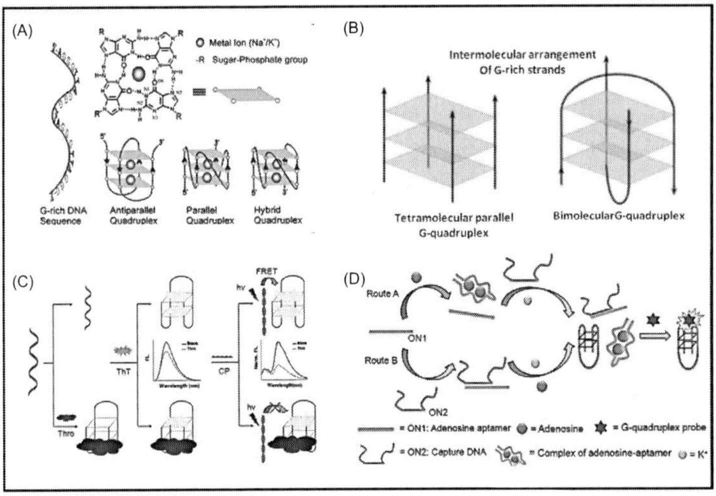

Figure 10.2 G-quadruplex. (A) Illustration of G-quartet/tetrad of G-quadruplex structure and their topology. (B) Structure of G-quadruplexes formed due to intermolecular or intramolecular arrangement of G-rich strands. (C) Thrombin detection using G-quadruplex-based thrombin aptamer. (D) Adenosine detection using G-quadruplex-forming adenosine aptamer [32]. (A) Adapted from Ref. [28] published by The Royal Society of Chemistry. (B) Adapted from Ref. [30]. (C) Adapted from Ref. [31]. Copyright (2015) American Chemical Society.

G-quadruplexes show a rich diversity in structural topologies owing to the number and length of the guanine tracts, the lengths of the intervening loop regions, and the character of the metal ion in solution. This extensive structural polymorphism presents them as a versatile sensing, diagnostic, and therapeutic platform. Liu et al. [31] demonstrated the detection of thrombin in the serum by Förster (or fluorescence) resonance energy transfer (FRET). Thrombin is a serine protease protein that converts fibrinogen into fibrin and has a major role in thrombosis. In this study, the authors used a thrombin-binding aptamer (TBA) sequence that can form the G-quadruplex after binding its target thrombin (for aptamers, refer to Section 10.4.2). Two dyes were used to demonstrate FRET, namely thioflavin (ThT) and a conjugated polymer (CP)-polyfluorene derivative doped with benzothiazole dyes. Initially, the addition of ThT induced the TBA to form into a G-quadruplex structure that gave a high-intensity fluorescence signal. The addition of target protein thrombin did not show a significant decrease in the fluorescence intensity. When the CP was added to the above system, an increase in the intensity of fluorescence in the blank was observed due to the FRET effect. However, when the CP was added to the thrombin-containing G-quadruplex solution, because of the huge size of thrombin, the distance between the CP and ThT was high, which resulted in less fluorescence intensity and less FRET effect. The difference in the FRET between the blank G-quadruplex and the thrombin-containing G-quadruplex gives the resultant protein levels comparable with the control [31] (Fig. 10.2C).

G-quadruplex-based devices can also be used as probes for a luminescent-based assay. The important factor for designing these devices is to develop probes that have a high affinity and selectivity for the G-quadruplex DNA. A great deal of interest has been shown in transition metal complexes in generating probes for biomolecules as these complexes have long lifetimes, large stroke shifts, simple synthesis methods, and variable excitation and emission in the visible region. Hence, iridium (III) octahedral complexes have been developed as G-quadruplex-based probes for luminescent assays. Lu et al. [32] developed a G-quadruplex-selective iridium (III) complex for the construction of an aptamer-based label-free luminescent switch-on detection device for adenosine. The working mechanism is as shown in the Fig. 10.2D (Route A). After adding adenosine, it reacts with ON1 to form an adenosine-aptamer complex. Then the capture DNA ON2 is added of the same concentration as that of ON1. The ON2 is partially hybridized with the ON1; the free ON2, which is not hybridized upon the addition of K^+, will form a G-quadruplex and the iridium probe captures the G-quadruplex, resulting in the fluorescence. The increased concentration of adenosine will lead to more hybridization of ON1, leaving free hybridized ON2, which helps in higher-intensity luminescent signals. In Fig. 10.2D, route B, ON1 and ON2 are hybridized to form an ON1 and ON2 complex. Once the adenosine is added to this complex, the ON1 detaches from the ON2, binding with the adenosine, leaving ON2 free. The free ON2 forms the G-quadruplex in the presence of K^+ ions; thereby the detection probe gives an increased luminescent response. When there is no adenosine or ON2, then the luminescence is not turned on, eventually leading to no signal [32].

10.2.2 I-motif-Based Devices

Like G-quadruplexes, which have a high G-content, i-motif nucleic acids are rich in cytosine content and form four-stranded DNA secondary structures called i-motifs. These i-motif structures are stable in acidic conditions, wherein protonation of the cytosine molecule helps in the formation of C-C^+ base pairing. They contain two parallel strands of DNA duplexes bound together in an antiparallel alignment by intercalated, cytosine-cytosine+ base pairing. An i-motif is also called an I-tetraplex. The stability of the i-motif decreases when the pH is above neutral. The pH dependency of i-motifs for folding helps in the development of the pH switches for usage in nanomedicine.

I-motif-based devices have been extensively studied for the detection of pH on the basis of fluorescence probe and quencher. As the microenvironment in the live organism is diverse and balanced dynamically, a small change in the physiological conditions can impact the metabolism of the organism, which can lead to the disease condition. One such change in the physiological system is pH, and it is a very important parameter for certain processes in the living system. Hence, Fu et al. have developed an i-motif-based device for the detection of pH using perylene-3,4,9,10 tetracarboxylic acid diamide (PTCDI) dyes (Fig. 10.3A). PTCDI has the ability to self-assemble linearly and form aggregates that quench the fluorescence intensity. When used in an aggregated form, the dye has a quenching effect on the fluorescence intensity because of the strong electrostatic interaction between the dye and the polyionic nucleic acid. Once the pH is slightly decreased, the nucleic acid strand tends to form the i-motif structure, breaking the aggregates. As a result, there are no binding sites on the nucleic acids for the dye, which leads to fluorescence. Decrease in the pH of the oligo in buffer increases the fluorescence intensity, and it is reversible with an increase in the pH of the system [33].

Ma et al. [34] have recently developed a ratiometric fluorescent probe with a high pH-sensitivity based on a hairpin consisting of an i-motif strand (I-strand) labeled with rhodamine green and black hole quencher-BHQ and a complementary C strand labeled with rhodamine red at the 5′-end (Fig. 10.3B). The I-strand and the C-strands are hybridized into a stable rigid duplex (I-C) at neutral pH, which quenches both rhodamine red and BHQ with concurrence.

This results in the shutdown of the fluorescence of rhodamine red and the fluorescence of the rhodamine green is "on." In the case of slightly acidic pH, the dehybridization of I-duplex happens once the I-strand forms the intramolecular i-motif. This turns on the fluorescence signal of rhodamine red, and the fluorescence of rhodamine green is reduced [34].

10.2.3 Molecular Walkers

A molecular walker belongs to a class of nucleic acid nanomachines where the nucleic acid–based walker moves on another nucleic acid–based track. The term "DNA walkers" was first coined by Reif in 2003 [35]. The molecular walker does consist of a motion that extends from linear to 2D and 3D that helps in carrying molecular cargo with high-velocity motion.

Molecular walkers are used for different applications, like DNA detection, small-molecule detection, and disease detection and diagnosis. Li et al. [36] have developed a catalytic-assembled enzyme-free 3D molecular walker for the detection of the smallpox gene C1 and the smallpox gene C2. The walker has the anchorage hairpin (H1-G-quadruplex sequence) and the smallpox gene C1 functionalized with a magnetic nanobead (MNB); the MNB works as a walking track. The walker has a swing arm consisting of the smallpox gene C2 and the blocker segment, which reduces the risk of hybridization with the other sequences. The function of the walker remains inactive if the swing arm and track are noninteractive. Once the target DNA is added, C2 recognizes the C1 and hybridization takes place, resulting in fixing of the swing arm with the track. The swing arms come near H1 and hybridize with H1, leaving blocker DNA. The resultant swing arm strand acts as the trigger for the H2 strand, and the repetition of this process will hybridize to form dsDNA on the track. The addition of N-methyl mesoporphrin IX (NMM) produces the fluorescence signal specific to a G-quadruplex assembly. This method has demonstrated that a single nucleic acid can catalyze the self-driven DNA walker by generating H1/H2 hybrid complexes and G-quadruplexes, which in turn gives a high-intensity fluorescence signal [36]. When there is no target DNA, the swing arm does not recognize the C1 and hybridization cannot take place because of the blocker DNA, which results in a low-intensity or no-fluorescence signal (Fig. 10.3C).

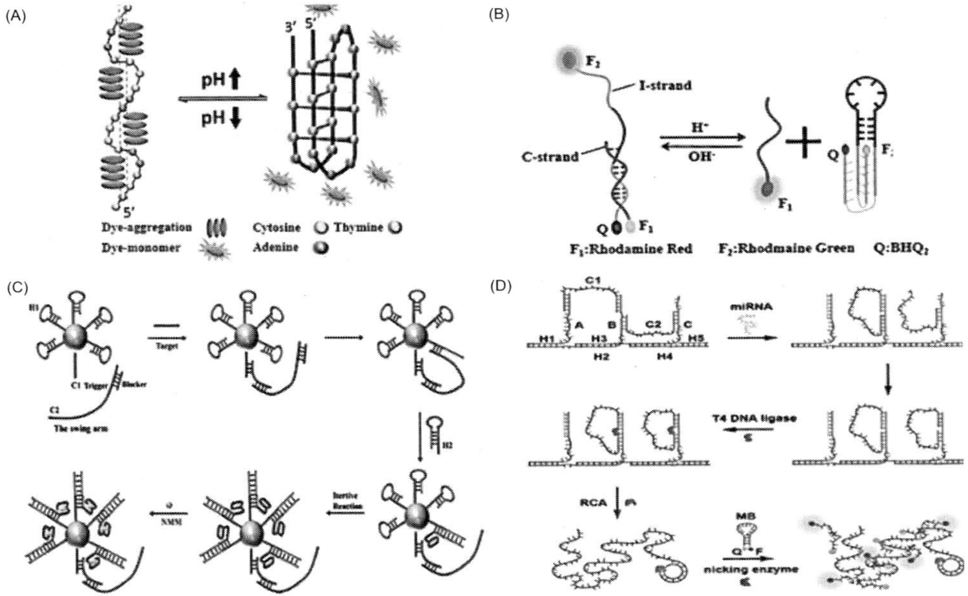

Figure 10.3 i-Motifs. (A) i-Motif-based pH detector using PTCDI dyes. (B) i-Motif-based pH sensor using dye and a quencher. (C) Schematic of catalytic assembled 3D DNA walker triggered by nucleic acids. (D) Molecular walker for miRNA detection. (A) Reprinted from Ref. [33] with permission of The Royal Society of Chemistry. (B) Adapted with permission from Ref. [34]. Copyright (2018) American Chemical Society. (C) Reprinted from Ref. [36] with permission of The Royal Society of Chemistry. (D) Reprinted from Ref. [37] published by The Royal Society of Chemistry.

DNA walkers are also used in the process of detecting the miRNAs, which are very important in biological systems as they are involved in the post-transcriptional regulations of certain gene expressions. miRNA-based therapies are now widely exploited as their importance in cancer treatment is high, especially in colon and lung cancers. Let7a is one such miRNA associated with many cancers. The detection of this miRNA is very important for cancer diagnosis. Wang et al. [37] have proposed a DNA walker–based system for the detection of miRNA, where miRNA is used as a "fuel" strand to initiate the walker strand to move along the track, enhancing the hybridization of the strand. Initially miRNA is bound to branch A by toehold-mediated strand displacement, moving the helices that are formed between the arm of strand C1 and the track to separate. The leftover arm of strand C1 walks over branch B, and the arm of strand C2 is moved from branch B to branch C, generating two padlock probes. Branch B and branch C then act as a ligation template in the presence of T4 DNA ligase, and the two padlock probes are circulated to create DNA circles C1 and C2. Rolling circle amplification (RCA) is performed, and the resultant product is hybridized with the molecular beacon (MB) having fluorophore and quencher. Finally, the nicking enzyme is added to catalyze the MB and to cleave at multiple sites of MB, leaving a fluorescence signal for the detection [37] (Fig. 10.3D).

10.2.4 Aptamer-Based Molecular Devices

Aptamers are small DNA/RNA oligonucleotide molecules that bind to a specific target molecule that can be a small molecule, a peptide, or a macromolecule (like proteins) (refer to Section 10.4.2). Aptamers have been used in clinical research and basic research in the field of drug delivery systems and target-based therapy. Aptamer-based molecular devices are designed for various applications, and the output of these devices Is mainly in the form of fluorescence, calorimetric, SPR, etc. Aptamer-based molecular devices will be discussed in detail in Section 10.4.2.

10.3 Nucleic Acid–Based Biosensors

Biosensors were first introduced by Clark and Lyons in 1962, with the development of the first glucose biosensor [38]. A biosensor is a small device utilizing biological recognition properties for a selective bioanalysis. These devices are based on the coupling of a biological recognition element with a signal transduction element, converting the biological signals into an electrical signal or other physically detectable signals (in the form of light, heat, pH, charge, or mass change), where these signals are proportional to the concentration of analytes. Biosensors are highly selective owing to the high affinity of the biomolecules toward their target. A typical biosensor consists of a recognition element, a transducer, and a processor. The sensing/recognition elements may be antibodies, enzymes, or nucleic acids. The biological recognition is very selective, where the target analyte and the recognition elements act like lock and key, leading to highly selective biosensors. These recognition elements are integrated with transducers via immobilization by adsorption, cross-linking, or covalent binding. The choice of the transducer element is based upon the parameters of measurement. On the basis of the choice of transducer, biosensors can be categorized as electronic biosensors (electrical or electrochemical) [39], optical biosensors (fluorescent, colorimetric, surface plasmon resonance [SPR], or Raman) [40–42], and piezoelectric biosensors (quartz crystal microbalance [QCM]) [43]. On the basis of the choice of the recognition element, biosensors can also be categorized as enzyme-based biosensors; immunological biosensors; and nucleic acid biosensors (NABs), or those based on DNA and RNA.

Nucleic acids have been extensively used for developing a range of biosensors due to their inherent physicochemical stability and biological activities. Biosensors built with nucleic acids are usually composed of a ssDNA, which can hybridize with the complementary strand with high efficiency and good specificity, facilitating the detection of the complementary DNA/RNA strand or analyte molecules. These DNA-based biosensors are also called "genosensors" [44, 45]. Unlike enzymes or antibodies, the advantages of nucleic acids lie in the simplicity of their synthesis, their suitability for structural modification, and their high selectivity and affinity. NABs can be more sensitive and specific when combined with polymerase chain reaction (PCR) methods [46]. NABs are developed by immobilizing nucleic acids to a solid support by adsorption, covalent bonding, or ionic interaction and coupled to a physicochemical transducer [47]. Various methods are available for immobilization, for example, adsorption on inert carriers, chemical cross-linking to macroscopic beads or magnetic particles, physical entrapment in a gel-like lattice, and microencapsulation in nanospheres. The immobilization process also facilitates probe orientation and easy accessibility to target element [48]. For the physicochemical method, different fabrication strategies are available for attaching a DNA probe to the transducer surface. Some of them include use of thiolated DNA for self-assembly onto gold-plated transducers (electrodes, piezoelectric crystals), covalent linkages to the gold surface by stabilizer molecules, use of biotinylated DNA to form conjugates with surface-immobilized avidin or streptavidin, and carbodiimide coupling to functional groups on carbon nanotubes (CNTs, single-walled carbon nanotubes [SWCNTs], and multiwalled carbon nanotubes [MWCNTs]) [49]. NABs can be designed with various types of transducers, like optical, electrochemical, and piezoelectric.

10.3.1 Optical Biosensors

Optical methods are the most commonly used for the detection of analytes. Nucleic acids themselves do not possess intrinsic properties that are functional in direct detection; therefore, many of the nucleic acid–based systems require a label for detection, such as radioactive isotopes and chemiluminescent or fluorescent molecules. The choice of label is based on stability, sensitivity, and convenience. Optical biosensors can be miniaturized by utilizing fiber-optics, semiconductor photodevices, SPR, and quantum dot (QD)-based detection modules. Krull and coworkers developed the first DNA optical biosensor for the rapid fluorometric detection of T/AT triplex formation in both parallel and antiparallel configurations [50]. These biosensors were based on fiber-optics, which are devices that carry light from one place to another through a series of internal reflections and transduce the emission signal of the fluorescent label. Decaadenylic acid oligonucleotides were covalently immobilized at the ends of the functionalized fused optical fibers, and their hybridization with complementary DNA (cDNA) from the solution phase was detected by fluorescence signals from the intercalated ethidium bromide at the fiber surface. Optical sensors have an advantage over standard fluorometric detection in terms of low detection limits, reusability, reliability of the device, and complete automation, thus negating the requirement of skilled technicians. Various optical NABs have been reported in the literature. However, we will briefly discuss optical biosensors based on MBs, QDs, and SPR detection modules.

An MB is an oligonucleotide composed of a stem-loop structure and can be labeled with a fluorophore at one end and a quencher at the other, where the loop is the target recognition region of about 15–30 bases. The formation of the stem-loop structure brings the quencher and the fluorophore into close proximity, resulting in quenched fluorescence. In the presence of a target DNA or RNA molecule, hybridization between the target and the loop sequence of the MB takes place and the stronger intermolecular hybridization opens the weaker stem helix, which causes an increase in the distance between fluorophore and quencher, resulting in restoration of the fluorescence [51]. Flexibility in designing MBs and various attachment platforms makes them potential candidates for biosensor application; for example, ssDNA MBs with a biotin tag can be surface-immobilized at a liquid-solid interface of an optical fiber probes through avidin-biotin binding for the detection of single-nucleotide polymorphism (SNP) in the submicromolar level.

Recently, a simple and highly selective amplification biosensor for miRNA detection was developed on the basis of G-quadruplex molecular beacons (G4MBs) and duplex-specific nuclease (DSN) (Fig. 10.4A). A G4MB with a G4 motif stem was used as a recognition probe that hybridized with the target miRNA in its presence and formed a G4MB-miRNA duplex. DSN subsequently cleaved the G4MB of the G4MB-miRNA duplex, recycling the target miRNA. This led to fluorescence signal amplification. When target miRNAs are not present, the DSN is not able to digest the stem of the G4MB because of the protection of the G4 motif, which eliminates the false positive signal, and produces a low-fluorescence background [52].

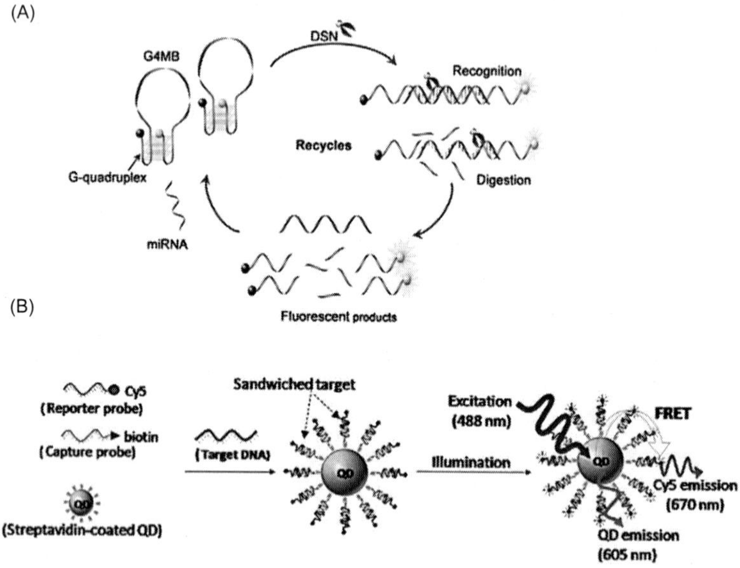

Figure 10.4 (A) Schematic illustration of the G4MB- and DSN-based signal amplification strategy for highly specific detection of miRNA. B) Principle of quantum dot–FRET nanosensor for sequence-specific DNA detection. (A) Reproduced from Ref. [52], Copyright (2017), with permission from Elsevier. (B) Adapted from Ref. [53], Copyright 2005, Springer Nature.

Another ultrasensitive nanosensor based on FRET utilizes QDs. These are the nanoparticles (NPs) used for the fluorescence tagging of probe biomolecules. QDs are fluorescent semiconductor nanocrystals. They are composed of materials from the elements in the periodic groups of II-VI, III-V, or IV-VI, for example, cadmium telluride (Cd from group II and Te from group VI) and indium phosphamide (In from group III and P from group V). Their diameters range from 2 to 10 nm. QDs are highly photostable, with broad absorption, narrow and symmetric emission spectra, slow excited-state decay rates, and broad absorption cross-sections owing to the effects of quantum confinement. Their emission color depends on their size, chemical composition, and surface chemistry and can be tuned from the ultraviolet to the visible and near-infrared (NIR) wavelengths. Biosensors utilizing QDs can detect very low concentrations of DNA. A QD-based hybrid FRET nanosensor was developed by using a CdSe-ZnS core-shell nanocrystal as the donor and Cy5 (fluorophore) as the acceptor [53]. This nanosensor had two target-specific oligonucleotide probes, which included a reporter probe labeled with a fluorophore and a capture probe labeled with biotin, in addition to a QD conjugated with several streptavidins (Fig. 10.4B). The QD functioned as both a FRET energy donor and a target concentrator. When target DNA was present, it was sandwiched by the two probes. Several sandwiched hybrids were then captured by a single QD through biotin-streptavidin binding. The resulting assembly brought the fluorophore acceptors and the QD donor into close proximity, which led to fluorescence emission from the acceptors by means of FRET on illumination of the donor. Detection of acceptor emission then indicated the presence of targets.

SPR provides another rapid, label-free, high-selectivity, and high-sensitivity assay method. Under total reflection conditions, when plane polarized light comes in contact with a metal film, SPR occurs. The biosensors based on SPR monitor changes in surface optical properties (change in the resonance angle due to a change in the interfacial refractive index), which result from the surface-binding reaction. The resonance conditions are influenced by the material adsorbed onto the thin metal film. A linear relationship is found between resonance energy and mass concentration of molecules, such as proteins, sugars, and DNA. The SPR signal, expressed in resonance units, is a measure of mass concentration at the sensor chip surface [54]. Alkane thiol–modified oligonucleotides, immobilized onto the gold surface, have been reported to detect DNA hybridization using SPR-based detection [55].

10.3.2 Electrochemical Biosensors

In electrochemical biosensors, DNA is immobilized onto an electrode, and the resulting change in electrical parameters due to hybridization reaction is measured. Electrical modes have been developed for the detection of both label-free (direct detection) and labeled (indirect detection) objects. Specific organic dyes, metal complexes, or enzymes are used for hybridization detection in label-based electrochemical biosensors [45]. Two commercialized DNA chips based on redox-active molecules have been commercialized in the molecular diagnosis market, with the trade names of eSensor^{TM} (produced by Motorola Life Sciences) and Genlyser^{TM} (by

Toshiba). In a label-free method, the immobilized probe recognizes a target complementary sequence in the sample. A label-free DNA biosensor has been developed based on a molecular sentinel (MS) immobilized on a plasmonic "nanowave" chip. This is called "metal film over nanosphere" (MFON). The sensing principle involves a decrease in the surface-enhanced Raman scattering (SERS) intensity when Raman label, which is tagged at one end of MS, is being removed from the MFON's surface due to DNA hybridization. The sensing process involves a single hybridization step between the target DNA sequences and the complementary MS probes. The potential application of this technique for medical diagnostics was demonstrated by the development of a biosensor for the detection of a common inflammation biomarker human radical S-adenosyl methionine domain containing the 2 (RSAD2) gene [56]. CNTs are also playing an important role in the development of electrochemical DNA-based biosensors. The unique electrical, thermal, chemical, mechanical, and 3D spatial properties of SWCNTs and MWCNTs have made them popular in the biosensor community [57].

10.3.3 Piezoelectric Biosensors

A piezoelectric biosensor is based on a quartz crystal that oscillates at a defined frequency upon application of an oscillating voltage, which allows high sensitivity. The increased mass, associated with the hybridization reaction, results in a decrease in the oscillating frequency, and a linear relationship exists between the two factors. The QCM is an extremely sensitive mass-measuring device allowing sensitive monitoring of hybridization events. Biosensors based on QCM hybridization have an oscillating crystal, with the DNA probe immobilized on its surface. The increased mass, associated with the hybridization reaction, results in a decrease in the oscillating frequency. The oligonucleotide QCM biosensor was developed for the microalgae *Alexandrium minutum*, by immobilizing a probe complementary strand of a partial sequence of the gene-encoding microalgae that produces neurotoxins responsible for paralytic shellfish poisoning on European and Asian coasts [58].

10.4 Functional Nucleic Acid–Based Biosensors

Nucleic acids can bind to their complementary strands with high specificity and are useful for nucleic acid detection. However, the detection of a broad range of analytes is facilitated by utilizing functional nucleic acids (FNAs) as probe molecules that are beyond the conventional role of nucleic acids. FNAs are broadly categorized into catalytic FNAs, such as nucleic acid enzymes (DNAzyme/RNAzymes), and noncatalytic FNAs, such as aptamers and aptazymes (combination of nucleic acid enzymes and aptamers).

10.4.1 DNAzyme- and RNAzyme-Based Biosensors

DNAzymes and RNAzymes (ribozymes) are the most popular nucleic acid scaffolds to be used in biosensors, for example, for metal ion sensing. These catalytic FNAs (RNAzymes and DNAzymes) first recognize their substrate sequence, followed by its cleavage. RNAzymes either occur naturally [59] or have been evolved by systematic evolution of ligand by exponential enrichment (SELEX)-type approaches [60]. DNAzymes however have been selected by artificial methods so far [61, 62]. The substrates for most of naturally occurring RNAzymes are other RNA strands, and they catalyze RNA splicing, cleavage, and peptide bond formation in ribosome machinery [63–65], whereas artificially derived RNAzymes that evolved through SELEX-type approaches are capable of catalyzing a range of reactions, such as RNA cleavage and ligation, and a vast array of chemical reactions, such as Diels–Alder reactions [66], Aldol reactions [67], Michael reactions [68], N-glycosidic bond formation [69], and acylation reactions [70]. Most RNAzymes and DNAzymes require metal ions for their activities, and some even show metal specificity, which makes them suitable for use in biosensors for metal ion sensing [71]. In the presence of the metal ion, DNAzyme/RNAzyme gets activated and cleaves a labeled substrate, which is observed through a change in the fluorescence or color. Various biosensor designs employ a DNAzyme/RNAzyme-substrate complex labeled with a fluorophore (orange) and quencher (black), where this complex exists in a fluorescently quenched form. When the specific metal ion (Mn^+) is present, it promotes cleavage in the DNAzyme/

RNAzyme-substrate complex. This dissociation causes the release of the shorter fragment carrying the fluorophore tag, resulting in an increase in the fluorescence. Metal ions that have been sensed using DNAzymes include Cu^{2+} [72], Pb^{2+} [73], Zn^{2+} [74], and UO_2^{2+} [75].

Another class of FNAs is riboswitches, which are regulatory, non-protein-coding elements residing within mRNA. They are able to directly bind and sense small metabolites, secondary messengers, vitamins, amino acids, ions, and other biomolecules. Apart from these, riboswitches are sensitive to changes in temperature and pH. A riboswitch primarily consists of a sensory domain (an aptamer; refer to Section 10.4.2) that specifically senses a metabolite and a regulatory domain (expression platform) that facilitates control over transcription termination or translation initiation by a structural rearrangement. When the sensory domain binds to the target ligand, it induces a conformational change in the expression platform, which affects translation or transcription of the downstream located gene. Because of the aforementioned properties, riboswitches offer a promising approach to the biosensor field.

There is a class of engineered riboswitches, known as aptazymes, that is generated in vitro by combining ribozymes and aptamers. The resulting aptazymes control gene expression by small molecule–induced self-cleavage of a ribozyme. The activity of these aptazymes can be controlled by the addition of the specific ligand of that aptamer. The first aptazyme was developed by the Breaker group by fusing an adenosine triphosphate (ATP)-binding aptamer with a minimal hammerhead ribozyme [76].

10.4.2 Aptamer-Based Biosensors

Aptamers are single-stranded DNA/RNA oligonucleotides that have the affinity to bind to a particular target with high selectivity and specificity. Their target can be virtually anything—small molecules, peptides, proteins, or even whole cells. Aptamers were discovered in 1990 by two independent groups, Tuerk and Gold [77] and Ellington and Szostak [78]. Aptamers can be selected from a large pool (library) of oligonucleotides (DNA or RNA) by an in vitro selection process known as SELEX. Aptamer binds with its target molecule with strong affinity and specificity, and the binding is determined by the aptamer's tertiary structure. The aptamer-based detection is more like antigen-antibody or receptor-ligand interactions, unlike other nucleic acid–based biosensors, where it's based upon sequence complementarity. Target recognition and binding involve 3D, shape-dependent interactions as well as hydrophobic interactions, base stacking, and intercalation.

Aptamers have been widely used as recognition elements, especially for the detection of proteins and small-molecule targets. Their binding with the target relies on specific conformations, such as loops, hairpins, junctions, and G-quadruplexes. The conformational variations before and after the formation of aptamer-ligand complexes offer a great possibility and feasibility for the construction of aptamer-based biosensors.

10.5 Design Strategies for Aptamer-Based Biosensors

As mentioned above, aptamers form various 3D conformations when they bind to their targets. Targets are divided into two groups on the basis of the relative spatial positions of both targets and aptamers in the target-aptamer complex. These are embedded groups and outside-binding groups. In an embedded group, the targets are generally buried within a small binding pocket formed by a special oligonucleotides sequence of aptamers. Examples include aptamers against ATP [79], cocaine [80], K^+, and theophylline [81, 82]. An outside-binding group includes targets with more complicated spatial structures or macromolecules/biomolecules, for example, proteins or even whole cells, bacteria, and viruses. Examples are aptamers against thrombin [83] and platelet-derived growth factor-BB (PDGF-BB) [84]. They have more than one aptamer-binding site offering diverse designs of sensing strategies. Aptamer-based biosensors have been constructed utilizing various methodologies, including electrochemical biosensors, optical biosensors, and piezoelectric biosensors, and these strategies can be divided into different modes, as follows:

10.5.1 Target-Induced Structure-Switching Mode

This mode fits both embedded groups and outside-binding groups of targets. In this case, the targets bind directly to the aptamer, leading to the conformational switch of aptamers to specific tertiary structures instead of random structures, which in turn leads to changes of detectable signals, which can be as follows: (i) the size or weight of aptamers, along with the formation of aptamer-target complexes, (ii) the position, quantity, or status of the signal moieties binding covalently to the end of aptamers or adsorbed on the aptamers via electrostatic force, stacking, hydrogen bond, etc., and (iii) other properties of aptamers, such as the ability to stabilize gold nanoparticles (Au NPs). Each of these change leads to a change of signal. On the basis of the target-induced structure-switching (TISS) mode, several aptamer-based biosensors have been designed. We will be discussing a few of them next.

10.5.1.1 Fluorescent aptamer-based biosensors

Fluorophores have been widely used as probes or labels/tracers for the detection and imaging of biomolecules. They possess properties like extinction coefficient, quantum yield (emission intensity), excitation/emission wavelengths, lifetime, anisotropy, and energy transfer, which enable them to be used rationally in the aptamer-based biosensor design. Fluorophores can be attached to nucleic acids through a number of interactions, like noncovalent binding, electrostatic or hydrophobic interactions, intercalation, and covalent attachment. Various fluorescent aptasensors are based on the FRET, a distance-dependent physical process involving nonradiative energy transfer from an excited donor fluorophore to another acceptor fluorophore.

Aptamer beacon approach: In aptamer beacon–based fluorescent detection, a fluorescence-quenching pair is used to detect changes in the conformation induced by ligand binding. On the basis of classic MB (refer to Section 10.3.1) designs, an antithrombin aptamer was constructed into an aptamer beacon by the addition of nucleotides to the 5′-end, complementary to nucleotides at the 3′-end of the aptamer [85A, B]. In the absence of thrombin, the added nucleotides form a duplex with the 3′-end, forcing the aptamer beacon into a stem-loop structure. In the presence of thrombin, the aptamer beacon forms the ligand-binding structure. This conformational change causes a change in the distance between a fluorophore attached to the 5′-end and a quencher attached to the 3′-end (Fig. 10.5A). In nonclassical MB design, aptamers are not forced to be in a stem-loop structure as such confinements could result in less binding affinity for some aptamers. Therefore, aptamer beacons were constructed that were based on the fluorophore/quencher approach but did not require the formation of a hairpin structure initially. For example, the cocaine aptamer was predicted to have a binding structure of a short hairpin flanked by two single-stranded regions [86]. The destabilization was constructed in one stem of a three-way junction (3WJ), which formed the cocaine-binding pocket, and the resulting short stem was end-labeled with a fluorophore and a quencher. Two stems remained open in the absence of cocaine; however, stems closed and formed a 3WJ in the presence of cocaine (Fig. 10.5B). Due to this structural change, fluorophore and quencher come close and signal the presence and concentration of ligand.

Split aptamers: Split aptamers are constructed by splitting aptamers into two fragments. In a study by Yamamoto-Fujita et al., the Tat protein aptamer was split into two halves. A fluorophore was labeled at one end and biotin on the other end, and it was immobilized on a streptavidin-coated surface. The surface lit up only in the presence of Tat. In the absence of Tat, the affinity between the two halves was very low because they were not complementary [87]. An anticocaine aptamer was split into two halves by Stojanovic et al. [88]. No DNA hairpin was designed, leading to an increased initial fluorescence signal. The two halves associated to bind the target in the presence of cocaine, resulting in quenched fluorescence. The detection limit for cocaine was from 10 to 1000 μM, with ~40% quenching. In 2011, utilizing the split aptamer approach, Sharma et al. reported DNA-templated reaction controlled by small-molecule binding rather than inherent WC affinity between the DNA strands. Two reactive groups, an azide and a strained alkyne, were placed at the ends of two aptamer fragments and the reaction was a cocaine-dependent split aptamer ligation using strain-promoted azide-alkyne cycloaddition. Cocaine directs assembly of the two nucleic acid strands of the split aptamer, bringing the azide and the cyclooctyne in close proximity, and thus promotes the ligation (Fig. 10.5C) [89].

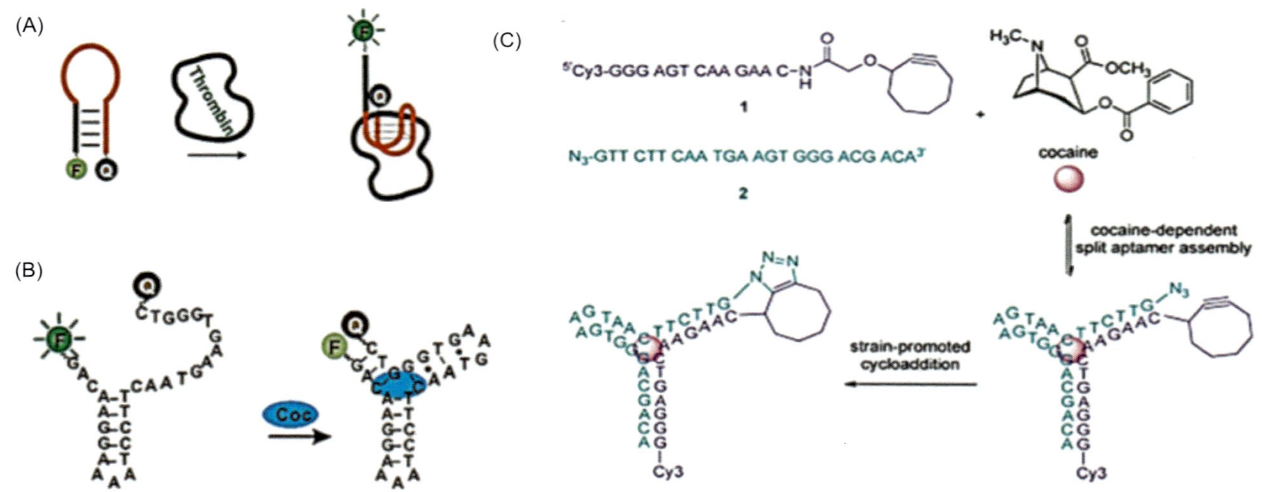

Figure 10.5 (A) Detection of thrombin with an aptamer beacon based on classical molecular beacon design. The aptamer contained an extension (in black), which formed the stem part of the hairpin. Binding to thrombin opened the hairpin and increased fluorescence. (B) Aptamer beacons with nonclassical molecular beacon design: An aptamer beacon for cocaine was constructed by labeling the two ends with a fluorophore and a quencher. (C) Cocaine-dependent split aptamer ligation using strain-promoted azide-alkyne cycloaddition. (A, B) Adapted with permission from Ref. [85B]. Copyright (2009) American Chemical Society. (C) Adapted with permission from Ref. [89]. Copyright (2011) American Chemical Society.

Structure-switching aptamers: An aptamer can associate either with a cDNA strand to form a duplex structure or with its target analyte, and there is a structure-switching process between the two states. In a study reported by Nutiu and Li, aptamer-based fluorescent sensors were reported that function by switching structures from the DNA/DNA duplex to the DNA/target complex (Fig. 10.6A) [90]. The duplex is formed between a fluorophore-labeled DNA aptamer and a small oligonucleotide, which is modified with a quenching moiety (denoted by QDNA). In the absence of the target, the aptamer binds to QDNA, bringing the fluorophore and the quencher close for maximum fluorescence quenching. However, when the target is introduced, the aptamer prefers to form an aptamer-target complex. The switch of the binding partners for the aptamer occurs in conjunction with the generation of a strong fluorescence signal owing to the dissociation of QDNA [91]. Sensors for detecting thrombin and ATP have been demonstrated, with detection limits of ~10 nM and 10 μM, respectively. In a different design by the Tang group, the two ends with fluorophore/quencher labels were linked by a polyethylene glycol (PEG) chain, forming a hairpin structure. In the presence of target molecules, increased fluorescence was observed [92]. In a different study by Rankin et al., an RNA aptamer for theophylline was hybridized to DNA labeled with fluorophore and quencher. High emission intensity was observed as fluorophore and quencher were far away from each other due to the rigidity of the short dsDNA, however; addition of theophylline released the DNA piece, which then formed a random coil structure in solution, resulting in quenched fluorescence [93]. In addition to using organic quenchers, Au NPs were also employed as quenchers by Zhao et al. for designing structure-switching aptamer sensors for protein detection [94]. A QD aptamer beacon was constructed by Ellington and coworkers on the basis of the structure-switching aptamer design [95]. A thrombin aptamer was attached to QDs, and a cDNA with a quencher was hybridized to the aptamer, resulting in quenched QD emission. Addition of thrombin removed the quencher and approximately 14-fold emission increase was observed. Structure-switching property of aptamers is not just significant in designing fluorescence sensors; it has also been applied to construct other types of sensors, including colorimetric sensors and electrochemical sensors.

10.5.1.2 Colorimetric aptamer-based biosensors

A change of color is observed in the presence of the target analyte in the case of colorimetric sensors. Since the native nucleic acids do not show absorption in the visible region, these colorimetric aptamer sensors require the introduction of color reporting groups. The most frequently used color reporting groups are small organic dyes, CPs, and metallic NPs. Metallic NPs exhibit strong distance-dependent optical properties and high extinction coefficients, which make them suitable for the construction of colorimetric sensors. Au NPs have been widely used due to their characteristic spectroscopic changes during target-induced aggregation or dispersion. Studies have been reported showing that adsorption of ssDNA on Au NPs can effectively stabilize the colloid against salt-induced aggregation while dsDNA has no such function [96, 97]. Later, in a study by Wang et al. [98], Au NPs were successfully employed as a colorimetric probe for both chemical and biological detection. When the target ATP was absent, the aptamer hybridized with its complementary sequence to form a rigid duplex, which could not prevent Au NPs from aggregating as salt was added into the system. However, as ATP was present, the duplex structure would switch to an ATP-aptamer complex together with the release of its complementary sequence, which could stabilize Au NPs and show high resistance to salt-induced aggregation, causing change in the solution color caused by Au NP aggregation, easily observed by the naked eye. The corresponding changes in ultraviolet (UV)-visible spectra could be used for quantitative assay of ATP (Fig. 10.6B) [98].

Sharma et al. reported the aptamer analog of the enzyme-linked immunosorbent assay (ELISA) for cocaine detection via colorimetric readout. The two fragments of the cocaine-split aptamer assembled in the presence of cocaine (Fig. 10.6C). One of the split aptamer fragments was immobilized on the microplate, and the second split aptamer fragment was in the test sample. In the presence of cocaine, split aptamer fragments assembled and promoted a chemical ligation between azide and cyclooctyne functional groups tagged to the termini of the split aptamer fragments. A colorimetric output was observed after ligation resulted in the covalent attachment of biotin to the microplate due to conjugation to streptavidin-horseradish peroxidase [99].

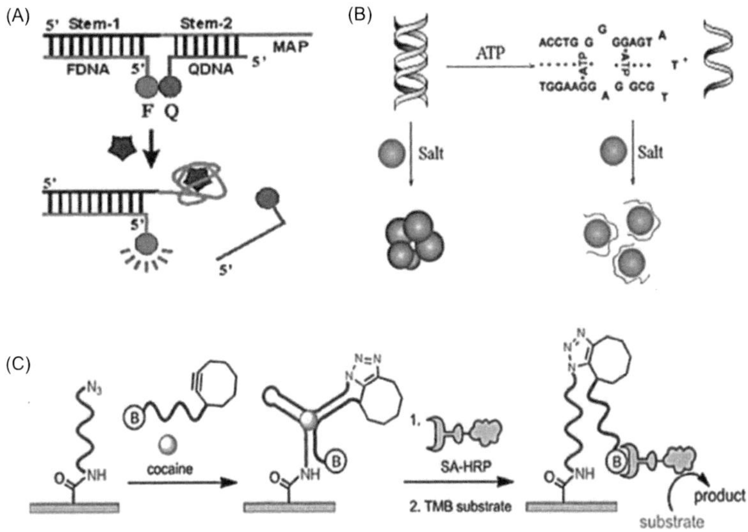

Figure 10.6 (A) Design of structure-switching signaling aptamers. (B) Target-induced structure switching and salt-induced AuNP aggregation. (C) Enzyme-linked cocaine detection using split aptamer ligation. (A) Adapted with permission from Ref. [90]. Copyright (2003) American Chemical Society. (B) Adapted with permission from Ref. [98]. Copyright (2007) John Wiley and Sons. (C) Adapted with permission from Ref. [99]. Copyright (2012) American Chemical Society.

10.5.1.3 Electrochemical aptamer-based biosensors

A redox probe-labeled, signal-off electrochemical biosensor was first developed for thrombin detection [83]. In this study, electroactive labels were covalently attached to the aptamers, in turn increasing the performance of the sensors. These sensors are called electronic aptamer-based sensors (E-AB) sensors. Thirty-one-base-pair-long DNA was modified with a 5′-thiol, and the 3′ terminal contained a thrombin aptamer, which was tagged with methylene blue. In the absence of thrombin, the MB could transfer electrons along the electrode surface due to the flexible conformation of the aptamer. In the presence of thrombin, the aptamer part folded into a G-quadruplex structure, inhibiting the electron transfer as only half of the DNA strand remained flexible. However, the negative part about this method is its "turn off" nature, that is, there is a decrease in the electron transfer when thrombin is present (Fig. 10.7A).

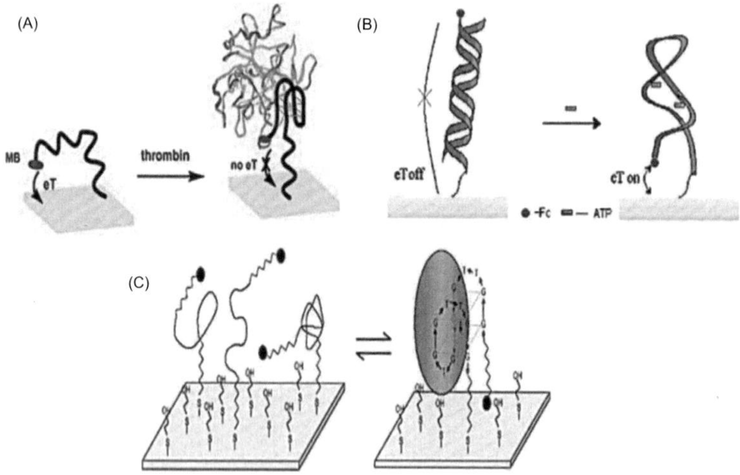

Figure 10.7 (A) Design of a signal-off electrochemical aptamer biosensor for thrombin detection. (B) TREAS strategy for ATP detection. (C) Design of a signal-on electrochemical aptamer biosensor for thrombin detection. (A) Adapted with permission from Ref. [83]. Copyright (2005) John Wiley and Sons. (B) Adapted with permission from Ref. [102]. Copyright (2007) American Chemical Society. (C) Adapted with permission from Ref. [100]. Copyright (2012) American Chemical Society.

To overcome this disadvantage, "turn on" electrochemical biosensors were developed. A 15-base-pair-long TBA was immobilized to the gold electrode and was attached with electroactive label ferrocene (Fc) at the other end. Addition of thrombin forced the aptamer to fold into a G-quartet to bind thrombin, and the Fc group was pointed down toward the surface, which resulted in an electron transfer, making it a turn-on sensor (Fig. 10.7C) [100]. Similar strategies have been applied to design a cocaine sensor [101].

In the turn-off design for the biosensor, the flexibility of the immobilized aptamer on the electrode leads to a high background signal. Zuo et al. reported a biosensor for the detection of ATP, where the anti-ATP aptamer was dually labeled with 3′-SH and 5′-Fc. It self-assembled on gold electrodes to form dsDNA [102]. In this conformation, the distance between the Fc and the electrode surface was large (~10 nm), which inhibited the efficient exchange of electrons with the underlying electrode, causing an eT OFF state. However, when target ATP was present, the tertiary aptamer structure was stabilized, which in turn denatured the duplex, liberating the cDNA. Because of this structural switch from duplex to tertiary aptamer structure, the distance between the Fc moiety and the electrode surface was much reduced and the electron transfer path was turned on (eT ON) (Fig. 10.7B). This biosensing strategy was called "target-responsive electrochemical aptamer switch" (TREAS). Here, the design of the complementary sequence of the aptamer is very significant.

10.5.1.4 Aptamer-based biosensors based on mass difference upon binding

In this case, signal transduction occurs through binding-induced mass change and the respective change of the resonance frequency of supporting materials. Usually, very small mass changes can be detected, resulting in high sensitivities.

SPR-based detection: As discussed in Section 10.3.1, this technology has been widely used for the study of interactions of biological molecules. Since aptamers can be used for targeting small molecules, the sensing surface can be fabricated for the detection of small molecules by SPR. In aptamer-based Au NP–enhanced SPR detection, Au NPs—the SPR signal enhancer—were immobilized via DNA hybridization with the aptamer sequence. The binding of ATP led to the formation of a tertiary structure and dehybridization between aptamer and the Au NP–tagged complementary ssDNA, which finally resulted in the decrease of the SPR signal [103]. Other examples include the study of RNA aptamers binding to aminoglycosides, S-adenosylhomocysteine, and HIV-1 Tat protein in the presence of Zn^{2+} and DNA aptamers binding to tubulin, adenosine, thrombin, and thyroid transcription factor 1 [104–108].

QCM-based detection: As discussed in Section 10.3.3, a QCM hybridization biosensor consists of an oscillating crystal with the DNA probe immobilized on its surface. In a study carried out by Liss and coworkers [109], a quartz crystal biosensor was designed to detect concentrations and ligand affinity parameters of free unlabeled proteins in real time. The crystal oscillated at its resonance frequency under constant flow conditions. In the presence of the target protein (IgE), its binding led to the increased mass of the crystal and thus decreased the resonating frequency. Several different attachment methods have been employed to immobilize the IgE aptamer onto the gold surface. Due to the high selectivity, the device could detect IgE in complex protein matrices, such as meat extract, bouillon, and milk. QCM has also been applied to aptamer-based detection of the HIV-1 Tat protein and thrombin [110, 111].

SAW-based detection: For the detection of change of mass, acoustic wave sensors have proven to be highly specific devices. When mass is loaded onto the surface of these sensors, there is a decrease in the propagation velocity of acoustic waves, which results in a reduction of resonance frequency or in the alteration of the phase shift between output and input signal. Surface acoustic waves (SAWs) have been used to detect analytes in the liquid phase, and antibody-based detections have been successfully carried out [112]. In addition, SAW devices have been used to detect aptamer-protein interactions. The sensor device was constructed by using interdigital transducers (IDTs) on a piezoelectric material (i.e., quartz) and sealed with SiO_2 to avoid direct contact with liquid [113]. IDTs are metal electrodes for creating and receiving acoustic waves, through which electric signals and acoustic waves can interconvert. The SiO_2 layer also served waveguide purposes. A gold layer was deposited onto the SiO_2 layer for electric shielding and for attaching thiol-modified aptamers. The device had a thrombin detection limit of 72 pg/cm^2. In a follow-up work, the SAW-based device was used to monitor complex formation in the blood-coagulation cascade with both DNA and RNA aptamers for thrombin [114].

10.5.2 Sandwich Mode

Sandwich-like assays, for example, ELISA, have been one of the most widely used diagnostic methods for many important analytes. In this assay, two different antibodies against the same target are involved where the target molecule acts as "a piece of meat" placed between "two slices of bread," which corresponds to two proteins, making up a sandwich and hence called "sandwich mode." Proteins like thrombin and PDGF have dual binding sites because of which they can bind two recognition molecules. In this assay, the aptamer is immobilized onto a surface that acts as a capture probe and another aptamer is modified with a label that acts as a signal probe. The target is captured on the surface of electrodes by binding to the capture aptamer and then forms a sandwich complex with the second aptamer. For the aptamer-based biosensors, two different formats for the sandwich complexes have been widely employed: aptamer-protein-aptamer and aptamer-protein-antibody.

An electrochemiluminescence (ECL) biosensor was constructed for the detection of thrombin based on aptamer-protein-aptamer assay format (Fig. 10.8A) [115]. The aptamers acting as capture probe were labeled with Au NPs, and the aptamers acting as signal probe were tagged with ECL labels. The capture aptamers were then immobilized onto the thio-silanized indium titanium oxide (ITO) electrode surface through strong Au-S bonds. In the presence of thrombin, signal aptamers attach to the assembled electrode surface, forming a Au NP–capture-aptamer/thrombin/ECL-tagged signal-aptamer sandwich. The electrode surface was treated with tri-n-propylamine (TPA), and after applying a swept potential, an ECL response was generated. Aptamer-protein-antibody format has also been utilized for the detection of thrombin [116].

For PDGF detection via sandwich mode, the aptamer is immobilized on the electrode surface through self-assembly. In the presence of the target, a PDGF-aptamer complex is formed and captured onto the interface. Then the negatively charged aptamer–modified Au NPs recognize the target and are bound to the electrode surface, forming a sandwich structure. Then, positively charged $[Ru(NH_3)_5Cl]^{2+}$ is adsorbed to the sandwich structure via electrostatic interaction and the obtained electrochemical signals are directly related with the concentration of PDGF (Fig. 10.8B) [117].

Sandwich mode can be adapted for the detection of the small-molecule targets (targets belonging to embedded group). These assays involve the manipulation and design of the aptamers and their complementary sequences. An electrochemical biosensor was developed for the detection of adenosine, where a thiolated capture probe was immobilized onto the Au NP–modified gold electrode. Then by utilizing the sandwich mode, a linker DNA containing an antiadenosine aptamer sequence and reporter DNA loaded on Au NPs was bound to the electrode through hybridization. When the target was present, the aptamer part of the linker DNA bound with adenosine, forming a target-aptamer complex structure. This resulted in the release of reporter probe–modified Au NPs together with $[Ru(NH_3)_6]^{3+}$, which in turn led to a decrease in the electrochemical signal (Fig. 10.8C) [118].

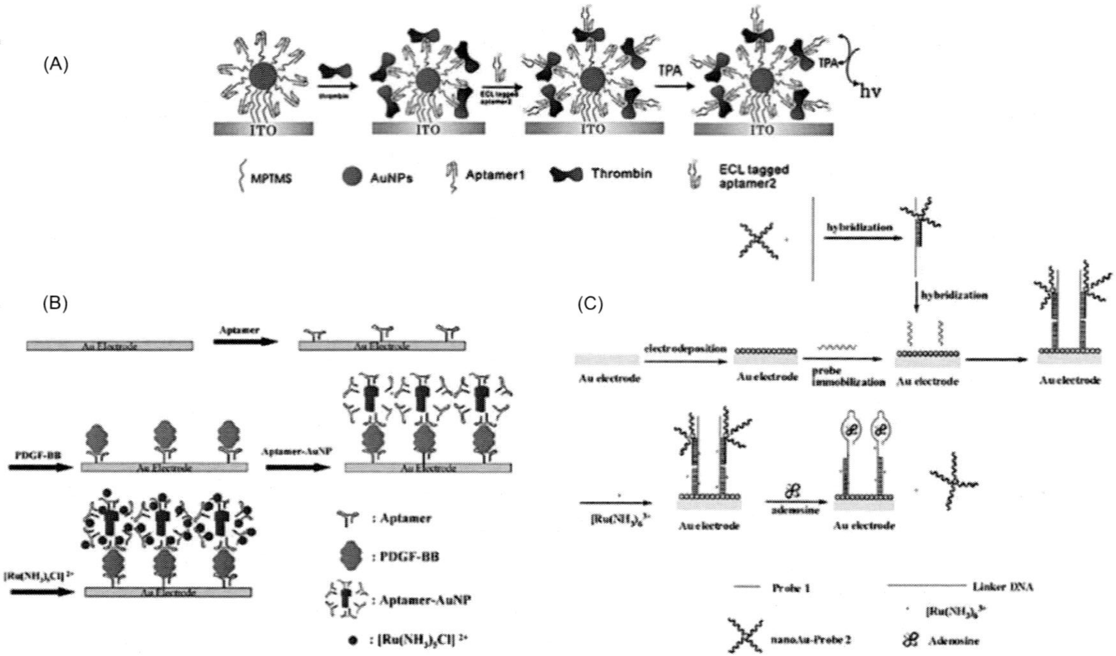

Figure 10.8 (A) Schematic diagram of the fabrication of an ECL biosensor. (B) Schematic for the formation of a sandwich structure and AuNP/electrochemical probe detection of PDGF. (C) Scheme for a sandwich-like structure for the detection of adenosine. (A) Adapted from Ref. [115], Copyright (2008), with permission from Elsevier. (B) Adapted from Ref. [117], Copyright (2008), with permission from Elsevier. (C) Adapted with permission from Ref. [118]. Copyright (2008) American Chemical Society.

10.5.3 Target-Induced Dissociation/Displacement Mode

Aptamer-based biosensors based on TISS mode and sandwich mode rely on the special conformations of the aptamers (G-quadruplex or hairpin conformation) and the specific structure of the targets or the target-aptamer complexes; however, they do not rely on a structure-dependent assay, as demonstrated by Han et al. [79]. The above-mentioned assays might involve heavy metals and also rely on the binding of the targets with the aptamers on an electrode surface. In target-induced dissociation/displacement mode (TID) mode, the complementary sequence of aptamers is employed as anchors to localize the aptamers. After incubation with targets, the target-aptamer complexes are liberated into the solution, leading to the changes of detectable signals. The TID mode further involves signal-off mode [119], signal-on mode [120], and label-free mode [121, 122].

In signal-off mode, the presence of targets can be determined from the signal suppression due to the release of the redox-modified aptamer from the aptamer-capture DNA duplex bound on an electrode. For example, in the case of ATP and thrombin biosensors [119] the redox-tagged aptamers were near the gold electrode surface through hybridization with their thiolated complementary sequence in the absence of targets, thereby giving electrochemical signals. In the presence of the target, the redox-tagged aptamers were released and the electrochemical signals were greatly reduced (Fig. 10.9A). Han et al. [79] reported an ATP biosensor based on a signal-on mode TID strategy. An aptamer hybridized with its thiolated complementary sequence was immobilized on the gold electrode via a Au-S bond. As the target ATP was added, the aptamer sequence was released from double strands to form a target-aptamer complex. Then the remaining complementary sequence was further hybridized with another Fc-moiety-modified ssDNA, resulting in the emergence of an electrochemical signal (Fig. 10.9B). In the label-free TID strategy, the release of the aptamer due to the formation of the target-aptamer complex decreased the interfacial electron transfer resistance, which was utilized to quantify the concentration of lysozyme through faradic impedance spectroscopy (FIS) [122].

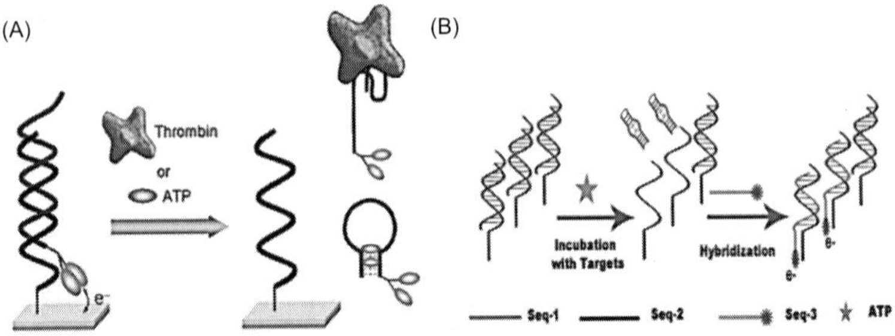

Figure 10.9 (A) Schematic representation of target-induced strand release from an aptamer–DNA duplex based on signal-off TID mode. (B) Schematic illustration of TID strategy. (A) Adapted from Ref. [119] with permission of The Royal Society of Chemistry. (B) Adapted from Ref. [122], Copyright (2008), with permission from Elsevier.

10.6 Nucleic Acids as Diagnostics

Nucleic acid–based diagnostic (NAD) systems are available in a variety of formats, ranging from simple nucleic acid probe hybridization systems to tests that incorporate amplification of a specific genomic target utilizing in vitro amplification technologies. DeLong et al. in 1989 reported fluorescence in situ hybridization (FISH), a direct detection method that involves the hybridization of the fluorescent probe with intact nucleic acid [123]. Another format involves the detection of nucleic acid following an in vitro amplification technique. PCR, real-time PCR (RT-PCR), nucleic acid sequence–based amplification (NASBA), and strand displacement amplification (SDA) are some common amplification techniques employed to increase the target nucleic acid prior to detection.

RT-PCR detection platforms utilize different detection formats where fluorescent technologies are employed. These are either nonspecific, using dyes such as SYBR Green I or SYBR Gold, which intercalate into the PCR product during the reaction, or specific, utilizing probes to detect specific sequence amplification in the PCR. TaqMan® (5′exonuclease) probes, HybProbes, MBs, and Scorpion probes are different fluorescent probe chemistries that have been employed in RT-PCR assays [124, 125]. An example of a commercial assay using MBs is the methicillin-resistant *Staphylococcus aureus*

(MRSA) assay from Cepheid (CA, USA). Various commercialized products have been developed for the detection of infectious pathogens in clinical samples utilizing nucleic acid–based biosensor technology [126]. On the basis of DNA hybridization, Dupont had developed a product named BAX® for the detection of numerous infectious agents, like *E. coli* O157:H7, *Listeria*, *Enterobacter*, *Campylobacter*, *Vibrio*, and *Salmonella*. ACCUPROBE′ and ANSR™ are DNA-based biosensors developed by Biomèrieux and Neogene, respectively, for monitoring *Campylobacter*, *Salmonella*, and *Listeria* in clinical samples. NABs have been utilized for the early detection and identification of various biomarkers for the diagnosis of bladder cancer [127], such as FGFR3 (fibroblast growth factor receptor 3) and HRAS (Harvey RAS), in urine samples using silicon resonator as an optical sensor. This biosensor employss specific interaction between the DNA probes with their mutant FGFR3 and HRAS targets [128].

Owing to various advantages of aptamers, for example, their small size, stable folding structures, ease of preparation, and economy, they are potential tools for clinical diagnosis of diseases.

10.6.1 Aptamers as Diagnostics

Proteins are a direct measure of health, and their evaluation is important in the understanding and accurate diagnosis of disease. Multiple types of diagnostic tools have been developed for the detection of disease-related proteins, small molecules, and diseased cells overexpressing particular protein biomarkers. Aptamers can detect very low amounts of these proteins or protein biomarkers on diseased cells, like cancer cells, for which they are selected. Aptamers bind to their targets with high affinity and specificity, where dissociation constants can range from micromolar to subpicomolar values. Aptamers are easy to label, and their ease of conjugation allows them to be combined with other advanced technologies, like flow cytometry, microfluidic cell separation, endogenous nucleic acid analysis, and NP-based sensing, presenting them as a potential tool for the diagnosis of different diseases.

Aptamers have been extensively tested as cancer diagnostic tools because of their high selectivity. The capability of aptamers to detect small differences between proteins with similar structures allows them to differentiate cancerous cells from normal cells on the basis of their unique cell-surface homology. Biomarkers are crucial in the diagnosis and treatment of cancer [129]. Biomarkers are distinctive and unique to the specific cells and are either not expressed or expressed in very low amounts elsewhere.

To facilitate the detection of various cancers in the early stages, various aptasensors have been developed to detect the interaction between a biomarker and the aptamers specific to it. MUC1 (Human mucin-1) and VEGF-165 (vascular endothelial growth factor-165) are important biomarkers for diagnosing cancer, and various aptasensors have been developed using these two biomarkers [130, 131]. Aptamers for MUC1 and VEGF-165 were used by Zhao et al. for the development of "signal-on" electrochemical aptasensors by using a Fc-labeled aptamer-cDNA as probe [132]. The Fc-labeled cDNA fragment was immobilized on a gold electrode, with a complementary binding site for MUC1 and VEGF165 aptamers (Fig. 10.10A). The aptamers hybridize with a cDNA fragment to form a long double strand in the absence of MUC1 and VEGF165 biomarkers, keeping Fc molecules away from the electrode surface, which results in no signal. However, in the presence of MUC1 and VEGF165, both aptamers bind to their respective targets, reducing the distance between the electrode and the Fc molecule due to stem loop formation by the ss-cDNA fragment, resulting in a signal.

Many high-affinity aptamers have been selected against the extracellular domain of prostate-specific membrane antigen (PSMA) [133], cell adhesion molecule P-selectin [134], and protein tyrosine phosphatase 1B (PTP1B) [135].

Nanomaterials, including metallic NPs, semiconductor nanocrystals (QDs), silica NPs, carbon nanomaterials, and other nanomaterials, such as liposomes, possess unique optical, electronic, and magnetic properties [136, 137]. These properties provide aptamers with more flexibility as cancer cell–targeting tools. A hydrazine–Au NP–aptamer bioconjugate was developed by Zhu and coworkers for the detection of both human epidermal growth factor receptor (HER-2) protein and SK-BR-3 breast cancer cells (human breast cancer cell line that overexpresses the HER-2) where the aptamer was specific toward HER-2 [138]. Approximately 10%–34% of breast cancer patients have tumors that show overexpression of HER-2. The sensor probe was fabricated by covalently immobilizing a monoclonal anti-HER2 antibody on the electrode, which is a nanocomposite layer that was composed of self-assembled 2,5-bis(2-thienyl)-1H-pyrrole-1-(p-benzoic acid) (DPB) on Au NPs. While detecting HER-2 overexpressing SK-BR-3 breast cancer cells, a sandwich is formed between the sensor probe, HER-2 expressing cells, and hydrazine–Au NP–aptamer (Fig. 10.10B). This was then treated with a silver nitrate solution where silver ions are reduced by hydrazine and are deposited onto the breast cancer cells. This deposited silver on breast cancer cells was observed under a microscope and analyzed using square wave stripping voltammetry to determine the number of HER-2 cells with a limit of detection of 26 cells/mL in human serum samples.

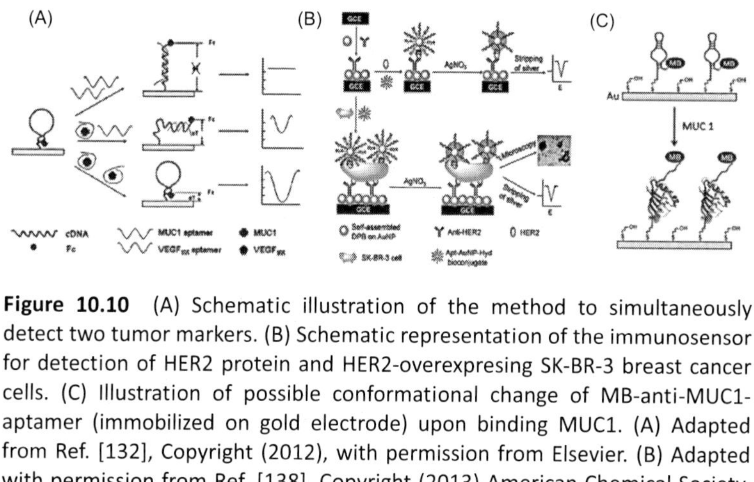

Figure 10.10 (A) Schematic illustration of the method to simultaneously detect two tumor markers. (B) Schematic representation of the immunosensor for detection of HER2 protein and HER2-overexpresing SK-BR-3 breast cancer cells. (C) Illustration of possible conformational change of MB-anti-MUC1-aptamer (immobilized on gold electrode) upon binding MUC1. (A) Adapted from Ref. [132], Copyright (2012), with permission from Elsevier. (B) Adapted with permission from Ref. [138]. Copyright (2013) American Chemical Society. (C) Adapted from Ref. [139], Copyright (2013), with permission from Elsevier.

Apart from the sandwich construction method, Ma et al. developed an aptasensor based on the conformation change. The advantage of this method lies in the fact that no antibody is used in this method, which reduces the cost and also increases the reproducibility. Modification steps are also fewer, which makes its suitable for its clinical application. An aptamer-modified gold electrode was developed, and MUC1 was chosen for the detection in the assay. By modifying the anti-MUC1 DNA aptamers on the surface of the gold electrode, these DNA single strands formed a hairpin structure without the MUC1. Therefore, electron transfer was facilitated between the MB and the gold electrode. However, in the presence of MUC1, the hairpin structured aptamer bonded to MUC1, and then the MB was moved away from the gold electrode, hindering the electron transfer (Fig. 10.10C) [139].

A number of other aptasensors have been developed for direct cancer cell detection. An aptamer-conjugated FRET silica NP assay for multiplexed detection of different cancer cells was reported by Chen et al. [140]. Three different cancer cells, including Toledo, CEM, and Ramos, were successfully distinguished by the aptamer-conjugated coded FRET silica NPs. Herr et al. modified magnetic NPs for extracting detected cancerous cells from the matrix [141]. CCRF-CEM cells were detected and extracted from complex mixtures, including whole blood samples, by using a combination of these two aptamer-modified NPs.

Epithelial cell adhesion molecule (EpCAM) is overexpressed in most adenocarcinomas, cancer stem cells, and metastases circulating tumor cells (CTCs) [142]. Song and coworkers screened DNA aptamers that selectively bind against human recombinant EpCAM proteins. These aptamers were able to detect live human cancer cells derived from breast, colorectal, and gastric cancers expressing EpCAM. The length of one of the selected aptamers SYL3 was optimized for improved binding ability, and the aptamer was able to detect cancerous cells from mixed cells in cell media [143].

Various aptasensors have been developed for metastatic hepatocellular carcinoma detection (selected aptamer LY-1) [144]: gastric carcinoma detection (selected aptamer S4a) [145], lung cancer detection [146], leukemia detection (selected aptamers KH1C12) [147], prostate cancer detection (RNA aptamer A10 and peptide aptamer DUP-1) [148], and glioblastoma detection (aptamers named GBM128 and GBM131) [149].

Apart from cancer, aptamers are also being used as a diagnostic tool for the detection of various diseases caused by bacteria and viruses. Various studies have reported diagnostic applications of aptamers in viral infections, for example, human immunodeficiency virus (HIV), hepatitis B virus (HBV), hepatitis C virus (HCV), severe acute respiratory syndrome (SARS), H5N1 avian influenza, and the recently spread Ebola. Hemagglutinin (HA), a well-known influenza protein, is a glycoprotein expressed in high amounts on the viral surface. It is responsible for the fusion of virus with the host cell. There are at least 18 different HA antigens; therefore, it could serve not only for infection diagnosis but also to distinguish current influenza types and subtypes. A DNA aptamer that binds to HA1 protein of the multiple influenza A virus subtype has been reported by Shiratori et al. [150]. In this study, H5N1, H1N1, and H3N2 subtypes of influenza A virus were detected. RNA aptamers P30-10-16 and A-20 that specifically bind type A and type B of HA, respectively, were generated by Gopinath et al. [151, 152]. Aptamers specific to HIV-1 Tat protein were generated by Tombelli et al. HIV-1 Tat protein was detected using biotin-streptavidin interaction, which allowed high specificity distinction between Tat and Rev (another HIV protein structurally similar to Tat) [153]. Enzyme-linked apto-sorbent assay (ELASA) was developed by Park et al. for the quantification of infectious HCV particles using aptamers against HCV E2 envelope protein [154]. An RNA aptamer specific to the surface antigen of hepatitis B virus (HBsAg) was reported by Liu and coworkers for the diagnosis of the virus [155]. Aptamer HBs-A22 was selected, which binds specifically to the HBsAg expressing hepatoma cell line HepG2.2.15 but does not bind to HBsAg-devoid HepG2 cells. Nonstructural protein 10 enzyme (nsP10) with NTPase/helicase activity is one of the few discovered proteins of coronavirus SARS. An RNA aptamer ES15 against nsP10 was generated by Jang et al. [156]. Viral enzyme activity was inhibited in a dose-dependent manner confirmed by a FRET test. Huang et al. [157] and Binning et al. [158] reported aptamers as a potential tool in Ebola diagnosis and treatment. Binning et al. constructed aptamers against VP35 protein (filoviral viral protein 35) of Ebola virus, one of the most important proteins in the virus replication complex. The selected aptamers were able to compete with viral dsRNA for binding to VP35 protein and by causing disruption in interactions between VP35 and Ebola nucleoprotein.

Rapid and sensitive diagnostic tests utilizing aptamers have been also developed for diagnosing bacterial infections. For the detection of a single cell of *Staphylococcus aureus*, Chang et al. developed a non-PCR-based method that measures the resonance light-scattering signal of aptamer-conjugated gold NPs [159]. An aptasensor with magnetic NPs was developed for the detection of *Salmonella typhimurium* and *Staphylococcus aureus* simultaneously through a luminescent bioassay. Two different aptamers were used and were immobilized on magnetic NPs for capturing the bacteria. This assay is highly sensitive to and selective for these two different types of bacteria [160]. Tuberculosis (TB) is a lethal disease spread worldwide. The 10 kDa culture filtrate protein (CFP-10) and the 6 kDa early secreted antigen target (ESAT-6) are potent T-cell antigens recognized in over 70% of TB patients. Rotherham et al. selected six aptamers against CFP-10 and ESAT-6 with K_d values in the nanomolar range [161]. Aptamers against these targets were also developed by Tang et al. [162].

10.7 Nucleic Acids in Imaging

Imaging of endogenous nucleic acids is very important for us to further our understanding of the complex biochemical processes involved in normal physiology and disease states and for opportunities for advancement in molecular biology, disease pathophysiology, drug discovery, and medical diagnostics. In this regard, RNA imaging is of utmost importance as it provides insight into the dynamics of RNA biosynthesis; in vivo RNA folding; and RNA localization, movement, processing, and cellular activities.

Among various approaches reported in the literature is fluorescent protein–based RNA imaging, where RNA-binding proteins (RBPs) tagged with optical reporters, such as green fluorescent protein (GFP), have been used to detect mRNA in living cells. The MS2 system is the prototypical RNA imaging system design introduced by Singer and colleagues [163]. However, the drawback of this approach lies in the difficulty in distinguishing the GFP-tagged protein bound to mRNA from unbound GFP causing high background fluorescence. Protein fragment complementation (PFC) is another method that has been used to improve the signal-to-background ratio of fluorescent protein–based imaging. In this approach, a marker protein is split into two inactive fragments, an N-terminal fragment and a C-terminal fragment, which cannot reassemble by themselves. However, if these fragments are fused to two proteins that interact, their binding forces the two fragments of the marker to stay in close proximity and regain functionality. The complementation of the two fluorescent protein fragments results in the regeneration of a fluorescent signal. In a study reported by Valencia-Burton et al., two RBPs were expressed in living cells, where each RBP was tagged with a fragment of a split GFP. Binding of the two RBPs to adjacent sequences of RNA brought two GFP fragments in close proximity, which led to the GFP complementation resulting in a fluorescence signal. In the absence of target mRNA, there was no fluorescent signal [164].

FISH has been the traditional method for visualizing RNA containing a known sequence. In this approach, RNA is required to be fixed in its native subcellular position through chemical cross-linking. A short exogenous fluorescent oligonucleotide is annealed to the complementary target cellular RNA sequence, and the target RNA/probe hybrid is visualized using fluorescence microscopy. FISH is highly sequence specific and can distinguish between RNA molecules differing by only a single base [165]. However, fixing the cells before hybridization followed by extensive washes of the unbound probes prevents monitoring the critical dynamic regulation aspects of RNA. Such steps were circumvented with the development of MBs (refer to Section 10.1). These probes are dually labeled at one end with a reporter fluorophore and at the other end with a quencher. In the absence of the complementary target, they form a stem-loop hairpin structure so that the fluorescence of the fluorophore is quenched. Hybridization with the target nucleic acid opens the hairpin and physically separates the fluorophore from the quencher, allowing a fluorescent signal to be emitted upon excitation. Fluorescently labeled linear oligonucleotide antisense probes have been used for imaging RNAs, where the antisense probe hybridizes to its target RNA, enabling imaging of the RNA by fluorescence microscopy. The antisense probes bind a particular RNA sequence and block the synthesis of diseased protein, thus acting as an antisense drug. These linear probes may lack detection specificity because a partial match between the probe and target sequences could induce probe hybridization to RNA molecules of multiple genes. However, backbone and/or base modifications can improve the oligonucleotide probe's affinity.

An emerging class of molecular probes based on RNA aptamers is being routinely used as a protein-free in vitro and in vivo optical imaging tool. These RNA aptamers are referred to as light-up aptamers and afford protein-free fluorescence detection of endogenous molecules. These aptamers bind with their cognate fluorogenic ligands and, in turn, activate their fluorescence. Major advantages offered by light-up aptamers over other imaging techniques (such as FISH, GFP, and MB) are that they have low-fluorescence backgrounds in the absence of RNA aptamers, small molecules are attached rather than proteins like GFP, and these aptamers possess better photobleaching properties. The malachite green aptamer (MGA) was the first RNA aptamer developed to confine the formation of radicals on RNA upon laser activation, leading to the rapid degradation of such RNA. This MGA had the ability to induce fluorescence upon binding to the otherwise nonfluorescent malachite green (MG) molecule [166]. However, the intrinsic property of MG to form radicals upon excitation makes it susceptible to cytotoxicity and has hindered the development of this aptamer for in vivo imaging [167]. In 2011, Jaffrey and coworkers reported RNA aptamer "Spinach" for an analog of the GFP fluorophore DFHBI (3,5-difluoro-4-hydroxybenzylidene imidazolinone) that could be attached to the 5S rRNA, enabling fluorescence-based visualization of this RNA in living cells. This aptamer has since been utilized to visualize small-molecule metabolites in bacterial cells and for in vitro labeling of RNA [168]. Tagging RNAs with Spinach has provided insights into the dynamics of RNA localization in cells. Because of the advantages offered by Spinach, other RNA aptamers with increased fluorescence, alternative sequences, and optimized folding properties for the cellular milieu have been identified, such as Mango and Broccoli RNA aptamers [169, 170]. These aptamers presented a significant advancement in RNA imaging technology. However, the noncovalent nature of the RNA–small molecule interaction can limit the imaging sensitivity. Sharma et al. reported an alternative approach, in which a ribozyme, selected using SELEX, is used as the recognition element. The ribozyme catalyzes self-alkylation with a fluorophore (fluorescein iodoacetamide) bearing an electrophilic reactive group. This results in a robust covalent linkage between the RNA and the fluorophore, negating the problem of signal loss from RNA fluorophore dissociation [171].

10.8 Nucleic Acid–Based Therapeutics

Nucleic acids (DNA/RNA) have been utilized for a broad range of gene therapy wherein a functional copy of a gene is added to the cells. It is based on the introduction of a gene that can produce the defective or lacking protein. On the other hand, antigene or antisense therapy involves the blocking of transcription or translation of a specific gene through hybridization with the gene of interest using short antigene/antisense oligonucleotides. Antigene oligonucleotides bind to the DNA, forming a triplex and thus inhibiting transcription, whereas an antisense oligonucleotide binds to the transcribed sense strand and silences the gene by inhibiting translation.

siRNAs and miRNAs belong to the class of antisense therapy, and they can be delivered to the target site by using a variety of different delivery vehicles, like aptamers and polymeric and metallic NPs. The basic principle behind nucleic acids as therapeutics is the inhibition of DNA or RNA expression, which works on the principle of sequence-specific recognition of nucleic acids through WC base pairing [172] that inhibits the production of abnormal protein related to the disease [173]. Nucleic acid therapeutics are used to treat diseases like inflammatory disorders, cancer, neurological disorders, and infectious diseases.

siRNAs and miRNAs are ncRNA structures. siRNAs can be used for the downregulation of disease-causing gene by RNA interference. When the dsRNA is introduced into the cytoplasm, it is cleaved by the Dicer, an RNAse III enzyme. The cleavage along the dsRNA breaks the RNA into short 21–23 nucleotide duplexes having 2–3 nucleotide overhang bases at the ends. The siRNA gets into the RNA-inducing silencing complex (RISC), which binds to the mRNA of interest. This stimulates mRNA degradation, which leads to the partial or full silencing of the gene of interest. siRNAs cannot integrate and interfere with the genome and hence are advantageous over other nucleic acid–based therapeutics, like DNAzymes and plasmid DNAs. siRNAs are chemically not very stable as they can be cleaved by the RNAse in vivo. Certain chemical modifications can be incorporated into the siRNA to make them chemically stable, which has been discussed in Section 10.3. On the other hand, miRNAs are ncRNAs (21–25 nucleotides) that take part in RNA silencing and post-transcriptional regulations. The genes responsible for specific miRNAs are transcribed to primary miRNA (pri-miRNA). The pri-miRNAs are converted into pre-miRNAs in the nucleus by RNAse III enzymes called Drosha and Pasha. These pre-miRNAs are transported to the cytoplasm by karyopherin exportin and a Ran guanosine triphosphate (Ran-GTP) complex. The Ran-GTP complex is bound to the karyopherin and forms a heterodimer with pri-miRNAs. The RNAse III enzyme Dicer processes these formed pri-miRNAs to form miRNAs. These miRNAs form an RISC and hence are involved in gene silencing due to RNA interference. miRNAs regulate the genes other than the ones that expressed them by inhibiting the translation of mRNA, whereas siRNAs regulate the same genes that expressed them, by cleaving mRNA.

Therapeutic siRNAs and miRNAs are delivered to the target site by a variety of modes, including DNA/RNA nanostructures, synthetic/metallic NPs, and aptamers. DNA/RNA-based NPs are used as the delivery vehicles as they can fold to form different 2D and 3D shapes in the nanoscale, which helps in drug delivery to the target site. More complex DNA origami nanostructures are constructed by using the M13mp18 genomic DNA sequence as the scaffold, which can be folded into different shapes via hybridization with the staple strands (Fig. 10.11). Jiang et al. have demonstrated the use of DNA nanostructures, triangle and tube-shaped nanostructures, and intercalated doxorubicin (DOX) (Fig. 10.11). Once incubated with DOX, the drug-loading nanostructures were studied in the breast cancer cell line MCF-7 cells. The study was conducted between the regular breast cancer cell line (reg-MCF-7) and the drug resistance breast cancer cell line (res-MCF-7). The study confirmed the increased drug activity with DNA origami, hence claiming that it overcame the drug resistance of the cancer cells [174]. RNA-based nanostructures are similar to DNA nanostructures, where RNA folds into nanoscale structures. RNA nanostructures have been extensively studied for their therapeutic use and diagnostic properties. Guo et al. have demonstrated constructing stable RNA multifunctional NPs, such as packaging RNA (pRNA) nanostructures, for drug delivery applications [175]. The building block of the RNA NPs are obtained from the bacteriophage phi29 DNA pRNA. This pRNA is reduced to the smaller 3WJ motif consisting of three individual RNA strands [26]. Haque et al. have used pRNA-3WJ for constructing a tetravalent NP having an MG aptamer, luciferase siRNA, folate, and the surviving siRNA. The RNA NP was tested in HT29 colon cancer cell lines. Due to the overexpression of folate on HT29 cells, the NPs were localized on the cell membrane and gene silencing was observed [176].

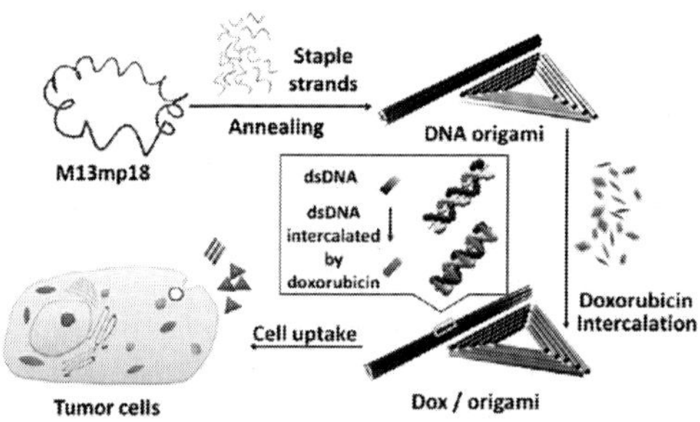

Figure 10.11 Nucleic acid–based therapeutics: Schematics of DNA nanostructure (triangle) for DOX delivery to cancer cells. Adapted with permission from Ref. [176]. Copyright (2012) American Chemical Society.

Other classes of drug delivery vehicles include metallic or polymeric NPs, which play an important role in the delivery of the therapeutic agents to the target site. Metallic NPs, like Au NPs, have been extensively used for delivering active pharmaceutical ingredients and other therapeutic molecules. The reason behind using gold is its inertness and nontoxic and biocompatible nature. Au NPs are easy to make; different sizes can be synthesized, ranging from 1 nm to 500 nm; they have a monodisperse nature; and the ideal size is maintained to be 10 nm to 100 nm for therapeutic applications. Au NPs have an affinity for the thiol group, and hence the modified siRNA aptamers can be easily attached to the Au NPs. Labala et al. have demonstrated the use of Au NPs for the codelivery of STAT3 siRNA along with imatinibmesylate (IM) for melanoma treatment. Treatment individually by STAT3 siRNA and IM showed less efficacy when compared with a combinational therapy with both. Significant reduction in the tumor volume was observed in the melanoma-bearing mice [177].

Aptamers have been extensively used as therapeutic agents for targeted drug delivery owing to their unique properties, such as low immunogenicity, high affinity, excellent specificity, thermal stability, no batch-to-batch variation, and wider range of targets. Different strategies are employed for aptamer-based therapeutics, where aptamers can act as antagonists, agonists, and aptamers as delivery agents. While acting as antagonists, aptamers can block the interaction of disease-associated targets, such as protein-protein or receptor-ligand interactions, and can disrupt the function of a pathological target protein. These inhibitory aptamers can be used directly as therapeutic agents. Many aptamers currently in clinical trials fall in this category, for example, aptamers against VEGF-165 and PDGF. Various aptamers have been developed against therapeutically relevant targets associated with several human diseases, such as cancer, degenerative neurological disease, ocular disorders, bleeding disorders, and infectious disease [178]. While acting as agonists, aptamers activate the function of target receptors, for example, RNA aptamers, against human epidermal growth factor receptor 3 (HER-3, also known as ERBB3) [179], OX40 [180], 4-1BB [181], and CD28 [182]. CD28, OX40, and 4-1BB are immune costimulatory receptors and have been engineered to act as receptor agonists for improved cancer immunotherapy [183].

While aptamers, acting as antagonists and agonists, can be used as therapeutic agents as such, aptamers can also be used as delivery vehicles to escort other therapeutic drugs to their targets. Since aptamers possess high affinity and selectivity against their targets, therapeutic compounds can be conjugated to them and can be targeted to specific cells for improved therapeutic efficacy. Many cell-specific aptamers have been conjugated with small-molecule drugs (e.g., DOX and Taxol) and oligotherapeutics (e.g., siRNAs and miRNAs). For aptamer-oligotherapeutic conjugates, McNamara et al. have reported PSMA aptamer-siRNA chimeras for cell-specific targeting and binding and delivery of functional siRNAs into cells [184]. Several variants of this chimera have also been developed to enhance cargo-loading capability [185]. For aptamer-drug conjugates, Boyacioglu et al. have reported a dimeric PSMA DNA aptamer conjugated with the antitumor chemotherapeutic drug DOX through a pH-sensitive linkage to selectively deliver DOX to PSMA-positive cells [186]. Xiang et al. have reported using the EpCAM aptamer to deliver DOX for colorectal cancer cells [187].

Apart from directly conjugating the drug to the aptamer, aptamers have also been used to decorate various nanocarriers carrying drug payload. In the case of aptamer-decorated nanomaterials as a drug delivery system, several components are integrated in one nanoscale carrier. Various nanomaterials, including Au NPs, iron oxide NPs, SWCNTs, QDs, liposomes, micelles, and dendrimers, have been assembled with cell-type-specific aptamers. One of the examples is aptamer-functionalized lipid nanoparticles (LNPs) for siRNA delivery [188].

Because of the small size of aptamers, they are susceptible to rapid excretion through renal filtration. Various aptamer derivatives have been developed to resist rapid renal clearance, in turn extending the circulation time for efficient therapeutic effect. Lee et al. have reported a cholesterol-oligonucleotide (cholODN) conjugate where the 5′-end of an aptamer was derivatized with cholesterol. The half-time of the resulting cholODN (9–11 min.) in plasma was considerably longer than the unmodified ODN (<1 min.) [189]. Derivatization of the aptamer with dialkyl lipids and *N*-hydroxysuccinimide (NHS)-ester-activated PEG has also been reported [190, 191]. Another approach for increasing the size of aptamers in order to resist fast renal clearance is to attach them with NPs of a suitable size. In other words, NPs can be decorated with the aptamers. Guo et al. reported a multifunctional RNA conjugate where they used a 3WJ RNA NP that was harbored with an epidermal growth factor receptor (EGFR) aptamer as a targeting agent, an anti-miR-21 sequence as a therapeutic agent, and a fluorescent dye (Alexa647) as an imaging agent [192].

Macugen is the only Food and Drug Administration (FDA)-approved therapeutic aptamer available. However, a number of novel aptamer-based therapeutics are currently undergoing clinical trials for treating diseases such as macular degeneration, choroidal neovascularization, intravascular thrombus, acute coronary syndrome, von Willebrand factor–related disorders, von Hippel–Lindau syndrome (VHL), angiomas, acute myeloid leukemia, renal cell carcinoma, non-small-cell lung cancer, and thrombotic thrombocytopenic purpura. Some of the aptamers that are in clinical trials are AS1411 (nucleolin) for renal cell carcinoma and leukemia, NU172 (thrombin) for heart disease, E10030 (PDGF) for age-related macular degeneration, and NOX-E36 (monocyte chemoattractant protein-1 [MCP-1]) for type 2 diabetes mellitus: albuminuria [193].

10.9 Conclusions

Nucleic acids are exquisite molecules, and apart from the central dogma, their role as FNAs has revolutionized the area of biosensing, diagnostics, medicine, and healthcare owing to their highly specific molecular recognition property, selective ligand binding, and catalysis. Their structural stability, programmability of sequences, and predictable self-assembly have made them a promising material for constructing various molecular devices and for application in theranostics. Specific interaction between complementary base pairs has also resulted in very versatile and flexible 2D and 3D nanostructures called DNA origami. The ability to engineer nucleic acid molecular devices and nanostructures to have diverse and addressable chemical compositions, shapes, sizes, and hollow or solid structures; simple loading methods for biomolecules; and biocompatibility and biodegradability has established them as new-generation drug delivery vehicles, contrast agents for imaging, and diagnostic devices.

Acknowledgment

The authors would like to thank the IISER, Tirupati, for its infrastructure and support. Deepti Sharma thanks SERB for the National Post-Doctoral Fellowship, and Prasanna Kumar Athyala thanks the IISER, Tirupati, for postdoctoral funding. The authors would like to thank the editor for giving them an opportunity to contribute to this volume.

References

1. Cech, T. R. and Steitz, J. A. (2014). The noncoding RNA revolution-trashing old rules to forge new ones, *Cell*, **157**, 77–94.
2. Slaby, O., Laga, R. and Sedlacek, O. (2017). Therapeutic targeting of non-coding RNAs in cancer, *Biochem. J.*, **474**, 4219–4251.
3. Seeman, N. C. (2003). DNA in a material world, *Nature*, **421**, 427–431.
4. Chelyapov, N., et al. (2004). DNA triangles and self-assembled hexagonal tilings, *J. Am. Chem. Soc.*, **126**, 13924–13925.
5. Seeman, N. C. (2005). Structural DNA nanotechnology: an overview, *Methods Mol. Biol.*, **303**, 143–166.
6. Guo, P. (2010). The emerging field of RNA nanotechnology, *Nat. Nanotechnol.*, **5**, 833–842.

7. Seeman, N. C. (2005). From genes to machines: DNA nanomechanical devices, *Trends Biochem. Sci.*, **30**, 119–125.

8. Szewczak, A. A., et al. (1998). A minor groove RNA triple helix within the catalytic core of a group I intron, *Nat. Struct. Biol.*, **5**, 1037–1042.

9. Boudvillain, M., de Lencastre, A. and Pyle, A. M. (2000). A tertiary interaction that links active-site domains to the 5′ splice site of a group II intron, *Nature*, **406**, 315–318.

10. Kim, N. K., et al. (2008). Solution structure and dynamics of the wild-type pseudoknot of human telomerase RNA, *J. Mol. Biol.*, **384**, 1249–1261.

11. Varani, G. and McClain, W. H. (2000). The G•U wobble base pair: a fundamental building block of RNA structure crucial to RNA function in diverse biological systems, *EMBO Rep.*, **1**, 18–23.

12. Hingerty, B., Brown, R. S. and Jack, A. (1978). Further refinement of the structure of yeast tRNAPhe, *J. Mol. Biol.*, **124**, 523–534.

13. Giege, R., Moras, D. and Thierry, J. C. (1977). Yeast transfer RNAasp: a new high-resolution X-ray diffracting crystal form of a transfer RNA, *J. Mol. Biol.*, **115**, 91–96.

14. Berg, P., et al. (1965). Competition between RNA polymerase and DNA polymerase for the DNA template, *Biochem. Biophys. Res. Commun.*, **18**, 932–942.

15. Kool, E. T. (2001). Hydrogen bonding, base stacking, and steric effects in DNA replication, *Annu. Rev. Biophys. Biomol. Struct.*, **30**, 1–22.

16. Shi, K., Pan, B. and Sundaralingam, M. (2003). The crystal structure of an alternating RNA heptamerr(GUAUACA) forming a six base-paired duplex with 3′-end adenine overhangs. *Nucleic Acids Res.*, **31**, 1392–1397.

17. Conte, M. R., et al. (1996). Hydration of the RNA duplex r(CGCAAAUUUGCG)2 determined by NMR. *Nucleic Acids Res.*, **24**, 3693–3699.

18. Watts, J. K., Deleavey, G. F. and Damha, M. J. (2008). Chemically modified siRNA: tools and applications, *Drug Discov. Today*, **13**, 842–855.

19. Shaw, B. R., et al. (2008). Boranophosphate siRNA-aptamer chimeras for tumor-specific downregulation of cancer receptors and modulators, *Nucleic Acids SympSer (Oxf)*, **52**, 655–656.

20. Mathé, C. and Périgaud, C. (2008). Recent approaches in the synthesis of conformationally restricted nucleoside analogues, *Eur. J. Org. Chem.*, **2008**, 1489–1505.

21. Boulmé, F., et al. (1998). Modified (PNA, 2′-O-methyl and phosphoramidate) anti-TAR antisense oligonucleotides as strong and specific inhibitors of in vitro HIV-1 reverse transcription, *Nucleic Acids Res.*, **26**, 5492–5500.

22. Madhuri, V. and Kumar, V. A. (2010). Design, synthesis and DNA/RNA binding studies of nucleic acids comprising stereoregular and acyclic polycarbamate backbone: polycarbamate nucleic acids (PCNA), *Org. Biomol. Chem.*, **8**, 3734–3741.

23. Hong, J., et al. (2010). Comprehensive analysis of sequence-specific stability of siRNA, *FASEB J.*, **24**, 4844–4855.

24. Liu, J., et al. (2011). Fabrication of stable and RNase-resistant RNA nanoparticles active in gearing the nanomotors for viral DNA packaging, *ACS Nano*, **5**, 237–246.

25. Haque, F., et al. (2012). Ultrastable synergistic tetravalent RNA nanoparticles for targeting to cancers, *Nano Today*, **7**, 245–257.

26. Shu, Y., et al. (2013). Fabrication of 14 different RNA nanoparticles for specific tumor targeting without accumulation in normal organs, *RNA*, **19**, 767–777.

27. Han, H. and Hurley, L. H. (2000). G-quadruplex DNA: a potential target for anti-cancer drug design, *Trends Pharmacol. Sci.*, **21**, 136–142.

28. Bhasikuttan, A. C. and Mohanty, J. (2015). Targeting G-quadruplex structures with extrinsic fluorogenic dyes: promising fluorescence sensors, *Chem. Commun. (Camb)*, **51**, 7581–7597.

29. Bochman, M. L., Paeschke, K. and Zakian, V. A. (2012). DNA secondary structures: stability and function of G-quadruplex structures, *Nat. Rev. Genet.*, **13**, 770–780.

30. Song, G. and Ren, J. [2010]. Recognition and regulation of unique nucleic acid structures by small molecules, *Chem. Commun.*, **46**, 7283–7294.

31. Liu, X., et al. (2015). Thioflavin T as an efficient G-quadruplex inducer for the highly sensitive detection of thrombin using a new Föster resonance energy transfer system, *ACS Appl. Mater. Interfaces*, **7**, 16458–16465.

32. Lu, L., et al. (2016). Development of a luminescent G-quadruplex-selective iridium(III) complex for the label-free detection of adenosine, *Sci. Rep.*, **6**, 19368.

33. Fu, B., et al. (2015). Label-free detection of pH based on the i-motif using an aggregation-caused quenching strategy, *Chem. Commun.*, **51**, 16960–16963.

34. Ma, W., et al. (2018). Hairpin-contained i-motif based fluorescent ratiometric probe for high-resolution and sensitive response of small pH variations, *Anal. Chem.*, **90**, 1889–1896.

35. Reif, J. H. (2003). The design of autonomous DNA nano-mechanical devices: walking and rolling DNA, *Nat. Comput.*, **2**, 439–461.

36. Li, W., Wang, L. and Jiang, W. (2017). A catalytic assembled enzyme-free three-dimensional DNA walker and its sensing application, *Chem. Commun.*, **53**, 5527–5530.

37. Wang, L., Deng, R. and Li, J. (2015). Target-fueled DNA walker for highly selective miRNA detection, *Chem. Sci.*, **6**, 6777–6782.

38. Clark, L. C. and Lyons, C. (1962). Electrode systems for continuous monitoring in cardiovascular surgery, *Ann. N.Y. Acad. Sci.*, **102**, 29–45.

39. Park, S. J., Taton, T. A. and Mirkin, C. A. (2002). Array-based electrical detection of DNA with nanoparticle probes, *Science*, **295**, 1503–1506.

40. Taton, T. A., Mirkin, C. A. and Letsinger, R. L. (2000). Scanometric DNA array detection with nanoparticle probes, *Science*, **289**, 1757–1760.

41. Gaylord, B. S., Heeger, A. J. and Bazan, G. C. (2002). DNA detection using water-soluble conjugated polymers and peptide nucleic acid probes, *Proc. Natl. Acad. Sci. U.S.A.*, **99**, 10954–10957.

42. Xu, H., et al. (2005). Magnetically assisted DNA assays: high selectivity using conjugated polymers for amplified fluorescent transduction, *Nucleic Acids Res.*, **33**, e83–e83.

43. Cooper, M. A., et al. (2001). Direct and sensitive detection of a human virus by rupture event scanning, *Nat. Biotechnol.*, **19**, 833–837.

44. Erdem, A., et al. (2000). Novel hybridization indicator methylene blue for the electrochemical detection of short DNA sequences related to the hepatitis B virus, *Anal. Chim. Acta*, **422**, 139–149.

45. Campbell, C. N., et al. (2002). Enzyme-amplified amperometric sandwich test for RNA and DNA, *Anal. Chem.*, **74**, 158–162.

46. Prabhakar, N., et al. (2007). DNA entrapped polypyrrole-polyvinyl sulfonate film for application to electrochemical biosensor, *Anal. Biochem.*, **366**, 71–79.

47. Wang, J. (2000). From DNA biosensors to gene chips, *Nucleic Acids Res.*, **28**, 3011–3016.

48. Herne, T. M. and Tarlov, M. J. (1997). Characterization of DNA probes immobilized on gold surfaces, *J. Am. Chem. Soc.*, **119**, 8916–8920.

49. Sperling, R. A., et al. (2006). Electrophoretic separation of nanoparticles with a discrete number of functional groups, *Adv. Funct. Mater.*, **16**, 943–948.

50. Piunno, P. A. E., et al. (1994). Fiber optic biosensor for fluorimetric detection of DNA hybridization, *Anal. Chim. Acta*, **288**, 205–214.

51. Tan, W., Wang, K. and Drake, T. J. (2004). Molecular beacons, *Curr. Opin. Chem. Biol.*, **8**, 547–553.

52. Zhou, H., Yang, C., Chen, H., Li, X., Li, Y. and Fan, X. (2017). G-quadruplex molecular beacon-based biosensor for highly selective detection of microRNA, *Biosens. Bioelectron.*, **87**, 552–557.

53. Zhang, C. Y., et al. (2005). Single-quantum-dot-based DNA nanosensor, *Nat. Mater.*, **4**, 826–831.

54. Wang, R., et al. (2004). Immobilisation of DNA probes for the development of SPR-based sensing, *Biosens. Bioelectron.*, **20**, 967–974.

55. Rella, R., et al. (2004). Liquid phase SPR imaging experiments for biosensors applications, *Biosens. Bioelectron.*, **20**, 1140–1148.

56. Ngo, H. T., et al. (2013). Label-free DNA biosensor based on SERS molecular sentinel on nanowave chip, *Anal. Chem.*, **85**, 6378–6383.

57. He, P., Xu, Y. and Fang, Y. (2005). Applications of carbon nanotubes in electrochemical DNA biosensors, *Microchim. Acta*, **152**, 175–186.

58. Lazerges, M., et al. (2006). Oligonucleotide quartz crystal microbalance sensor for the microalgae Alexandriumminutum (Dinophyceae), *Biosens. Bioelectron.*, **21**, 1355–1358.

59. Doudna, J. A. and Cech, T. R. (2002). The chemical repertoire of natural ribozymes, *Nature*, **418**, 222–228.

60. Wilson, D. S. and Szostak, J. W. (1999). In vitro selection of functional nucleic acids, *Annu. Rev. Biochem.*, **68**, 611–647.

61. Achenbach, J. C., et al. (2004). DNAzymes: from creation in vitro to application in vivo, *Curr. Pharm. Biotechnol.*, **5**, 321–336.

62. Silverman, S. K. (2008). Catalytic DNA (deoxyribozymes) for synthetic applications-current abilities and future prospects, *Chem. Commun.*, **30**, 3467–3485.

63. Szostak, J. W. (1986). Enzymatic activity of the conserved core of a group I self-splicing intron, *Nature*, **322**, 83–86.

64. Symons, R. H. (1992). Small catalytic RNAs, *Annu. Rev. Biochem.*, **61**, 641–671.

65. Nissen, P., et al. (2000). The structural basis of ribosome activity in peptide bond synthesis, *Science*, **289**, 920–930.

66. Seelig, B. and Jaschke, A. (1999). A small catalytic RNA motif with Diels-Alderase activity, *Chem. Biol.*, **6**, 167–176.

67. Fusz, S., et al. (2005). A ribozyme for the aldol reaction, *Chem. Biol.*, **12**, 941–950.

68. Sengle, G., et al. (2001). Novel RNA catalysts for the Michael reaction, *Chem. Biol.*, **8**, 459–473.

69. Unrau, P. J. and Bartel, D. P. (1998). RNA-catalysed nucleotide synthesis, *Nature*, **395**, 260–263.

70. Lee, N., et al. (2000). Ribozyme-catalyzed tRNAaminoacylation, *Nat. Struct. Biol.*, **7**, 28–33.

71. Willner, I., et al. (2008). DNAzymes for sensing, nanobiotechnology and logic gate applications, *Chem. Soc. Rev.*, **37**, 1153–1165.

72. Carmi, N., Balkhi, S. R. and Breaker, R. R. (1998). Cleaving DNA with DNA, *Proc. Natl. Acad. Sci. U.S.A.*, **95**, 2233–2237.

73. Li, J. and Lu, Y. (2000). A highly sensitive and selective catalytic DNA biosensor for lead ions, *J. Am. Chem. Soc.*, **122**, 10466–10467.

74. Santoro, S. W., et al. (2000). RNA cleavage by a DNA enzyme with extended chemical functionality, *J. Am. Chem. Soc.*, **122**, 2433–2439.

75. Liu, J., et al. (2007). A catalytic beacon sensor for uranium with parts-per-trillion sensitivity and millionfold selectivity, *Proc. Natl. Acad. Sci. U.S.A.*, **104**, 2056–2061.

76. Tang, J. and Breaker, R. R. (1997). Rational design of allosteric ribozymes, *Chem. Biol.*, **4**, 453–459.

77. Tuerk, C. and Gold, L. (1990). Systematic evolution of ligands by exponential enrichment: RNA ligands to bacteriophage T4 DNA polymerase, *Science*, **249**, 505–510.

78. Ellington, A. D. and Szostak, J. W. (1990). In vitro selection of RNA molecules that bind specific ligands, *Nature*, **346**, 818–822.

79. Han, K., et al. (2009). Target induced dissociation (TID) strategy for the development of electrochemical aptamer-based biosensor, *Electrochem. Commun.*, **11**, 157–160.

80. Liu, J. and Lu, Y. (2006). Fast colorimetric sensing of adenosine and cocaine based on a general sensor design involving aptamers and nanoparticles, *Angew. Chem. Int. Ed.*, **45**, 90–94.

81. Zhao, W., et al. (2008). DNA aptamer folding on gold nanoparticles: from colloid chemistry to biosensors, *J. Am. Chem. Soc.*, **130**, 3610–3618.

82. Ferapontova, E. E., Olsen, E. M. and Gothelf, K. V. (2008). An RNA aptamer-based electrochemical biosensor for detection of theophylline in serum, *J. Am. Chem. Soc.*, **130**, 4256–4258.

83. Xiao, Y., et al. (2005). Label-free electronic detection of thrombin in blood serum by using an aptamer-based sensor, *Angew. Chem. Int. Ed.*, **44**, 5456–5459.

84. Wang, J., et al. (2009). Combination of aptamer with gold nanoparticles for electrochemical signal amplification: Application to sensitive detection of platelet-derived growth factor, *Biosens. Bioelectron.*, **24**, 1598–1602.

85. (A) Hamaguchi, N., Ellington, A. and Stanton, M. (2001). Aptamer beacons for the direct detection of proteins, *Anal. Biochem.*, **294**, 126–131. (B) Liu J., Cao Z. and Lu Y. (2009). Functional nucleic acid sensors, *Chem. Rev.*, **109**, 1948–1999.

86. Stojanovic, M. N., de Prada, P. and Landry, D. W. (2001). Aptamer-based folding fluorescent sensor for cocaine, *J. Am. Chem. Soc.*, **123**, 4928–4931.

87. Yamamoto-Fujita, R. and Kumar, P. K. (2005). Aptamer-derived nucleic acid oligos: applications to develop nucleic acid chips to analyze proteins and small ligands, *Anal. Chem.*, **77**, 5460–5466.

88. Stojanovic, M. N., de Prada, P. and Landry, D. W. (2000). Fluorescent sensors based on aptamer self-assembly, *J. Am. Chem. Soc.*, **122**, 11547–11548.

89. Sharma, A. K. and Heemstra, J. M. (2011). Small-molecule-dependent split aptamer ligation, *J. Am. Chem. Soc.*, **133**, 12426–12429.

90. Nutiu, R. and Li, Y. (2003). Structure-switching signaling aptamers, *J. Am. Chem. Soc.*, **125**, 4771–4778.

91. Nutiu, R. and Li, Y. (2005). In vitro selection of structure-switching signaling aptamers, *Angew. Chem. Int. Ed. Engl.*, **44**, 1061–1065.

92. Tang, Z., et al. (2008). Aptamer switch probe based on intramolecular displacement, *J. Am. Chem. Soc.*, **130**, 11268–11269.

93. Rankin, C. J., et al. (2006). A simple fluorescent biosensor for theophylline based on its RNA aptamer, *Nucleosides Nucleotides Nucleic Acids*, **25**, 1407–1424.

94. Wang, W., et al. (2008). Aptamer biosensor for protein detection using gold nanoparticles, *Anal. Biochem.*, **373**, 213–219.

95. Levy, M., Cater, S. F. and Ellington, A. D. (2005). Quantum-dot aptamer beacons for the detection of proteins, *Chem. Bio. Chem.*, **6**, 2163–2166.

96. Li, H. and Rothberg, L. (2004). Colorimetric detection of DNA sequences based on electrostatic interactions with unmodified gold nanoparticles, *Proc. Natl. Acad. Sci. U.S.A.*, **101**, 14036–14039.

97. Li, H. and Rothberg, L. J. (2004). Label-free colorimetric detection of specific sequences in genomic DNA amplified by the polymerase chain reaction, *J. Am. Chem. Soc.*, **126**, 10958–10961.

98. Wang, J., et al. (2007). A gold nanoparticle-based aptamer target binding readout for ATP assay, *Adv. Mater.*, **19**, 3943–3946.

99. Sharma, A. K., Kent, A. D. and Heemstra, J. M. (2012). Enzyme-linked small-molecule detection using split aptamer ligation, *Anal. Chem.*, **84**, 6104–6109.

100. Radi, A. E., et al. (2006). Reagentless, reusable, ultrasensitive electrochemical molecular beacon aptasensor, *J. Am. Chem. Soc.*, **128**, 117–124.

101. Baker, B. R., et al. (2006). An electronic, aptamer-based small-molecule sensor for the rapid, label-free detection of cocaine in adulterated samples and biological fluids, *J. Am. Chem. Soc.*, **128**, 3138–3139.

102. Zuo, X., et al. (2007). A target-responsive electrochemical aptamer switch (TREAS) for reagentless detection of nanomolar ATP, *J. Am. Chem. Soc.*, **129**, 1042–1043.

103. Wang, J. and Zhou, H. S. (2008). Aptamer-based Au nanoparticles-enhanced surface plasmon resonance detection of small molecules, *Anal. Chem.*, **80**, 7174–7178.

104. Verhelst, S. H. L., et al. (2004). Surface plasmon resonance evaluation of various aminoglycoside–RNA hairpin interactions reveals low degree of selectivity, *Chem. Bio. Chem.*, **5**, 937–942.

105. Gebhardt, K., et al. (2000). RNA aptamers to S-adenosylhomocysteine: kinetic properties, divalent cation dependency, and comparison with anti-S-adenosylhomocysteine antibody, *Biochemistry*, **39**, 7255–7265.

106. Kawakami, J., et al. (2000). In vitro selection of aptamers that act with Zn^{2+}, *J. Inorg. Biochem.*, **82**, 197–206.

107. Fukusaki, E., et al. (2001). SELEX for tubulin affords specific T-rich DNA aptamers. Systematic evolution of ligands by exponential enrichment, *Bioorg. Med. Chem. Lett.*, **11**, 2927–2930.

108. Murphy, M. B., et al. (2003). An improved method for the in vitro evolution of aptamers and applications in protein detection and purification, *Nucleic Acids Res.*, **31**, e110.

109. Liss, M., et al. (2002). An aptamer-based quartz crystal protein biosensor, *Anal. Chem.*, **74**, 4488–4495.

110. Minunni, M., et al. (2004). Development of biosensors with aptamers as bio-recognition element: the case of HIV-1 Tat protein, *Biosens. Bioelectron.*, **20**, 1149–1156.

111. Hianik, T., et al. (2005). Detection of aptamer-protein interactions using QCM and electrochemical indicator methods, *Bioorg. Med. Chem. Lett.*, **15**, 291–295.

112. Josse, F., Bender, F. and Cernosek, R. W. (2001). Guided shear horizontal surface acoustic wave sensors for chemical and biochemical detection in liquids, *Anal. Chem.*, **73**, 5937–5944.

113. Schlensog, M. D., et al. (2004). A love-wave biosensor using nucleic acids as ligands, *Sens. Actuators, B*, **101**, 308–315.

114. Gronewold, T. M. A., et al. (2005). Monitoring complex formation in the blood-coagulation cascade using aptamer-coated SAW sensors, *Biosens. Bioelectron.*, **20**, 2044–2052.

115. Fang, L., et al. (2008). A electrochemiluminescence aptasensor for detection of thrombin incorporating the capture aptamer labeled with gold nanoparticles immobilized onto the thio-silanized ITO electrode, *Anal. Chim. Acta*, **628**, 80–86.

116. Kang, Y., et al. (2008). Electrochemical detection of thrombin by sandwich approach using antibody and aptamer, *Bioelectrochemistry*, **73**, 76–81.

117. Wang, J., et al. (2009). Combination of aptamer with gold nanoparticles for electrochemical signal amplification: application to sensitive detection of platelet-derived growth factor, *Biosens. Bioelectron.*, **24**, 1598–1602.

118. Zhang, S., Xia, J. and Li, X. (2008). Electrochemical biosensor for detection of adenosine based on structure-switching aptamer and amplification with reporter probe DNA modified Au nanoparticles, *Anal. Chem.*, **80**, 8382–8388.

119. Yoshizumi, J., et al. (2008). Target-induced strand release (TISR) from aptamer-DNA duplex: a general strategy for electronic detection of biomolecules ranging from a small molecule to a large protein, *Analyst*, **133**, 323–325.

120. Fan, H., et al. (2008). An electrochemical aptasensor for detection of thrombin based on target protein-induced strand displacement, *Electroanalysis*, **20**, 2113–2117.

121. Lu, Y., et al. (2008). Aptamer-based electrochemical sensors with aptamer-complementary DNA oligonucleotides as probe, *Anal. Chem.*, **80**, 1883–1890.

122. Han, K., et al. (2009). Target induced dissociation (TID) strategy for the development of electrochemical aptamer based biosensor, *Electrochem. Commun.*, **11**, 157–160.

123. DeLong, E. F., Wickham, G. S., Pace, N. R. (1989). Phylogenetic stains: ribosomal RNA-based probes for the identification of single cells, *Science*, **243**, 1360–1363.

124. Bustin, S. (2002). Quantification of mRNA using real-time reverse transcription PCR (RT-PCR): trends and problems, *J. Mol. Endocrinol.*, **29**, 23–39.

125. McKillip, J. L. and Drake, M. (2004). Real-time nucleic acid-based detection methods for pathogenic bacteria in food, *J. Food Prot.*, **67**, 823–832.

126. Singh, A., Poshtiban, S. and Evoy, S. (2013). Recent advances in bacteriophage based biosensors for food-borne pathogen detection, *Sensors (Basel)*, **13**, 1763–1786.

127. Proctor, I., Stoeber, K. and Williams, G. H. (2010). Biomarkers in bladder cancer, *Histopathology*, **57**, 1–13.

128. Shin, Y., Perera, A. P. and Park, M. K. (2013). Label-free DNA sensor for detection of bladder cancer biomarkers in urine, *Sens. Actuators, B*, **178**, 200–206.

129. Shangguan, D., et al. (2008). Cell-specific aptamer probes for membrane protein elucidation in cancer cells, *J. Proteome Res.*, **7**, 2133–2139.

130. Mukhopadhyay, P., et al. (2011). Mucins in the pathogenesis of breast cancer: implications in diagnosis, prognosis and therapy, *Biochim. Biophys. Acta*, **1815**, 224–240.

131. Ghosh, S., et al. (2008). High levels of vascular endothelial growth factor and its receptors (VEGFR-1, VEGFR-2, neuropilin-1) are associated with worse outcome in breast cancer, *Hum. Pathol.*, **39**, 1835–1843.

132. Zhao, J., et al. (2012). A "signal-on" electrochemical aptasensor for simultaneous detection of two tumor markers, *Biosens. Bioelectron.*, **34**, 249–252.

133. Lupold, S. E., et al. (2002). Identification and characterization of nuclease-stabilized RNA molecules that bind human prostate cancer cells via the prostate-specific membrane antigen, *Cancer Res.*, **62**, 4029–4033.

134. Gutsaeva, D. R., et al. (2011). Inhibition of cell adhesion by anti-P-selectin aptamer: a new potential therapeutic agent for sickle cell disease, *Blood*, **117**, 727–735.

135. Townshend, B., et al. (2010). An RNA aptamer that selectively inhibits the enzymatic activity of protein tyrosine phosphatase 1B in vitro, *ChemBioChem*, **11**, 1583–1593.

136. Wang, Z. and Lu, Y. (2009). Functional DNA directed assembly of nanomaterials for biosensing, *J. Mater. Chem.*, **19**, 1788–1798.

137. Chen, T., et al. (2011). Aptamer-conjugated nanomaterials for bioanalysis and biotechnology applications, *Nanoscale*, **3**, 546–556.

138. Zhu, Y., Chandra, P. and Shim, Y. B. (2013). Ultrasensitive and selective electrochemical diagnosis of breast cancer based on a hydrazine–Au nanoparticle–aptamer bioconjugate, *Anal. Chem.*, **85**, 1058–1064.

139. Ma, F., et al. (2013). Immobilization of redox-labeled hairpin DNA aptamers on gold: electrochemical quantitation of epithelial tumor marker mucin 1, *Electrochim. Acta*, **110**, 139–145.

140. Chen, X., et al. (2009). Using aptamer-conjugated fluorescence resonance energy transfer nanoparticles for multiplexed cancer cell monitoring, *Anal. Chem.*, **81**, 7009–7014.

141. Herr, J. K., et al. (2006). Aptamer-conjugated nanoparticles for selective collection and detection of cancer cells, *Anal. Chem.*, **78**, 2918–2924.

142. Bellone, S., et al. (2009). Overexpression of epithelial cell adhesion molecule in primary, metastatic, and recurrent/chemotherapy-resistant epithelial ovarian cancer: implications for epithelial cell adhesion molecule-specific immunotherapy, *Int. J. Gynecol. Cancer*, **19**, 860–866.

143. Song, Y., et al. (2013). Selection of DNA aptamers against epithelial cell adhesion molecule for cancer cell imaging and circulating tumor cell capture, *Anal. Chem.*, **85**, 4141–4149.

144. Wang, F.-B., et al. (2013). Recognition and capture of metastatic hepatocellular carcinoma cells using aptamer-conjugated quantum dots and magnetic particles, *Biomaterials*, **34**, 3816–3827.

145. Ding, F., et al. (2015). Diagnostic applications of gastric carcinoma cell aptamers in vitro and in vivo, *Talanta*, **134**, 30–36.

146. Kunii, T., et al. (2011). Selection of DNA aptamers recognizing small cell lung cancer using living cell-SELEX, *Analyst*, **136**, 1310–1312.

147. Sefah, K., et al. (2009). Molecular recognition of acute myeloid leukemia using aptamers, *Leukemia*, **23**, 235–244.

148. Min, K., et al. (2010). Simultaneous electrochemical detection of both PSMA (+) and PSMA (-) prostate cancer cells using an RNA/peptide dual-aptamer probe, *Chem. Commun. (Camb)*, **46**, 5566–5568.

149. Kang, D., et al. (2012). Selection of DNA aptamers against glioblastoma cells with high affinity and specificity, *PLoS One*, **7**, e42731.

150. Shiratori, I., et al. (2014). Selection of DNA aptamers that bind to influenza A viruses with high affinity and broad subtype specificity, *Biochem. Biophys. Res. Commun.*, **443**, 37–41.

151. Gopinath, S. C., Kawasaki, K. and Kumar, P. K. (2005). Selection of RNA-aptamer against human influenza B virus, *Nucleic Acids SympSer (Oxf)*, **49**, 85–86.

152. Gopinath, S. C., et al. (2006). An RNA aptamer that distinguishes between closely related human influenza viruses and inhibits haemagglutinin-mediated membrane fusion, *J. Gen. Virol.*, **87**, 479–487.

153. Tombelli, S., et al. (2005). Aptamer-based biosensors for the detection of HIV-1 Tat protein, *Bioelectrochemistry*, **67**, 135–141.

154. Park, J. H., et al. (2013). Infectivity of hepatitis C virus correlates with the amount of envelope protein E2: development of a new aptamer-based assay system suitable for measuring the infectious titer of HCV, *Virology*, **439**, 13–22.

155. Liu, J., et al. (2010). Development of HBsAg-binding aptamers that bind HepG2.2.15 cells via HBV surface antigen, *Virol. Sin.*, **25**, 27–35.

156. Jang, K. J., et al. (2008). Isolation of inhibitory RNA aptamers against severe acute respiratory syndrome (SARS) coronavirus NTPase/Helicase, *Biochem. Biophys. Res. Commun.*, **366**, 738–744.

157. Huang, Z., Wang, X. and Gao, G. (2010). Analyses of SELEX-derived ZAP-binding RNA aptamers suggest that the binding specificity is determined by both structure and sequence of the RNA, *Protein Cell.*, **1**, 752–759.

158. Binning, J. M., et al. (2013). Development of RNA aptamers targeting Ebola virus VP35, *Biochemistry*, **52**, 8406–8419.

159. Chang, Y. C., et al. (2013). Rapid single cell detection of Staphylococcus aureus by aptamer-conjugated gold nanoparticles, *Sci. Rep.*, **3**, 1863.

160. Duan, N., et al. (2012). Dual-color upconversion fluorescence and aptamer-functionalized magnetic nanoparticles-based bioassay for the simultaneous detection of Salmonella Typhimurium and Staphylococcus aureus, *Anal. Chim. Acta*, **723**, 1–6.

161. Rotherham, L. S., et al. (2012). Selection and application of ssDNA aptamers to detect active TB from sputum samples, *PLoS One*, **7**, e46862.

162. Tang, X. L., et al. (2014). CFP10 and ESAT6 aptamers as effective Mycobacterial antigen diagnostic reagents, *J. Infect.*, **69**, 569–580.

163. Bertrand, E., et al. (1998). Localization of ASH1 mRNA particles in living yeast, *Mol. Cell.*, **2**, 437–445.

164. Valencia-Burton, M., et al. (2007). RNA visualization in live bacterial cells using fluorescent protein complementation, *Nat. Methods*, **4**, 421–427.

165. Larsson, C., et al. (2010). In situ detection and genotyping of individual mRNA molecules, *Nat. Methods*, **7**, 395–397.

166. Grate, D. and Wilson, C (1999). Laser-mediated, site-specific inactivation of RNA transcripts, *Proc. Natl. Acad. Sci. U.S.A.*, **96**, 6131.

167. Kraus, G. A., et al. (2008). Fluorinated analogs of malachite green: synthesis and toxicity, *Molecules*, **13**, 986–994.

168. Paige, J. S., Wu, K. Y. and Jaffrey, S. R. (2011). RNA mimics of green fluorescent protein, *Science*, **333**, 642–646.

169. Dolgosheina, E. V., et al. (2014). RNA mango aptamer-fluorophore: a bright, high-affinity complex for RNA labeling and tracking, *ACS Chem. Biol.*, **9**, 2412–2420.

170. Filonov, G. S., et al. (2014). Broccoli: rapid selection of an RNA mimic of green fluorescent protein by fluorescence-based selection and directed evolution, *J. Am. Chem. Soc.*, **136**, 16299–16308.

171. Sharma, A. K., et al. (2014). Fluorescent RNA labeling using self-alkylating ribozymes, *ACS Chem. Biol.*, **9**, 1680–1684.

172. Cavagnari, B. M. (2011). Gene therapy: nucleic acids as drugs. Action mechanisms and delivery into the cell, *Arch. Argent. Pediatr.*, **109**, 237–244.

173. Wraight, C. J. and White, P. J. (2001). Antisense oligonucleotides in cutaneous therapy, *Pharmacol. Ther.*, **90**, 89–104.

174. Jiang, Q., et al. (2012). DNA origami as a carrier for circumvention of drug resistance, *J. Am. Chem. Soc.*, **134**, 13396–13403.

175. Shu, D., et al. (2011). Thermodynamically stable RNA three-way junctions as platform for constructing multi-functional nanoparticles for delivery of therapeutics, *Nat. Nanotechnol.*, **6**, 658–667.

176. Haque, F., et al. (2012). Ultrastable synergistic tetravalent RNA nanoparticles for targeting to cancers, *Nano Today*, **7**, 245–257.

177. Labala, S., et al. (2017). Effective melanoma cancer suppression by iontophoretic co-delivery of STAT3 siRNA and imatinib using gold nanoparticles, *Int. J. Pharm.*, **525**, 407–417.

178. Shum, K. T., Zhou, J. and Rossi, J. J. (2013). Aptamer-based therapeutics: new approaches to combat human viral diseases, *Pharmaceuticals (Basel)*, **6**, 1507–1542.

179. Chen, C. H., et al. (2003). Inhibition of heregulin signaling by an aptamer that preferentially binds to the oligomeric form of human epidermal growth factor receptor-3, *Proc. Natl. Acad. Sci. U.S.A.*, **100**, 9226–9231.

180. Pratico, E. D., Sullenger, B. A. and Nair, S. K. (2013). Identification and characterization of an agonistic aptamer against the T cell costimulatory receptor, OX40, *Nucleic Acid Ther.*, **23**, 35–43.

181. McNamara, J. O., et al. (2008). Multivalent 4-1BB binding aptamerscostimulate CD8+ T cells and inhibit tumor growth in mice, *J. Clin. Invest.*, **118**, 376–386.

182. Pastor, F., et al. (2013). CD28 aptamers as powerful immune response modulators, *Mol. Ther. Nucleic Acids*, **2**, e98.

183. Gilboa, E., McNamara, J. and Pastor, F. (2013). Use of oligonucleotide aptamer ligands to modulate the function of immune receptors, *Clin. Cancer Res.*, **19**, 1054–1062.

184. McNamara Ii, J. O., et al. (2006). Cell type–specific delivery of siRNAs with aptamer-siRNA chimeras, *Nat. Biotechnol.*, **24**, 1005.

185. Dassie, J. P., et al. (2009). Systemic administration of optimized aptamer-siRNA chimeras promotes regression of PSMA-expressing tumors, *Nat. Biotechnol.*, **27**, 839–849.

186. Boyacioglu, O., et al. (2013). Dimeric DNA aptamer complexes for high-capacity-targeted drug delivery using pH-sensitive covalent linkages, *Mol. Ther. Nucleic Acids*, **2**, e107.

187. Xiang, D., et al. (2017). Transforming doxorubicin into a cancer stem cell killer via EpCAM aptamer-mediated delivery, *Theranostics*, **7**, 4071–4086.

188. Liang, C., et al. (2015). Aptamer-functionalized lipid nanoparticles targeting osteoblasts as a novel RNA interference–based bone anabolic strategy, *Nat. Med.*, **21**, 288.

189. Lee, C. H., et al. (2015). Pharmacokinetics of a cholesterol-conjugated aptamer against the hepatitis C virus (HCV) NS5B protein, *Mol. Ther. Nucleic Acids*, **4**, e254.

190. Green, L. S., et al. (1995). Nuclease-resistant nucleic acid ligands to vascular permeability factor/vascular endothelial growth factor, *Chem. Biol.*, **2**, 683–695.

191. Hoffmann, S., et al. (2011). RNA aptamers and spiegelmers: synthesis, purification, and post-synthetic PEG conjugation, *Curr. Protoc. Nucleic Acid Chem.*, Chapter 4:Unit 4.46.1–30.

192. Shu, D., et al. (2015). Systemic delivery of anti-miRNA for suppression of triple negative breast cancer utilizing RNA nanotechnology, *ACS Nano*, **9**, 9731–9740.

193. Sundaram, P., et al. (2013). Therapeutic RNA aptamers in clinical trials, *Eur. J. Pharm. Sci.*, **48**, 259–271.

Chapter 11

DNA-Based Nanoswitches and Devices

Bappaditya Roy,* Madhu Ramesh,* and Thimmaiah Govindaraju
Bioorganic Chemistry Laboratory, New Chemistry Unit,
Jawaharlal Nehru Centre for Advanced Scientific Research,
Jakkur PO, Bengaluru 560064, Karnataka, India
tgraju@jncasr.ac.in

11.1 Introduction

The unique molecular recognition of DNA strands through complementary nucleobases enables the programmable self-assembly to create diverse DNA architectures. In this context, the concept of DNA nanotechnology, introduced by Seeman, deals with developing diverse structural architectures and systems, materials, and devices with at least one dimension in the nanoregime (1 to 100 nm) [1, 2]. The physical properties of single strands of DNA, unlike other natural or synthetic polymers, are very interesting and more flexible than those of the double-stranded DNA (dsDNA).

*These authors contributed equally.

Templated DNA Nanotechnology: Functional DNA Nanoarchitectonics
Edited by Thimmaiah Govindaraju

ISBN 978-981-4800-21-1 (Hardcover), 978-0-429-42866-1 (eBook)
www.panstanford.com

The characteristic base pairing and stacking and the phosphate groups in the backbone make DNA one of the most highly charged biomacromolecules with a linear and regular structure. Over the past few decades, DNA has become a versatile building block for bottom-up nanofabrication to construct 2D [3] and 3D lattices [4], nanoscale objects [5, 6] and complex wireframe structures [7] with potential applications in chemistry, biology, medicine, and materials science. The structural aspects of DNA nanoarchitectures combined with their ability to respond to stimuli have led to the creation of cost-effective and sensitive biosensors that are functional under a wide range of biologically relevant conditions [8–10].

A sensor platform or device contains at least two components required for target recognition and signal transduction. DNA molecules can provide multivalent interactions, and in addition, have free termini that are allowed to functionalize with specific functional moieties. The unique synthesis and structural and functional features make DNA easily accessible, versatile, and a practically viable tool for studying multimolecular interactions with several biologically relevant targets, viz ions, small molecules (SMs), biomacromolecules, and biological events, in real time. Therefore, DNA can be explored in developing smart sensory platforms or devices for detecting different analytes/targets and applicable in delivery, diagnostics, therapeutics, etc.

11.1.1 DNA Nanoswitches and Devices

DNA switches and devices are the assembly states corresponding to distinct interchangeable conformations (typically referred as "off" and "on" and vice versa) in response to specific molecular inputs or external stimuli [2, 11–13]. In living systems, switching between different states involving shape or conformation of biomolecules (viz protein conformations, motor proteins, gene expression regulation switches, riboswitches, receptors, and transporters) is ubiquitous, which occurs in response to environmental cues (e.g., different types of phobia) to perform biological functions and maintain total homeostasis. Artificial molecular switches, on the other hand, are typically programmed molecular interactions either to generate an easily measurable signal (e.g., fluorescence [FL] or electrochemical output) or to trigger a precise analyte activity, including drug release, in response to a specific analyte or stimuli. For instance, new systems and methods based on DNA can be developed for drug delivery that are cost effective and allow optimal and controlled cargo release to the target.

In one of the earliest reports Seeman and coworkers described a DNA-based mechanical device as shown in Fig. 11.1 [11]. They constructed the molecular device by employing two rigid DNA "double-crossover" (DX, chromatids from two homologous chromosomes come in contact at two points) molecules connected by 4.5 double-helical turns. One domain of each DX molecule is attached to the other by the connecting helix. The transition between B (right-handed double helix) and Z (left-handed double helix) forms of DNA was applied to position the two unconnected domains of the DX molecules to lie on the same side (B form) or on the opposite side (Z form) of the central helix (Fig. 11.1). This relative switching was detected by means of fluorescence resonance energy transfer (FRET) spectroscopy, which measures the relative proximity of two dye molecules (FRET pair fluorescein [green] and Cy3 [magenta] as donor and acceptors) attached to the free ends of the DX molecules. This idea has enabled DNA nanotechnology to rationally design a number of 2D and 3D architectures (switches and devices) with a molecular precision and homogeneity that cannot be achieved by any other known technology on the nanometer scale.

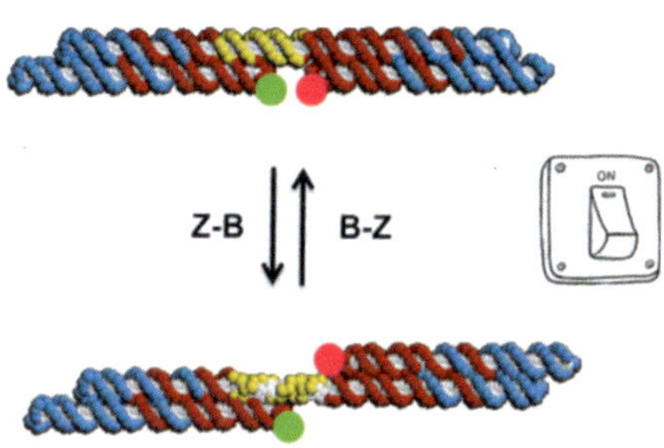

Figure 11.1 Design of a DNA nanodevice reported by Seeman. Reprinted by permission from Springer Customer Service Centre GmbH: Springer Nature, *Nature*, Ref. [11], Copyright (1999).

DNA with flexible strands and sticky ends are most important in the development of DNA nanotechnology. The concept of DNA switches and devices is an extension of DNA nanotechnology useful in developing biosensor platforms based on reversible switching between shapes or conformations that rely on specific recognition events as triggers or stimuli evoked by a target analyte (viz ions, SMs, biomacromolecules, and biological events). In addition, the nanoscale dimensions of these DNA architectures provide large surface-to-volume ratios, thus resulting in large signal changes on the target-binding event. The ease of design and synthesis of short DNA sequences or aptamers (40–50 base pair [bp] units) in large quantities, the cost effectiveness, and the prospect of introducing diverse chemical modifications to enhance their functionality to suit a specific application drive the development of a wide range of DNA nanoswitches and devices [14]. In this chapter, we mainly focus on the DNA switches and devices that have been developed to detect different biologically important targets, for example, ions, pH, SMs, proteins, and antibodies, and are applicable in drug delivery and cancer treatment.

11.2 pH Sensing

Intracellular pH is one of the important parameters that control numerous physiological functions in living organisms. It regulates a wide range of biological functions, viz enzymatic catalysis, protein folding and degradation, membrane function, and cell proliferation and apoptosis [15]. Interestingly, pH varies across compartments and organelles within the cellular milieu, regulated by both secretory and endocytic pathways. For instance, different compartments of the cell, such as lysosomes and endosomes, are normally acidic in nature, with pH ranging from 6.5 to 4.5. This pH is optimal for the functioning and activity of many enzymes, such as hydrolases. Moreover, cells under abnormal conditions, viz cancer cells, are also found to exhibit acidic pH. Therefore, it is of great importance to develop effective tools and methods to detect the intracellular pH with high sensitivity to monitor the slightest changes in the pH. Conventional techniques, including indicators and electrodes, are often used for pH measurement, and their obvious drawbacks, such as sensitivity and instrument size, limit their in cellulo and in vivo applications. The DNA structure is highly sensitive to pH as the protonation and deprotonation of nucleobases drive the reversible conformational/structural transformations, which can serve as the basis for the design of DNA-based nanoswitches and devices for accurate sensing of pH. The conformational changes, depending on the specific protonation of nucleobases, have been demonstrated in the formation of noncanonical DNA structures, including i-motifs [16, 17], duplex to triplexes [18–20], A-motifs [21–23], DNA tweezers [24], and switches and machines [25], among other architectures.

DNA motifs are short sequences and segments in DNA that are presumed to have numerous biological functions [26]. The reversible switching pattern of DNA motifs can be used for selective detection and monitoring of biological events and targets. Recently, we reported a SM–single stranded DNA ensemble of dB_n (oligonucleotides, B = nucleobases) and adenine (A)-functionalized perylene dye (**APA**) [23]. We found pH-dependent transformations of the dA_n:(**APA**)$_n$:dA_n ensemble structure due to protonation of the adenine unit in acidic media (pH = 2–4). This facilitates the formation of homoduplexes (A-motifs) of the type dA_n:dA_n by triggering the self-complementary base pairing [d(AH$^+$):d(AH$^+$)] through reverse Hoogsteen hydrogen-bonding interactions [21]. The A-motif is further stabilized by electrostatic interactions by the negatively charged backbone phosphates. Notably, a well-defined and characteristic circular dichroism (CD) signal for the self-complementary A-motif was observed [22].

DNA nanoswitches designed by taking advantage of the pH dependency of DNA motifs have become an important class of sensors. The variation and choice of DNA sequences can help in fine-tuning the performance of the nanoswitches over a wide range of pH. The protonation patterns are the common mechanism of pH sensing by the nanoswitches constructed from short DNA sequences. The DNA nanoswitches can be used for in vitro and in vivo pH sensing in real time. We designed a molecular beacon (MB)-based DNA nanoswitch (Fig. 11.2a) that can transform between the closed state (MB) and the A-motif (open state) under an acidic window of pH 5 to 3 [27]. The MB with an A-rich loop and a mixed sequence stem was used for the design of a DNA nanoswitch, which changes its structure to A-motif, a right-handed parallel duplex, upon protonation of As. The actual DNA nanoswitch consisted of 24 nucleobases (UMB, 5′-GACGCCAAAAAAAAAAAACGCGTC, and LMB, Cy3-5′-GACGCCAAAAAAAAAAAACGCGTC-3′-Cy5) and adopted a MB (closed) structure. Out of the 24 nucleobases, 12 consecutive A-bases are in the loop region and two stretches of 5 complementary bps each at 5′- and 3′-ends, forming the hairpin-like duplex (MB) structure. The 5′- and 3′-ends of the DNA nanoswitch (in LMB) were labeled with Cy3 (energy donor, λ_{em} = 565 nm) and Cy5 (energy acceptor, λ_{ex} = 650 nm) dyes of high quantum yields. Under normal physiological conditions (pH = 7), the DNA nanoswitch exists in a closed state (MB) where the predisposed close proximity of Cy3 and Cy5 dyes is appropriately placed for efficient FRET. At relatively low pH, the A-bases undergo protonation and the DNA nanoswitch transforms from MB into an open state (A-motif) through reverse Hoogsteen [AH^+–H^+A] hydrogen-bonding interactions that suppress the effective FRET (Fig. 11.2a). The A-motif structure further stabilizes by electrostatic interactions between the AH^+ and the negatively charged backbone phosphate, which results in the suppression of the FRET process. The DNA nanoswitch was found to perform reversibly between pH 7 and 3. Further, the DNA nanoswitch was demonstrated to assess the feasibility of switching action in response to acidic pH in an artificial vesicle, which mimics a cell-like environment. An efficient FRET signal in FL microscopy was observed in the LMB-encapsulated vesicle in the pH range of 5 to 3, with sensitivity corresponding to small step sizes of 0.2– 0.3 units. The experiment was extended to a cell-based study to validate the DNA nanoswitch in live HeLa cells. Under physiological conditions (pH 7.4), the HeLa cells were incubated with the LMB. The results demonstrate successful cellular uptake of LMB without the need for an external transfection agent. The successful transformation of LMB from the closed state to the opened state inside live HeLa cells was monitored by the ratio of change in FRET between Cy3 and Cy5. This provides the real-time, rapid, and sensitive sensing of change in intracellular acidic pH.

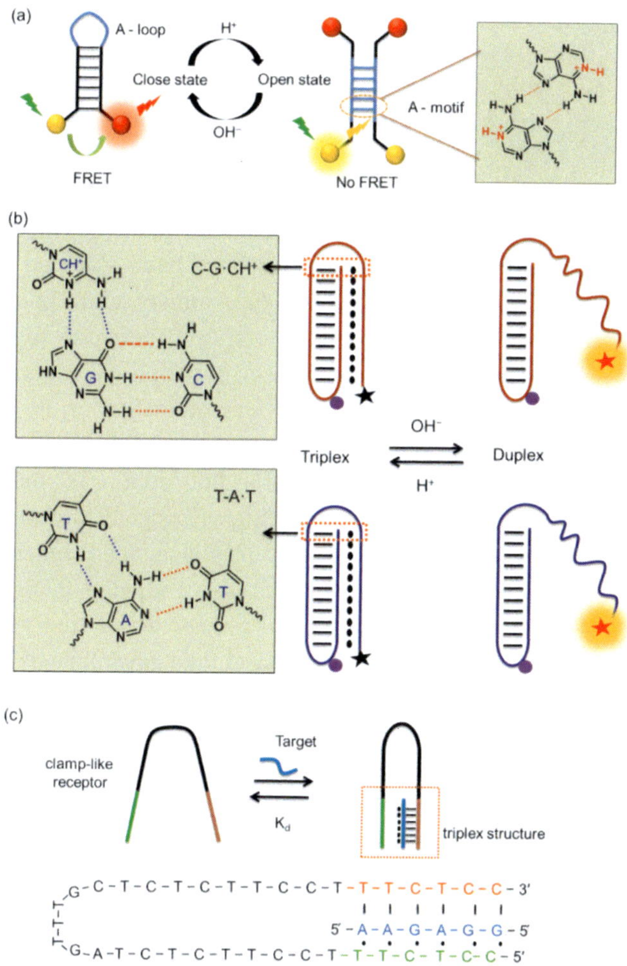

Figure 11.2 (a) Schematic representation of a thoughtfully designed pH-triggered DNA nanosensor under a narrow acidic pH window studied by means of the FRET mechanism, (b) programmable DNA nanoswitches triggered by triplex structure formation via Watson–Crick (WC) (dashed) and a pH-sensitive Hoogsteen (dots) interaction, and (c) clamp-like triplex forming a DNA nanoswitch: a recognition element is linked through a random loop (black) to a triplex-forming portion (green) capable of recognizing the dsDNA and to form a triplex structure composed of both WC and parallel Hoogsteen interactions. (a) Reproduced from Ref. [27] with permission of The Royal Society of Chemistry. (b) Reprinted with permission from Ref. [18]. Copyright (2014) American Chemical Society. (c) Reprinted with permission from Ref. [20]. Copyright (2017) American Chemical Society.

Ricci and coworkers reported a programmable DNA nanoswitch that can function in a specific pH window (Fig. 11.2b) [18]. Triple-helix DNA with typical Watson-Crick (WC) base pairing and Hoogsteen pairing was used to construct the device. The weak Hoogsteen pairing was opened and modulated with the change in pH. The two ends of the nanoswitch were labeled with fluorophore and quencher to monitor the on/off signal in response to the proton gradient. Specific CGC/TAT sequences have been included in the Hoogsteen base pairing sequence of the nanoswitch to create the pH tunable structural folding and unfolding from triplex to duplex structures due to protonation and deprotonation of certain bases depending on the pH. The FL signal was quenched in the triplex state (folded) as both the fluorophore and the quencher come closer, while in the duplex state (unfolded) the fluorophore is positioned far from the quencher and shows a high FL signal intensity. The nanoswitches work in the time span of milliseconds, which helps in the real-time sensing and monitoring of change in pH. This is one of the interesting examples of nanoswitches that form an intramolecular triplex DNA structure and take advantage of the pH dependence of parallel Hoogsteen bp interactions [28–30]. Further, the authors combined two or more switches working in different pH windows simply by changing the relative content of CGC/TAT triplets in the sequence to improve the pH sensing range up to 5.5. The DNA switches with a high CGC content (≥50%) are opened in the acidic pH window of 5.5 ≈ 6.0, while those with a low CGC content (<50%) opened at a relatively more basic pH window of 9–11. Switches that do not contain CGC triplets (and contain TAT triplets) unfolded only at pH above 10.0.

A parallel DNA triplex helix-based clamp-like nanoswitch was developed for better and rapid sensing of pH. Incorporation of the triplex-forming sequence in a DNA nanoswitch can trigger conformational adjustment as a function of change in pH and serves as a simple two-state clamp-like nanoswitch (Fig. 11.2c) [20]. The actual nanoswitch design consists of triplex clamp-switch (5′-*CCTTT*-TCCTTCTCTCGTTTGCTCTCTTCCT-**TTCTCC**-3′) and linear probe (5′-CTCTCTTCCT-**TTCTCC**-3′) with common stretches of six bases. These sequences are labeled with Alexafluor 680 (AF680) and Black Hole Quencher 2 (BHQ-2) at the 5′- and 3′-ends. Bases in bold represent duplex-forming sequences, whereas those in italic represent the triplex-forming sequences. A six-base target sequence (5′-GGAGAA-3′) was used to form the triplex (Fig. 11.2c). The duplex (WC base pairing)-forming portion was linked to a triplex (Hoogsteen base pairing)-forming portion through the random loop. The triplex formation requires TAT and a C^+GC triplet between the clamp switch and its complementary six-base DNA strand. The C^+GC triplet formation is favorable when the N3 of cytosine (C) is protonated and hence the triplex state was found stable in acidic pH. In contrast, TAT is stabilized at higher pH due to deprotonation of thymine (T). Hence, the nanoswitch adjusts the closed triplex and open unfold states in lower and higher pH, which correspond to a decrease and an increase in the FL signal, respectively. This DNA nanoswitch operates between pH 5 to 8 in a reversible manner and over several cycles without any perturbation. The simulation method was attempted to characterize and describe the switching mechanism at the atomistic level of the DNA nanoswitch, and results are shown to be in good agreement with experimental results.

The pH sensitivity window of the DNA nanoswitch by Ricci and coworkers discussed above was further improved by Yu and coworkers (Fig. 11.3a) [31]. Two sensitive probes with different TAT contents (50% and 80%) were used in the triplex DNA structure formation. These DNA probes in different conformations were shown to have different affinities to graphene oxide (GO). In addition, the probes showed selective adsorption on GO, which also resulted in efficient change in the FL signal. Under acidic conditions (pH = 5.0), probes fold into a triplex structure that resists adsorption on to the GO surface and becomes free, thereby switching on the FL signal. In basic condition (pH = 9.0), probes formed single-stranded tail structures that adsorb on to GO and thereby quench the FL signal. Two switches with variable TAT contents responded to and resulted in sensing of the relatively large range of pH. These DNA nanoswitches showed a high signal-to-background ratio and fast response time in the pH range of 5 to 8 and thus have the potential for application as a sensitive pH sensor for diagnostic purposes.

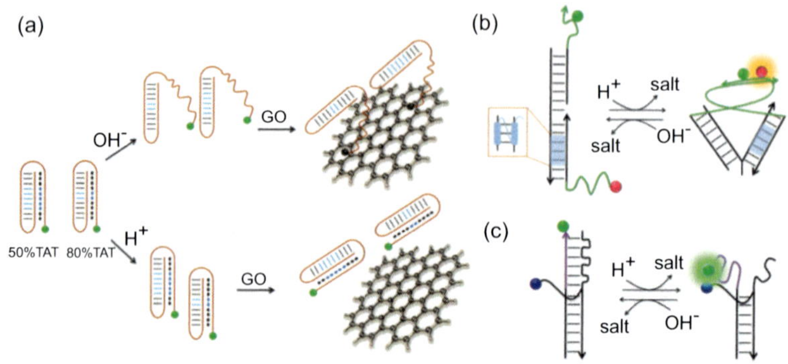

Figure 11.3 (a) Schematic presentation of the detection of pH on the basis of the conformational changes of fluorophore-tagged DNA molecules on the GO surface. (b) pH-dependent DNA nanomachine labeled with Alexa 546 (magenta circle) and Alexa 647 (green circle) on the DNA scaffold. The dsDNA domain on IFu (blue) acts as a recognition element for a recombinant antibody (scFv, blue cylinders). (c) Intramolecular i-motif formation at low pH leads to a high FRET. (a) Reproduced from Ref. [31] with permission of The Royal Society of Chemistry. (b, c) Reprinted by permission from Springer Customer Service Centre GmbH: Springer Nature, *Nature Nanotechnology*, Ref. [32], Copyright (2013).

Krishnan and coworkers developed two distinct DNA nanomachines to monitor the pH simultaneously in two intersecting pathways inside the cells to map pH gradients [32]. The two molecularly programmed nanodevices with suitably chosen FRET pairs having minimal cross talk were coupled to distinct organelles' pathways. These FRET-based nanomachines (Fig. 11.3b,c) were programmed to deliver them to different endosomal pathways. The authors chose the furin retrograde endocytic pathway and transferrin recycling pathways, which lead to trans-Golgi network (TGN) and perinuclear recycling endosome (RE), respectively, via distinct pathways. The first nanomachine was optimized with an extra eight-bp sequence that was used for the conjugation of the protein that localizes the nanodevice to the furin pathway. Another nanomachine that forms intrastrand i-motif as a response to pH change was employed to track the transferrin pathway. Transferrin was conjugated with the nanodevice and targeted to the transferrin pathway via receptor endocytosis. The devised nanomachines are efficient in the pH range of 5 to 6.5 and are suitable for studying typical intraorganelle pH. In cellulo studies in HeLa cells showed that the nanodevices are internalized and localized to the target compartments. The FL readout of the nanodevice was successfully used to map pH profiles of subcellular organelles. The nanodevices were successfully used in simultaneous monitoring of the pH in two intersecting organelle-compartment pathways without any cross talk. This type of DNA machine and pH sensing technology can be explored for compartment mixing studies and different fission-fusion studies.

11.3 Ion Sensing

Ions (cation and anion) play an important role in the fundamental processes of biological systems, including osmotic regulation, catalysis, metabolism, biomineralization, and signaling [33, 34]. The concentration of cellular metal ions (M^{n+}) is highly regulated in different parts of the cells, as both deficiency and surplus of M^{n+} can disrupt normal functions [35–38]. Such imbalance in the concentration levels of metal ions is the causative factor for a variety of disease conditions. Therefore, monitoring of concentration levels of metal ions in the biological milieu is of prime importance, as it provides a better understanding of the pathophysiological conditions. Metal ions such as Na^+, K^+, Ca^{2+}, Mg^{2+}, Ag^+, Zn^{2+}, Cu^{2+}, and Co^{2+} are of biological relevance and involved in various biological functions. However, imbalance in the concentration levels of these biometals leads to adverse health effects. On other hand, heavy metal ions, such as Pb^{2+} and Hg^{2+}, are extremely toxic at very low concentrations and cause serious health disorders. Na^+ is one of the most abundant metal ions in intracellular fluid [39] and plays a key role in numerous cellular processes by activating many signal transduction pathways as well as influencing the actions of hormones [40]. Ag^+ is useful in radioimmunotherapy and photographic technology [41]. It inhibits the growth of bacteria (antibacterial activity) and fungi (antifungal activity) and reduces the risk of bacterial and fungal infection and also finds use in cancer immunotherapy. Heavy metals ions, viz Pb^{2+} and Hg^{2+}, are considered as extremely toxic and cause environmental and health hazards. A wide variety of adverse health symptoms are observed upon exposure to these metals, including digestive, kidney, and especially neurological disorders [42].

Among the anions, chloride (Cl^-), fluoride (F^-), and cyanide (CN^-) are the notable ones that are essential for maintaining the electrolyte balance in living bodies and therefore are significantly important in clinical science [43]. F^- is found in drinking water, toothpaste, etc., and its accumulation in the human body is responsible for osteoporosis. Intracellular Cl^- transport is critical to organelle function. In general, anions are required for several critical biological functions, while in excess, they become toxic to cells and cause numerous disorders. This underlines the need to develop simple, cost-effective, and sensitive devices for in vitro as well as in vivo detection of aforementioned cations and anions. As a versatile molecule, DNA has been exploited in the form of aptamers and DNA nanodevices for the detection of cations and anions. DNA-based nanodevices and aptamers are designed with selective and specific recognition capability to achieve sensitive detection of target ions. The structural robustness, predictability, stability, and versatility of DNA make these nanodevices suitable for in vitro and in-vivo applications. The DNA-based sensor system can be coupled to a variety of signal readout strategies, like colorimetry, electrochemical, and FL, to improve the sensitivity of analyte detection. Furthermore, DNA nanodevices are also most suitable for potential biological applications due to improved sensitivity and simplicity.

11.3.1 Sodium Ion Sensing

Lu and coworkers reported in vitro detection (>10,000-fold relative signal increment) of Na^+ by employing an RNA-cleaving deoxyribozyme (DNAzyme) with high selectivity over other competing metal ions [44]. This detection system showed a wide dynamic range covering the physiological concentration range of Na^+ (0.135–50 mM) and exhibited a fast catalytic rate (k_{obs}, ~0.1 $min.^{-1}$). The metal-selective DNAzymes can be obtained without the prior knowledge of metal ion–binding sites or specific metal-DNA interaction, which makes them more advantageous compared to other techniques [45]. The enzyme strand was labeled with a quencher (Q) at the 3′-end and the DNA substrate strand with a fluorophore (F) and a quencher (Q) at the 5′- and 3′-ends, respectively, transforming the DNAzyme into a fluorescent sensor for Na^+ (Fig. 11.4a). This Na^+-specific DNAzyme in combination with efficient cationic polypeptide-based delivery method was used for imaging intracellular Na^+ in living cells. In addition, a photocaging strategy was employed to allow controlled activation of the probe inside cells.

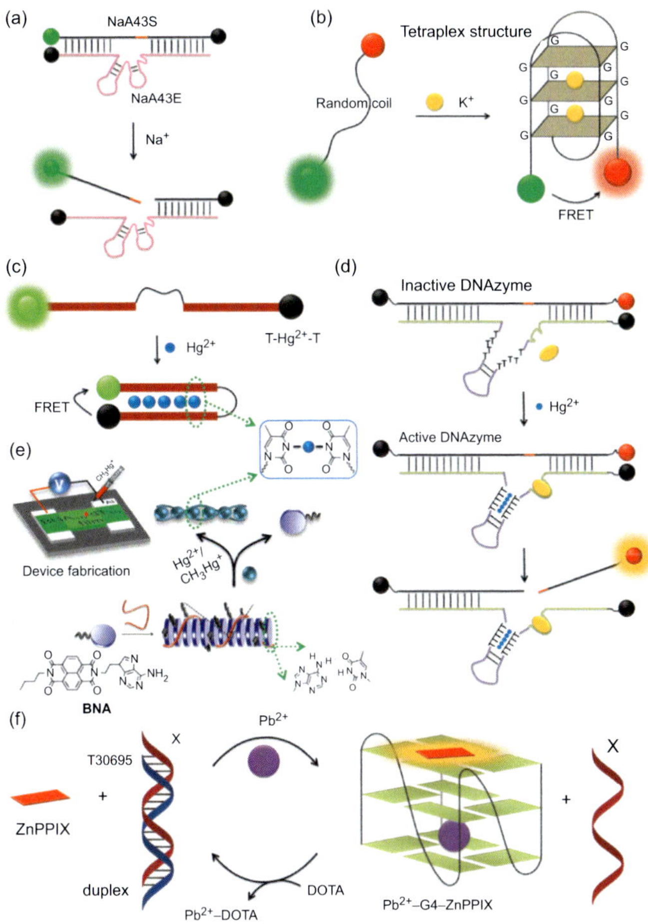

Figure 11.4 (a) Design of the DNAzyme beacon using a fluorophore (green) and quencher (black) for Na^+ [44]; (b) design strategy of a oligonucleotide-based tetraplex structure formation in the presence of K^+ ions, resulting in FRET; (c) hairpin structure formation induced by Hg^{2+} ion-mediated T–Hg–T pair formation, which results in the quenching of fluorescence from F; (d) rational design strategy of turn-on allosteric DNAzyme catalytic beacons for Hg^{2+}; (e) Hg^{2+}-mediated displacement of *N*-benzyl-2-methyl-4-nitroaniline (BNA) from a BNA_n-dT_n ensemble and subsequent formation of a metallo-DNA duplex $[dT\text{-}Hg\text{-}dT]_n$; and (f) Pb^{2+}-driven DNA nanodevice moduled on a duplex-quadruplex conformational change. (b) Reprinted with permission from Ref. [55]. Copyright (2002) American Chemical Society. (c) Reprinted with permission from Ref. [61]. Copyright 2004, John Wiley and Sons. (d) Reprinted with permission from Ref. [66]. Copyright 2016, John Wiley and Sons. (e) Reprinted with permission from Ref. [8]. Copyright (2016) American Chemical Society. (f) Reprinted with permission from Ref. [68]. Copyright (2010) American Chemical Society.

11.3.2 Potassium Ion Sensing

K^+ is one of the physiologically important metal ions and plays a key role in the maintenance of homeostasis [46]. Monitoring the in cellulo and in vivo concentration levels of K^+ is of prime importance as it has been implicated in various disease conditions, including diabetes, cardiovascular disease, kidney problems, Addison's disease, high blood pressure, and stroke [47]. In recent times, the G-quadruplex, one of the DNA polymorphic forms, has been well studied for its structure, biological significance, and molecular platform to be used for sensing different biological targets [48–51]. The G-quadruplexes are highly ordered structures composed of guanine quartets (G-quartets) wherein four G bases are held together by Hoogsteen base pairing. Interestingly, the G-quartet/G-quadruplex structure is stabilized by K^+ through coordination at the core of the complex, as shown in Fig. 11.4b [52–54]. Takenaka and coworkers [55] reported a novel potassium-sensing oligonucleotide (PSO) probe for the facile detection of K^+. PSO (Sigma Genosys) carries a part of the human telomere sequence, d(GGGTTAGGGTTAGGGTTAGGG), with two fluorophores, 6-carboxyfluorescein (6-FAM) and 6-carboxytetramethylrhodamine (6-TAMRA), at the 5′-terminus and the 3′-terminus of the oligonucleotide as the donor and acceptor, respectively (Fig. 11.4b). The two chromophores appear in close proximity to show FRET in the presence of K^+. The distance between the termini of the PSO-carrying telomere sequence changes dramatically upon formation of a tetraplex from a random-coil structure, and this transformation is strongly dependent on the concentration of K^+.

The PSO probe was found useful for the detection of K^+ in the presence of an excess of Na^+ ions. Later, they designed 15-mer oligonucleotide with the sequence d(GGTTGGTGTGGTTGG), known as a thrombin-binding aptamer (TBA) [56], for the detection of K^+ [57]. Unlike the FRET-based system discussed above [55], in this work, pyrene was employed as fluorophore, which is expected to produce efficient π-stacked excimer emission to report the conformational changes of the DNA sequence induced by the guest analyte (K^+). The spatial orientation of the PSO-py/K^+ complex emits only excimer FL, whereas in the absence of K^+ the random-coil structure of PSO-py gives only monomer emission. The dynamics of the FL response and the reproducibility of excimer emission were established by sequential dilution of solution and addition of KCl to maintain the K^+ concentration between 5 and 6 mM. Therefore, the PSO-py fluorescent probe can be potentially exploited for the real-time monitoring of K^+ under extracellular conditions.

11.3.3 Mercury Ion Sensing

Mercury (Hg) is one of the most predominant heavy metals in the environment, with both natural (volcanic and oceanic emissions) and anthropogenic (major industrial sources, such as coal and gold mining and fossil fuel combustion) sources. *It is toxic in all forms:* elemental (Hg^0), inorganic salts (Hg^{2+}), and organometallic compounds (CH3Hg$^+$ and CH_3HgCH_3). Hg^{2+} is known to bind with thymine bases to form a T-Hg(II)-T complex in a DNA duplex [58, 59]. The binding of Hg^{2+} to a T-T pair is highly selective, and the T-Hg(II)-T complex is thermally stable. There are several interesting literature reports where thymine-containing DNA sequences functionalized with fluorophores or fluorophore displacement systems were developed for the selective detection of Hg^{2+} in aqueous solution [60–65]. For example, Ono and coworkers reported an oligodeoxyribonucleotide (ODN)-based Hg^{2+} sensor, where the ODN was functionalized with a fluorophore (fluorescein, F) and a quencher moiety (dabcyl, D) at the 3′-terminus and the 5′-terminus, respectively (Fig. 11.4c) [61]. The ODN sequence is divided into two parts, the thymine-rich mercury-binding sequence and the linker sequence. In the presence of Hg^{2+}, T-Hg-T metal-base pairing leads to the formation of a hairpin-like structure. This brought together the termini of the ODN, and the close proximity of F and D moieties resulted in FL quenching. Hence, the change in the FL intensity of the hairpin structure relative to the random coil was used to detect Hg^{2+}, and a detection limit of 40 nM was reported [61].

Lu and coworkers reported that the T-Hg(II)-T interaction can be used to modulate DNAzyme activity through allosteric interactions [66]. In Fig. 11.4d, the secondary structure of the DNAzyme contains a substrate strand (39S) and an enzyme strand (39E). 39S has a single RNA linkage (rA) that serves as the cleavage site. 39E binds with 39S through two substrate-binding arms. The catalytic core in 39E contains a stem loop (blue) and an eight-nucleotide bulge (green). The substrate was labeled with FAM fluorophore at the 5′-end and the quencher (Q) at the 3′-end, and in addition the enzyme strand was labeled with Q at the 3′-end (Fig. 11.4d). Such a dual-quencher labeling method was devised to achieve very low background FL, which allowed high signal enhancement. The DNA nanodevice was then activated by mixing with UO_2^{2+}. The nanodevice showed Hg^+ detection only in the presence of UO_2^{2+}. In the absence of Hg^{2+}, the DNAzyme was unable to bind with UO_2^{2+} because the active secondary structure cannot be formed. However, the addition of Hg^{2+} quickly restored the stem-loop structure and activated the DNAzyme to cleave the substrate, releasing the fluorophore-labeled piece and resulting in increased FL. Overall, the activated nanodevice functions following allosteric regulation, which resulted in a detection limit of 2.4 nM.

In a recent study, our group reported SM-templated DNA nanoarchitectonics using an adenine-appended naphthalenedimide (BNA) and oligothymidine (dT_n) (Fig. 11.4e) [8]. The self-assembled organic semiconductor BNA acted as a template for the assembly of dT_n to form mutually templated chiral and conducting hybrid SM-DNA ensembles BNA_n-dT_n. The BNA_n-dT_n sensor platform facilitated the ultrasensitive detection of inorganic and organic mercury (Hg^{2+} or CH_3Hg^+) with high selectivity in water. This is one of the first designs of DNA-based sensor platforms that employ homo-thymidine sequences to enhance the detection sensitivity, and the signal out was in the form of change in the chirality of the supramolecular system and conductivity measurement from the field-effect transistor (FET) device. The effective displacement of BNA from BNA_n-dT_n by the Hg^{2+} to form the metallo-DNA duplex of the type (dT_n-Hg_n-dT_n) [67] is the sensor mechanism that results in reliable and quantifiable changes in induced chirality and conductivity (refer to Chapter 2 for a detailed account of this work). This design strategy described the SM-templated assembly of oligothymidine (dT_n), which acts as a highly selective and ultrasensitive (0.1 nM) sensor platform for mercury, a highly toxic and global environmental pollutant. In this design, we have overcome the drawbacks of previous reports of Hg(II) detection using DNA (or aptamer) containing a few T units.

11.3.4 Lead Ion Sensing

Lead (Pb) is one of the environment pollutants, a cause for numerous human health hazards, and hence has drawn immense interest from researchers. Recently, Wang and coworkers reported a Pb^{2+}-sensitive DNA nanodevice [68]. This nanodevice was developed on the basis of the concept of DNA "duplex-quadruplex" conformation interchange and utilized for selective and sensitive detection of Pb^{2+}. They established in the previous study that Pb^{2+} stabilizes the single-stranded DNA (T30695), in contrast to duplex DNA, and can interact with the zinc-protoporphyrinIX (ZnPPIX) complex, which enhances its FL intensity [69]. For device fabrication, they prepared a T30695-X duplex (X, complementary strand) and combined it with ZnPPIX, which serves as a source of FL signal output. The binding of Pb^{2+} to the duplex DNA resulted in the formation of the G-quadruplex from the T30695 strand and displaced the complementary X strand. The ZnPPIX further stabilizes the Pb^{2+}-bound G-quadruplex structure of T30695 and shows a FL turn-on response (Fig. 11.4f). This device can be reused by the addition of a strong Pb^{2+} chelator, like DOTA (1,4,7,10-tetraazacyclododecane-1,4,7,10-tetraacetic acid). The strong complexation of DOTA to the Pb^{2+} regenerates the duplex DNA structure of the nanodevice, which is subsequently available for the next detection cycle. The FL intensity or signal output of ZnPPIX can be tuned by its binding to and release from the G4 structure, and this structural change of a DNA nanodevice is directly reflected in the increase or decrease of FL in the output signal.

11.3.5 Chloride Ion Sensing

The Cl^- concentration greatly varies across different compartments of the cell. Under normal physiological conditions, the extracellular chloride concentration (120 mM) is higher than the intracellular chloride concentration (4–60 mM). In healthy cells no appreciable chloride influx takes place unless biological ion channels or transporters are activated [70–72]. Chloride-sensitive reporter proteins have been employed for sensing its concentration in different compartments of the cell. However, these protein-based sensors have limited applications due to a small working range and pH sensitivity. Krishnan and coworkers developed a DNA nanodevice to overcome the short falls of the existing Cl^- reporters [73]. This DNA nanodevice was found suitable for quantifying Cl^- transport in organelles of living cells. The pH independence of the nanodevice and FL readout makes the device useful for in vivo monitoring of variations in the concentration levels of Cl^-. The DNA nanodevice consists of three segments: (i) a sensing segment (P), which is a dodecamer peptide nucleic acid (PNA) conjugated to a chloride-sensitive FL molecule 10,10′-bis[3-carboxypropyl]-9,9′-biacridinium dinitrate (BAC), (ii) a 38-mer DNA sequence conjugated to Alexa 647 (A647) fluorophore (D2), which is insensitive to chloride, and (iii) a 26-mer targeting DNA sequence (D1) (Fig. 11.5). The P and D1 hybridize to D2 to develop the nanodevice. The double-stranded domain of the nanodevice, formed by the D1 and D2, acts as a negatively charged ligand for receptors (ALBRs), which will be trafficked to the endolysosomal pathway [74]. FL of A647 is Cl^- independent; however, BAC undergoes collisional quenching in the presence of Cl^-. This ratiometric approach ($I_{Alexa647}/I_{BAC}$) enables the quantification of physiological Cl^- at endolysosomal compartments. The DNA nanodevice shows linearity for the detection of Cl^- over 0–200 mM concentration in the pH range of 5 to 7.4. The device was further used in the drosophila hemocytes titrate with the Cl^- in a buffer solution. Overall, this DNA device provided good sensing of the Cl^- in different trafficking pathways.

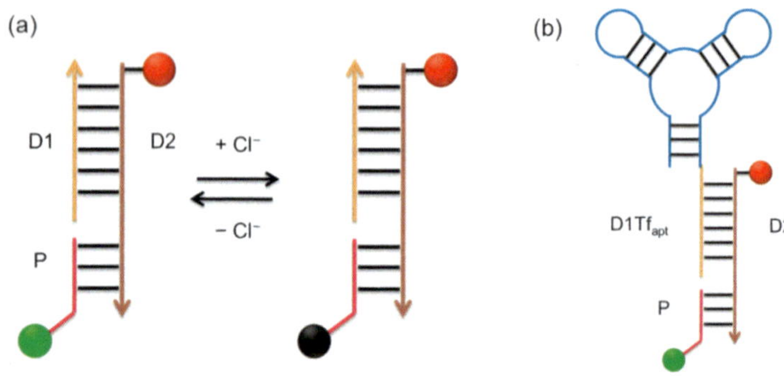

Figure 11.5 DNA nanodevice for Cl^- sensing. P, sensing module (pink line) containing a Cl^--sensitive fluorophore, BAC (green filled circle); D2, normalizing module (red line) containing a Cl^--insensitive fluorophore, Alexa 647 (red-filled circle); D1, targeting module (orange line). Reprinted by permission from Springer Customer Service Centre GmbH: Springer Nature, *Nature Nanotechnology*, Ref. [73], Copyright (2015).

11.4 Biomolecule Detection

Detection and monitoring of biomolecules are important facets in analytical chemistry and clinical diagnosis [75, 76]. Diagnosis of diseases on the basis of the presence of certain biomolecules or their concentration levels requires assays that can detect these target biomarkers with utmost selectivity and sensitivity. There are several instrumentation techniques and methods available that can detect and quantify biomolecules in physiological conditions. However, simple, rapid, and cost-effective detection platforms with required selectivity and sensitivity for the target biomolecules are necessary for reliable and practical applications. The advancements in the field of DNA nanotechnology lead to the development of oligonucleotide-based binders or aptamer nanoswitches and devices capable of detecting the target biomolecules with high specificity and sensitivity [77]. In this context, aptamers with a high selectivity and affinity can be developed for SMs, biopolymers, surfaces, or even whole cells. Similarly, modular DNA switches and devices with distinct mechanistic or working principles can be designed and employed with specific and reliable signal readouts, such as colorimetric, electrochemical, and FL responses [77, 78].

11.4.1 Small Molecules

In one of the earliest works, Landry and coworkers hypothesized a DNA nanodevice with a single-chain aptamer that might be rationally destabilized such that the stem formation would depend on ligand-binding interaction [79]. Such a biomolecular recognition event would promote a rapid response time and enable operation of the device in complex media. For this, a double-end-labeled aptameric nanodevice (MNS-7.9) with ligand-induced stem formation that permits optical detection of the ligand (cocaine) through FL signal was devised (Fig. 11.6a). The aptamer was coupled with fluorescent dye F (fluorescein) and a quencher Q (dabcyl) at 5′ and 3′ positions, respectively, and expected to form two stem loops with a three-way (S1, S2, and S3) junction.

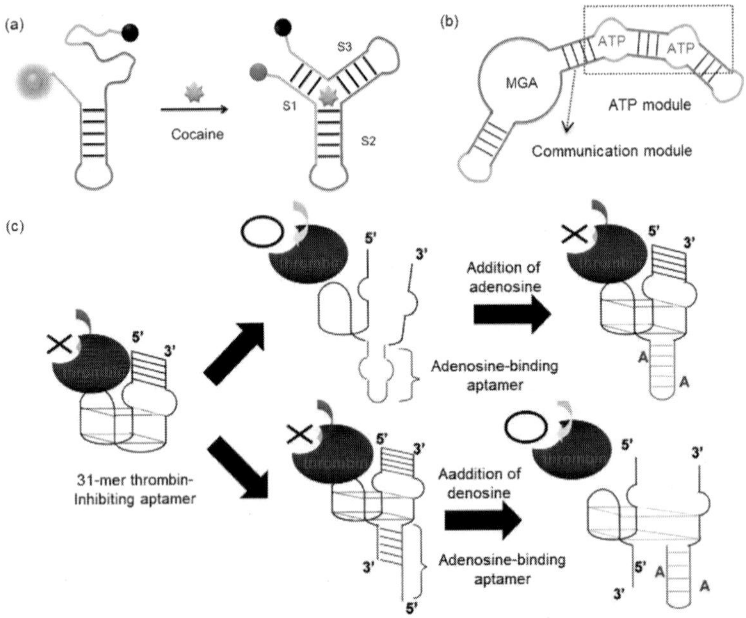

Figure 11.6 (a) Conformational change of a DNA nanodevice in the presence of cocaine. (b) A DNA-based allosteric sensor was designed by combining ATP aptamer with malachite green aptamer for recognition of different molecules, like ATP, FMN, and theophylline. (c) Two adenosine-sensing schemes: A thrombin-inhibiting aptamer is fused with an adenosine-binding aptamer. The thrombin-inhibiting aptamer folds into the G-quartet structure, which plays an important role in its inhibitory activity and sensing. (a) Reprinted with permission from Ref. [79]. Copyright (2001) American Chemical Society. (b) Reprinted with permission from Ref. [80]. Copyright (2004) American Chemical Society. (c) Reprinted with permission from Ref. [82]. Copyright (2006) American Chemical Society.

The pocket at the junction fits to bind with cocaine and stabilizes the closed-state compared to its free state. The FL quenching due to close proximity of the fluorophore and the quencher was found proportional to the concentration of the cocaine. The appropriate calibration curve aided the quantification of the target and showed a good affinity for cocaine with $K_d = 20\ \mu M$. This nanodevice displayed similar detection ability in solution and in a complex mixture, like serum, which perhaps suggests its applicability in a biological environment.

Most of the aptamer-based DNA nanosensors are not suitable for intracellular use as they need to be labeled with a fluorescent dye and delivered exogenously. Kolpashchikov and Stojanovic developed a DNA nanodevice by using a malachite green (MG)-binding RNA aptamer (MGA) through a ligand displacement mechanism (Fig. 11.6b) [80]. The idea of chimeric construction avoids trivial chemical modifications and shows FL changes upon simultaneously binding MG and target analytes, adenosine triphosphate (ATP), flavin mononucleotide (FMN), and theophylline (TH). This nanodevice was constructed of three domains: (i) a signaling domain (MG aptamer), (ii) a recognition domain, and (iii) a connecting stem for transducing the binding of targets, which results in FL change. The recognition domain has the capability to bind different molecules, like ATP, FMN, and TH. This generates a modular allosteric sensor that can be used for in vitro sensing of biomolecules. The designed DNA nanodevice can efficiently respond to ATP over the concentration range of 10 μM to 2 mM, with a fivefold signal amplification. It showed the maximum specificity for ATP, with minimal response for uracil triphosphate (UTP). This design strategy was extended to develop nanoswitches for FMN and TH. The TH sensor worked with sensitivity in the range of 2 to 250 nM and with an eightfold increase in the FL. The FMN sensor showed a 30- to 50-fold increase in FL at saturation concentration with a robust signal output. Kolpashchikov applied a similar design strategy to develop a nanodevice for the detection of nucleic acids [81]. Similarly, Ikebukuro and coworkers designed allosteric aptamers, as shown in Fig. 11.6c, wherein thrombin binding was enhanced by the binding of adenosine at an allosteric site [82]. Szostak and Huizenga reported a DNA aptamer nanodevice with MG as the reporter for the detection of ATP [83].

Li and coworkers developed a DNA nanoswitch as a sensor system for the detection of nucleotides, for example, ATP (Fig. 11.7). The nanoswitch showed conformational switching between off and on states upon addition/removal of the stimuli via toehold-mediated strand displacement reactions [84]. This is in contrast to other previous studies where ATP was typically utilized as a target to trigger DNA strand displacement reactions [85–87]. In another report, thioflavin T (ThT) was used as a fluorescent indicator for an ATP-binding aptamer [88]. It served as a competitive reporter for the binding and release of ATP and reported ATP binding by a decrease in the FL response. For the construction of a DNA nanodevice, a contractile three-way structure [89] was strategically adopted and functionalized by incorporating three 27-mer ATP aptamer sequences into single-stranded motifs at the center of the three-way junction (Fig. 11.7). The DNA nanodevice acted as a specific ATP-responsive three-way nanosensor (ARTWN) whose structural integrity was confirmed by the native polyacrylamide gel electrophoresis (PAGE).

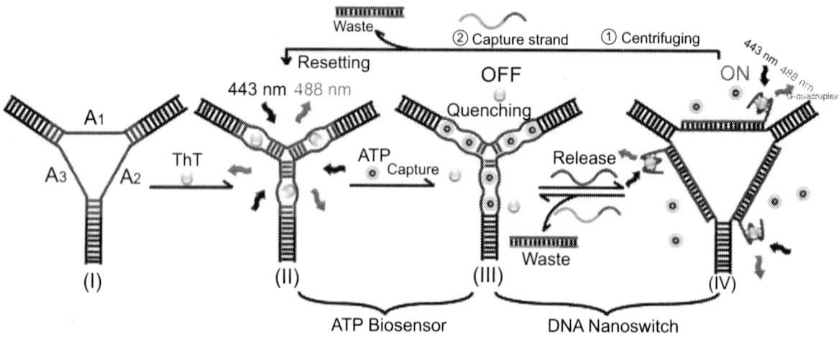

Figure 11.7 Schematic presentation of a DNA-based reversible nanosensor. (I) The nanostructure of ARTWN is self-assembled from the equivalent of three component strands: A1, A2, and A3. (II) The initial state of the nanosensor. ThT molecules intercalate into the two adjacent embedded ATP aptamers intermolecularly, giving rise to a fluorescence increase of ThT. (III) The off state of the switch cycle. The bound ThT molecules are displaced by ATP, accompanied by a sharp fluorescence decrease. (IV) The on state of the switch cycle. Upon introduction of the release strand, the central ATP aptamers are fully hybridized, thereby releasing the ATP molecules. The free ThT molecules bound to G-quadruplexes serve as a fluorescence indicator for the switch cycle. Adapted from Ref. [84]. Copyright (2016) Oxford University Press.

SM detection has benefited from the development of the concept of aptamers and DNAzymes that are based on novel methods of molecular recognition [90–96]. However, detection of nucleic acid remains dominated by base pairing, which relies fundamentally on differences in bonding enthalpies between probes and matched/mismatched targets. On this basis, Ghazal and coworkers developed a nanoswitch [97] that was structurally based on Holiday junction (HJ), a four-way DNA junction with a central branch point [98]. The switch was constructed of two DNA strands, (i) a labeled probe strand that comprises the majority of the structure and contains two unpaired binding arms and (ii) a target strand whose hybridization to the probe completes the assembly of the DNA switch. The hybridization of the target DNA strand completes the HJ, which brings the donor and the acceptor closer to each other and ensures effective FRET (Fig. 11.8a). HJ stabilized mostly by electrostatic interactions; therefore, an optimum ionic strength of cations like Mg^{+2} is required to specifically detect the target sequence. The FL readout in the solution and gel (FRET) was shown to detect the target with good specificity and sensitivity.

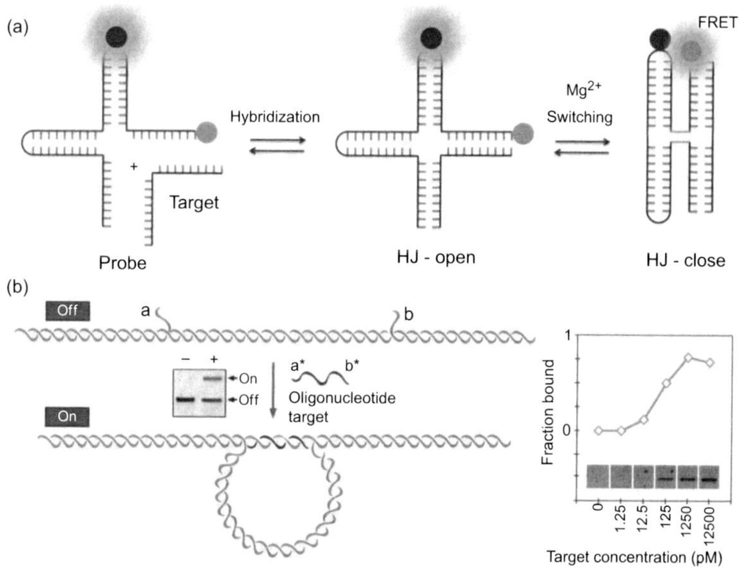

Figure 11.8 (a) DNA nanodevice based on Holiday junction with probe and target hybridization and Mg^{2+}-dependent conformational switching. (b) Two-state DNA nanoswitch showing conformational change by addition of the key oligonucleotide that binds the overhangs of the two detector regions (a) and (b), thereby forming a loop. Agarose gel shows the sequence-specific activation of the nanoswitch. (a) Reprinted with permission from Ref. [97]. Copyright (2007) American Chemical Society. (b) Reprinted with permission from Ref. [99]. Copyright (2016) American Chemical Society.

A programmable nanoswitch with a structural change between the linear and loop states was designed to detect the target DNA sequences by Halvorsen and coworkers (Fig. 11.8b) [99]. A simple technique, such as gel electrophoresis, is used to confirm the change in the structure of the nanoswitch, which makes this method easy to adopt and useful for biomedical applications involving DNA detection. The nanoswitch was designed as a linear double helix with one linear single strand and two short complementary sequences (a and b) hybridize to both ends of the main backbone. The short overhangs in the linear single strand are complementary to parts of the target sequence. In the presence of the target sequence the two overhangs hybridize and bring the two parts of the backbone to form a loop structure (Fig. 11.8b). The migration of the two states, linear and loop structures, is different, and hence that can be detected by a simple gel electrophoresis technique. The DNA nanoswitch can detect fragments as short as 20 bps, with sensitivity as low as picomolar range. In addition, the nanoswitch detects the target sequences without any cross reaction with other sequences.

11.4.2 Large Molecules

DNA nanodevices are equally important state-of-the-art analytical tools for the fast and selective detection of large molecules, like proteins and nuclei acids. For example, Wong and coworkers developed a nanoswitch-linked immunosorbent assay (NLISA) for a rapid and sensitive detection of a protein target (Fig. 11.9a) [100]. The DNA nanoswitch changes the structure upon binding to the target protein, and the readout was through gel electrophoresis. The nanoswitch consisted of a long 7.2 kb single-stranded DNA scaffold and antibody-oligonucleotide conjugates for hybridization on to the scaffold. In the presence of the target protein, the DNA backbone forms a loop through an antibody-protein-antibody sandwich. The conformational change, in turn, affects the migration on the agarose gel, and the intensity of the looped band on the gel measured the quantity of the target protein. This type of assay with high sensitivity in the femto-molar range can potentially replace the traditional ELISA technique [101]. This technique was further evaluated for the detection of streptavidin and prostate-specific antigen in biological fluids.

In another work, Ricci and coworkers reported a DNA nanoswitch for a rapid, single-step detection of antibodies in bodily fluids (Fig. 11.9b) [102].

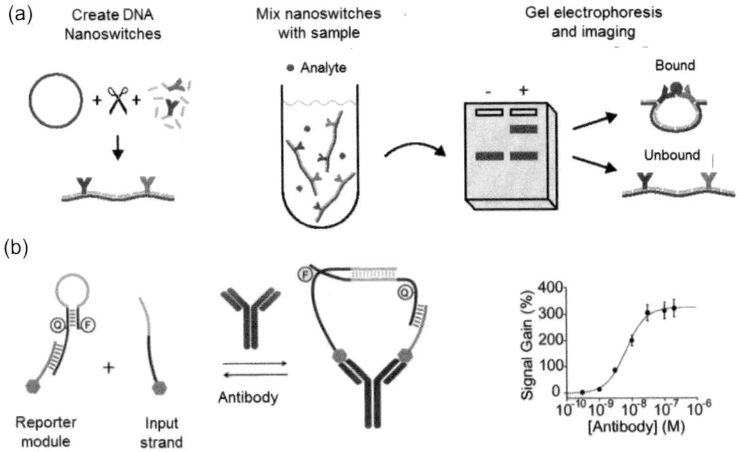

Figure 11.9 (a) DNA nanoswitches developed by hybridizing antibody–oligonucleotide conjugates and tiling oligonucleotides onto a linearized M13 scaffold. Agarose gel was used to monitor the target-binding process by the intensity of the slower migrating band that corresponds to looped nanoswitches [100]. (b) Representation of programmable DNA-based nanoswitches for anti-DNP antibody detection. Reprinted with permission from Ref. [102]. Copyright (2018) American Chemical Society.

Antibody detection plays a pivotal role in the diagnosis of pathogens and monitoring of the success of the vaccine immunization [103, 104]. The DNA nanoswitch is composed of three synthetic nucleic acid strands: (i) strand I, a fluorophore/quencher-modified DNA stem loop flanked by a 15-bp single-stranded tail, (ii) strand II, a synthetic DNA strand conjugated to an appropriate recognition element (e.g., antibody), and (iii) strand III, a DNA strand conjugated with another copy of the recognition element and containing a portion complementary to the loop sequence of strand I of the reporter module. This programmable DNA nanoswitch is activated by the antibody-induced colocalization. In the presence of the target antibody the reporter module and the input strand are colocalized in a confined volume. This brings the stem loops into close proximity, thereby increasing their local concentrations and allowing their efficient hybridization. The authors claimed that the sensing principle can be coupled with other output signals, such as electrochemical or ultraviolet-visual absorption/colorimetric, thus making the platform suitable for point-of-care applications and large-scale analysis [102].

11.5 Drug Delivery

Drug delivery is one of the challenging tasks in the fields of drug development, pharmaceutical sciences, and medicine, to achieve target-specific delivery and minimize the toxicity or side effects. There is a great demand for the development of targeted delivery methods to effectively treat major diseases, like cancers. Although organic and inorganic carriers effectively increase the water solubility, circulation time, and cellular uptake of drugs, they are well known for the inherent toxicity. In developing drug delivery systems, we need to consider the following points: (i) quantitatively loading individual or several drug molecules for synergic therapy and (ii) intelligently controlling drug release in space and time. In this context, stimuli-responsive DNA nanoswitches/devices can serve as efficient drug delivery vehicles with control on the release kinetics and targeted area. In general, DNA nanotechnology/DNA aptamer techniques can be used to develop drug delivery systems with a high affinity for the pathological location, exploiting external or internal stimulus [105]. The specific binding ability of aptamers can be explored for targeted delivery to improve the bioavailability of the drugs with significantly reduced toxicity. Therefore, stimuli-responsive DNA nanoswitches/devices and aptamers offer nanotherapy—novel drug delivery and treatment with numerous advantages.

As discussed earlier, DNA conformation can be manipulated by pH, and therefore protonation and deprotonation are the simplest ways to control its conformational changes. For example, Gueron and coworkers developed a proton (H^+)-fueled DNA nanomachine (PF-DNA) using a single-stranded DNA with four stretches of cytosine-rich sequences (i-motif domain) (Fig. 11.10b) [106]. It showed highly reversible conformational switching between a four-stranded i-motif structure at weakly acidic pH (<6.0) and random coil at higher pH (>6.4) [107]. The conformational switching of C-rich i-motif DNA was used successfully for different delivery purposes [108, 109]. In another work, Simmel and Liedl showed that the pH oscillations in a nonequilibrium chemical reaction (redox reaction) can be used to switch a pH-sensitive DNA structure between two distinct conformations: open and coiled states (Fig. 11.10a) [110].

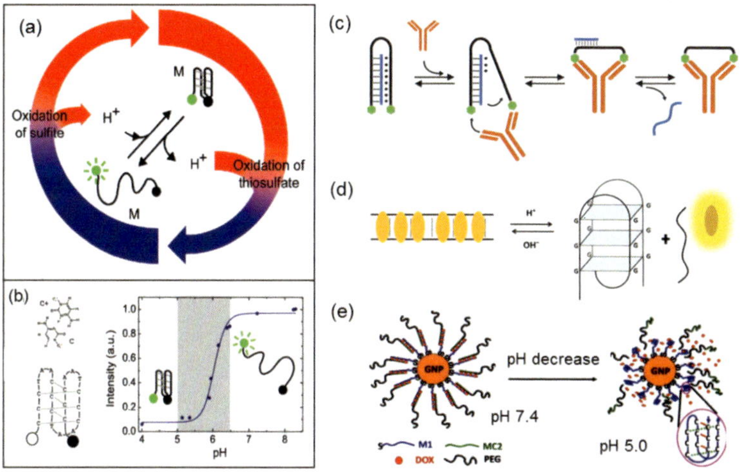

Figure 11.10 (a) Operation cycle for a redox-dependent DNA nanoswitch. (b) Scheme of how a DNA strand M-folds into the i-motif upon formation of six C^+–C pairs at pH > 6.4. (c) A DNA strand (black) labeled with two antigens (green hexagons) can load a nucleic acid strand (blue) through a clamp-like triplex-forming mechanism. The binding of an antibody to antigens causes a conformational change that reduces the stability of the triplex complex with the consequent release of the loaded strand [113]. (d) pH-dependent conformational change and release of a fluorophore-loaded DNA strand [113]. (e) The gold-nanoparticle-based naonodevice for pH-triggered drug delivery at normal physiological pH (7.4). (a, b) Reprinted with permission from Ref. [110]. Copyright (2005) American Chemical Society. (d, e) Reprinted with permission from Ref. [108], Copyright 2013, John Wiley and Sons.

Allostery is a mechanism through which binding of an effector at one site of the functional receptor causes a conformational change that modulates (activates or inhibits) its activity toward interaction/binding with another ligand [111]. Ricci and coworkers used this concept to rationally design a stem-loop DNA nanoswitch that adopts two mutually exclusive conformations, a "Load" conformation containing a doxorubicin (DOX)-intercalating domain and a "Release" conformation containing a duplex portion recognized by a specific transcription-factor [112, 113]. This strategy might allow the design and application of DNA nanoswitches in combination with different biological inputs (such as antibodies and proteins) and thus open the door for the next generation of DNA-based drug delivery systems [113]. In this study, the authors adopted a modular strategy that finds an optimal thermodynamic trade-off required for meeting the following conditions (Fig. 11.10c). First, there should be a strong difference in the stabilities between the triplex (both WC and Hoogsteen interactions) and duplex (only WC base pairings) conformations; second, the duplex conformation under the chosen experimental conditions should be unstable enough to allow release of the cargo DNA; finally, the triplex conformation should not be too stable, allowing bidentate binding to the nanomachine by the antibody. In the actual study, a cargo DNA of 12 bp labeled with a fluorophore/quencher pair was chosen to easily monitor its loading onto or release from the nanomachine. In the triplex state, the DNA was stretched and showed a maximum FL signal. The release of the cargo DNA was controlled by the amount of antibody added. The release could be triggered specifically, and again, the rate was found limited by the added antibody.

Utilization of nanoparticles (NPs) as carriers of drug molecules and to achieve controlled release to the targeted sites offers unprecedented opportunities in the field of DNA nanotechnology [108, 114–116]. For example, Jiang and coworkers recently reported an intelligent pH-responsive carrier and release system based on DNA nanoswitches. The nanoswitch was developed by the organization of gold nanoparticles (Au NPs) on mesoporous silica (MS) [117]. The DNA nanoswitch was modulated with a single-stranded DNA-2 containing four stretches of the C-rich domains, which can be efficiently switched between a folded i-motif conformation and the extended random conformation, simply by changing the solution pH from neutral to alkaline conditions, respectively [96, 118, 119]. In this DNA nanoswitch, DNA-2 was first conjugated with Au NP and then the conjugated system was further hybridized with the partially complementary DNA-1 that attached to the MS surface to form a rigid double-helical form. In mildly acidic conditions, the C-rich domain of DNA-2 can form closely packed, folded i-motif conformation through protonated cytosine-cytosine ($C{:}C^+$) base pairing interactions (Fig. 11.10b). The pH transition can thus be used to control the closing/opening of the pores of MS particles through the duplex-i-motif interconversion of this DNA nanoswitch. To prove the efficacy of the proposed system for controlled release, the pores of MS particles were filled with rhodamine B and the on-off switching was repeated several times by sequential adjustment of pH between 8 and 5. This DNA nanoswitch is a promising system for potential in vitro release of therapeutic agents.

Stimuli-responsive DNA Au NPs can be developed and used for biodiagnosis and nanomedicine. Zhou and coworkers used Au NPs in the construction of a DNA nanomachine and effectively delivered DOX into cancer cells with high cytotoxicity (Fig. 11.10d,e) [108]. In the design strategy, the authors used a PF-DNA made of a single-stranded DNA of four stretches of C-rich sequences (i-motif domain) [106]. The pH-dependent binding/release of DOX using PF-DNA duplex MN/MC2 (CCC TAA CCC TAA CCC TAA CCC/GTG TTA GG T TTA GGG TTA GGG) was developed, where MN is the PF-DNA strand made of the i-motif domain and MC2 has two mismatched bases to MN to tune the duplex stability. This ensures the formation of i-motif by MN, which can effectively displace MC2 and release intercalated DOX.

There are several other interesting reports that utilize different stimuli-responsive DNA conformational change and nanoarchitectures to design DNA nanodevices for targeted drug delivery. For example, Takakura and coworkers employed a Y-shaped oligodeoxynucleotide (Y-ODN) containing three strands partially complementary to each other to bear the CpG motifs [120]. Fan and coworkers assembled a series of DNA-origami-like nanoribbons with rolling circle amplification (RCA, isothermal nucleic acids amplification strategy)-based nanoarchitectures to carry CpG motifs into mammalian cells [121]. Tan and coworkers reported aptamer-initiated DNA nanotrains (aptNTrs) that self-assembled from two short DNA into locomotive-like guiding nanotrains [122]. Recently, Ricci and coworkers designed a stem-loop DNA nanoswitch guided by the antibody-stimulated conformational change and utilized for DOX release [112]. Our group developed a DNA nanoswitch based on pH-induced structural transition of a simple and highly efficient MB in physiological conditions [23]. This pH-dependent DNA nanoswitch showed efficient cellular uptake without the need for any transfection and may have implications in drug delivery applications. Overall, all these design principles indeed highlight the potential of DNA nanoswitches and devices as tool kits for cellular uptake and smart drug release regulated by different chemical and biological inputs.

11.6 Therapeutics and Diagnostics

Now, it is established that DNA nanocarriers are useful for targeted delivery of the drugs and biologicals (e.g., SMs and antibodies), and the concept has been intensively studied by leading pharmaceutical companies in order to achieve targeted therapy [123–126]. Although the concept of nanocarrier-drug conjugates appears simple, the development of efficient products for reliable and practical applications often represents a considerable challenge. The unique characteristics of DNA aptamers make them possible alternatives to SM/NP-based therapeutic tools for a variety of disease conditions. A number of proteins are overexpressed under cancer conditions, and suitably designed aptamers with a high affinity toward these proteins can be employed for the effective treatment of cancers [127]. The aptamer-based therapeutic option has advantages of targeted action and reduced side effects when compared to currently practiced chemotherapeutics. DNA nanodevices can be specifically designed using aptamers and conjugated with drug molecules for targeted treatment. The US Food and Drug Administration (FDA) has approved DNA molecules for therapeutic treatment and endorsed the scope of DNA nanodevices in the development.

AS1411 is a known 26-bp G-rich DNA aptamer that was previously used for the treatment of cancer [128]. This successfully passed the phase II clinical trial for the treatment of acute myeloid leukemia (AML) and renal cell carcinoma (RCC) [128, 129]. Again, with a G-rich strand that can form a G-quadruplex structure in certain biological conditions, AS1411 can be reproduced in laboratory conditions. For this, Liu and coworkers designed an aptamer-based DNA nanodevice that is useful in achieving targeted therapy (Fig. 11.11a) [130]. Studies have shown that AS1411 binds to plasma membrane nucleolin and induces tumor cell apoptosis and AS1411 arrests cell growth in the G2/M check point [131–133]. The aptamer was used to construct the DNA-drug adduct (folded) by cross-linking DOX with formaldehyde, which can be used as a therapeutic agent. This DNA-drug adduct binds to nucleolin, which is aberrantly expressed in cancer cells, and inhibits cancer progression. It has been also reported that AS1411 conjugation reduces the toxic effects of DOX due to targeted delivery.

Cancer is one of the leading diseases in the world, and there are many emerging techniques available to detect the circulating cancer cells in the bloodstream [134–136]. However, most of them are not sensitive enough for a reliable detection of circulating cancer cells. The conventional techniques employed for this purpose also include flow cytometric sorting and immunomagnetic enrichment, which are expensive and inefficient. To improve the sensitivity, DNA nanodevices conjugated with aptamers that bind strongly to the surface markers of cancer cells have been developed. Karp and coworker reported a DNA nanodevice in conjugation with the microfluidic system to capture and detect cancer cells with a high sensitivity (Fig. 11.11b) [137]. The nanodevice was found promising for the development of cancer diagnostics. In this work, the authors employed microfluidic devices with monovalent capturing agents like monoclonal antibodies. However, monoclonal antibodies are extended to a very small distance and fail to capture cells and bear the shear force. Therefore, aptamers with multiple binding sites were developed for the capture of cancer cells in microfluidic channels.

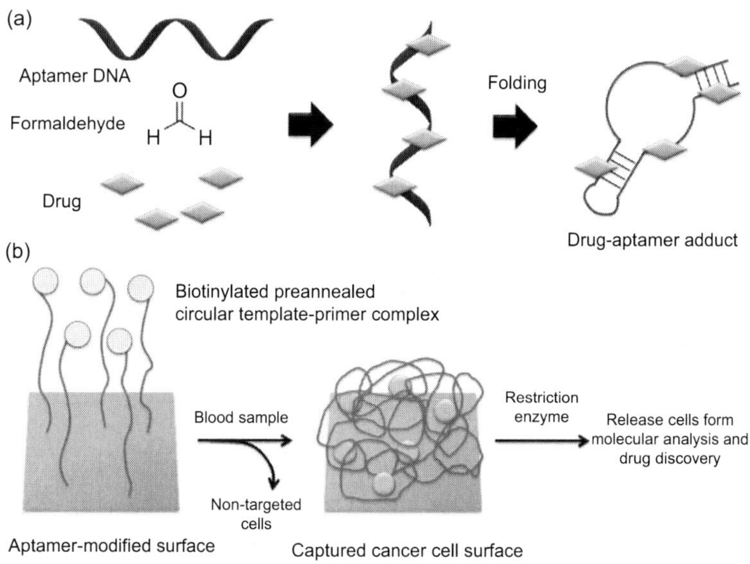

Figure 11.11 (a) DNA nanodevice developed using an aptamer to construct the DNA–drug adduct [130]. (b) Isolation and detection of cancer cells in whole blood using a long multivalent DNA aptamer–based microfluidic device [137].

The DNA nanodevice, with long aptamers, binds to the cancer marker protein tyrosine kinase-7 (PTK7), which is overexpressed on the surface of many human cancer cells. Multiple aptamers were hypothesized to bind cooperatively to the target cancer cells and increase the effective capture and sensitivity of detection/diagnosis. Extension of the aptamer chain lengths further increases the efficiency and the ability to bear the shear force. In particular, this nanodevice was reported to be efficient in the capture and diagnosis of leukemia cells.

11.7 Conclusion

In this chapter, we concisely but meticulously described the expansive amount of work undertaken to exploit the unique characteristics of flexible single-stranded DNA, rigid duplex DNA, highly predictable molecular recognition of nucleobases, and associated reversible conformational transformations of canonical and noncanonical DNA structures for developing nanoswitches and devices with numerous potential applications. This further reiterates the fact that DNA not only is an important component in modern molecular biology and allied areas of biological sciences in general but also plays an increasingly important role as a building block for nanoscale architectures, materials, and devices. We must categorically mention and acknowledge the fact that nanotechnology has made a significant contribution to analytical chemistry and diagnostics. In particular, DNA nanotechnology has provided new directions and prospects to the analyte or target recognition, capture, detection, delivery, and release-related aspects, with promising applications in the areas of bioimaging, biomedical sciences, laboratory/clinical diagnostics, drug delivery, and therapy. DNA nanoswitches and devices have been explored from the mechanistic point of view to understand the working principle, as well as their applications as sensors for ions; pH; and biomolecules, including nucleic acids, antibodies, proteins, etc. In addition, recent developments have shown several possibilities for DNA nanoarchitectures and nanodevices in the domain of applications related public health, viz development of diagnostic and therapeutic tools for cancer treatment, among other diseases.

Future developments in this area will build upon the current knowledge and modern design and technology to construct complex DNA-based nanoswitches and devices for reliable and practical applications. In designing the next generation of DNA nanoswitches and devices the emphasis must be on cost effectiveness; ease of use or operation; robustness; and performance in terms of time, site, and target. Recent advances encourage RNA-based nanodevices, despite their inherent instability, for potential application as sensor platforms, delivery vehicles, diagnosis, and therapy, due to their structural diversity. Furthermore, a detailed understanding of reversible and accurate conformational transformations between canonical and noncanonical nucleic acid (DNA/RNA) structures, a thorough knowledge of recognition/active sites, different synthetic schemes for backbone or side chain modifications with signaling moieties, mode of (bio)analysis, etc., are of utmost importance in the successful design and development of the next generation of DNA nanoswitches and devices. Above all, researchers from different fields, including chemistry, biology, material sciences, engineering, and clinicians, must work together to push the technological limits to realize the potential of DNA nanoarchitectures, nanoswitches, and devices in a wide range of biomedical applications. Overall, we expect that newer and more efficient DNA-based nanoswitches and devices will be developed for use in (bio)imaging, (bio)sensors, molecular operations in living systems, diagnostics, drug delivery, therapy, and theranostics, which will have a significant impact on the science and human life.

Acknowledgments

We thank Professor C. N. R. Rao FRS for constant support and encouragement and Jawaharlal Nehru Centre for Advanced Scientific Research (JNCASR), the DST-Nanomission (grant: DST/SMS/4428 or SR/NM/TP-25/2016), the Government of India, and Sheikh Saqr Laboratory (SSL), ICMS-JNCASR, for financial support.

References

1. Seeman, N. C. and Sleiman, H. F. (2017). DNA nanotechnology, *Nat. Rev. Mater.*, **3**, 17068.

2. Seeman, N. C. (2003). DNA in a material world, *Nature*, **421**, 427–431.
3. Winfree, E., Liu, F., Wenzler, L. A. and Seeman, N. C. (1998). Design and self-assembly of two-dimensional DNA crystals, *Nature*, **394**, 539–544.
4. Douglas, S. M., Dietz, H., Lied, T., Högberg, B., Graf, F. and Shih, W. M. (2009). Self-assembly of DNA into nanoscale three-dimensional shapes, *Nature*, **459**, 414–418.
5. Shen, B., Linko, V., Tapio, K., Kostiainen, M. A. and Toppari, J. J. (2015). Custom-shaped metal nanostructures based on DNA origami silhouettes, *Nanoscale*, **7**, 11267–11272.
6. Li, H., Carter, J. D. and LaBean, T. H. (2009). Nanofabrication by DNA self-assembly, *Mater. Today*, **12**, 24–32.
7. Simmel, S. S., Nickels, P. C. and Lied, T. (2014). Wireframe and tensegrity DNA nanostructures, *Acc. Chem. Res.*, **47**, 1691–1699.
8. Pandeeswar, M., Senanayak, S. P. and Govindaraju, T. (2016). Nanoarchitectonics of small molecule and DNA for ultrasensitive detection of mercury, *ACS Appl. Mater. Interfaces*, **8**, 30362–30371.
9. Lanier, L. A. and Bermudez, H. (2015). DNA nanostructures: a shift from assembly to applications, *Curr. Opin. Chem. Eng.*, **7**, 93–100.
10. Lo, P. K., Metera, K. L. and Sleiman, H. F. (2010). Self-assembly of three-dimensional DNA nanostructures and potential biological applications, *Curr. Opin. Chem. Biol.*, **14**, 597–607.
11. Mao, C., Sun, W., Shen, Z. and Seeman, N. C. (1999). A nanomechanical device based on the B-Z transition of DNA, *Nature*, **397**, 144–146.
12. Bath, J. and Turberfield, A. J. (2007). DNA nanomachines, *Nat. Nanotechnol.*, **2**, 275–284.
13. Simmel, F. C. and Dittmer, W. U. (2005). DNA nanodevices, *Small*, **1**, 284–299.
14. Stoltenburg, R., Reinemann, C. and Strehlitz, B. (2007). SELEX—a (r) evolutionary method to generate high-affinity nucleic acid ligands, *Biomol. Eng.*, **24**, 381–403.
15. Casey, J., Grinstein, R. S. and Orlowski, J. (2010). Sensors and regulators of intracellular pH, *Nat. Rev. Mol. Cell Biol.*, **11**, 50–61.
16. Gehring, K., Leroy, J. L. and Gueron, M. (1993). A tetrameric DNA structure with protonated cytosine-cytosine base pairs, *Nature*, **363**, 561–565.
17. Dong, Y., Yang, Z. and Liu, D. (2014). DNA nanotechnology based on *i*-motif structures, *Acc. Chem. Res.*, **47**, 1853–1860.

18. Idili, A., Vallée-Bélisle, A. and Ricci, F. (2014). Programmable pH-triggered DNA nanoswitches, *J. Am. Chem. Soc.*, **136**, 5836–5839.

19. Yang, M., Zhang, X., Liu, H., Kang, H., Zhu, Z., Yang, W. and Tan, W. (2015). Stable DNA nanomachine based on duplex–triplex transition for ratiometric imaging instantaneous pH changes in living cells, *Anal. Chem.*, **87**, 5854–5859.

20. Iacovelli, F., Idili, A., Benincasa, A., Mariottini, D., Ottaviani, A., Falconi, M., Ricci, F. and Desideri, A. (2017). Simulative and experimental characterization of a pH-Dependent clamp-like DNA triple-helix nanoswitch, *J. Am. Chem. Soc.*, **139**, 5321–5329.

21. Rich, A., Davies, D. R., Crick, F. H. C. and Watson, J. D. (1961). The molecular structure of polyadenylic acid, *J. Mol. Biol.*, **3**, 71–86.

22. Chakraborty, K., Sharma, S., Maiti, P. K. and Krishnan, Y. (2009). The poly dA helix: a new structural motif for high performance DNA-based molecular switches, *Nucleic Acids Res.*, **97**, 2810–2817.

23. Narayanaswamy, N., Suresh, G., Priyakumar, U. D. and Govindaraju, T. (2015). Double zipper helical assembly of deoxyoligonucleotides: mutual templating and chiral imprinting to form hybrid DNA ensembles, *Chem. Commun.*, **51**, 5493–5496.

24. Elbaz, J., Want, Z. G., Orbach, R. and Willner, I. (2009). pH-stimulated concurrent mechanical activation of two DNA "Tweezers". A "SET–RESET" logic gate system, *Nano Lett.*, **9**, 4510–4514.

25. Liu, X., Lu, C. H. and Willner, I. (2014). Switchable reconfiguration of nucleic acid nanostructures by stimuli-responsive DNA machines, *Acc. Chem. Res.*, **47**, 1673–1680.

26. D'haeseleer, P. (2006). What are DNA sequence motifs?, *Nat. Biotech.*, **24**, 423–425.

27. Narayanaswamy, N., Nair, R. R., Suseela, Y. V., Saini, D. K. and Govindaraju, T. (2016). A molecular beacon-based DNA switch for reversible pH sensing in vesicles and live cells, *Chem. Commun.*, **52**, 8741–8744.

28. Ohmichi, T., Kawamoto, Y., Wu, P., Miyoshi, D., Karimata, H. and Sugimoto, N. (2005). DNA-based biosensor for monitoring pH in vitro and in living cells, *Biochemistry*, **44**, 7125–7130.

29. Leitner, D., Schroder, W. and Weisz, K. (2000). Influence of sequence-dependent cytosine protonation and methylation on DNA triplex stability, *Biochemistry*, **39**, 5886–5892.

30. Sugimoto, N., Wu, P., Hara, H. and Kawamoto, Y. (2001). pH and cation effects on the properties of parallel pyrimidine motif DNA triplexes, *Biochemistry*, **40**, 9396–9405.

31. Luo, F., Xi, Q., Jiang, J. and Yu, R. (2016). Graphene oxide based DNA nanoswitches as a programmable pH-responsive biosensor, *Anal. Methods*, **8**, 6982–6985.
32. Modi, S., Nizak, C., Surana, S., Halder, S. and Krishnan, Y. (2013). Two DNA nanomachines map pH changes along intersecting endocytic pathways inside the same cell, *Nat. Nanotechnol.*, **8**, 459–467.
33. Carter, K. P., Young, A. M. and Palmer, A. E. (2014). Fluorescent sensors for measuring metal ions in living systems, *Chem. Rev.*, **114**, 4564–4601.
34. Zhou, W., Saran, R. and Liu, J. (2017). Metal sensing by DNA, *Chem. Rev.*, **117**, 8272–8325.
35. Guengerich, F. P. (2009). Thematic series: metals in biology, *J. Biol. Chem.*, **284**, 709.
36. Domaille, D. W., Que, E. L. and Chang, C. J. (2008). Synthetic fluorescent sensors for studying the cell biology of metals, *Nat. Chem. Biol.*, **4**, 168–175.
37. Fu, D. and Finney, L. (2014). Metalloproteomics: challenges and prospective for clinical research applications, *Expert Rev. Proteomics*, **11**, 13–19.
38. Gray, H. B. (2003). Biological inorganic chemistry at the beginning of the 21st century, *Proc. Natl. Acad. Sci. U.S.A.*, **100**, 3563–3568.
39. Ferré-D'Amaré, A. R. and Winkler, W. C. (2011). The roles of metal ions in regulation by riboswitches, *Met. Ions Life Sci.*, **9**, 141–173.
40. Jaitovich, A. and Bertorello, A. M. (2010). Intracellular sodium sensing: SIK1 network, hormone action and high blood pressure, *Biochim. Biophys. Acta*, **1802**, 1140–1149.
41. Elgqvist, J. (2017). Nanoparticles as theranostic vehicles in experimental and clinical applications – focus on prostate and breast cancer, *Int. J. Mol. Sci.*, **18**, 1102.
42. Jennrich, P. (2013). The influence of arsenic, lead and mercury on the development of cardiovascular diseases, *ISRN Hypertension*, **2013**, 234034.
43. Gale, P. A. and Caltagirone, C. (2015). Anion sensing by small molecules and molecular ensembles, *Chem. Soc. Rev.*, **44**, 4212–4227.
44. Torabi, S. F., Wu, P., McGhee, C. E., Chen, L., Hwang, K., Zheng, N., Cheng, J. and Lu, Y. (2015). In vitro selection of a sodium-specific DNAzyme and its application in intracellular sensing, *Proc. Natl. Acad. Sci. U.S.A.*, **112**, 5903–5908.

45. Santoro, S. W. and Joyce, G. F. (1997). A general purpose RNA-cleaving DNA enzyme, *Proc. Natl. Acad. Sci. U.S.A.*, **94**, 4262–4266.

46. Alberts, B., Johnson, A., Lewis, J., Raff, M., Roberts, K. and Walter, P. (2002). *Molecular Biology of the Cell*, 4th ed., Garland Science, New York.

47. Teixeira, M. F. S., Freitas, B. H., Seraphim, P. M., Salmazo, L. O., Nobre, M. A. and Lanfredi, S. (2009). Development of an electrochemical sensor for potassium ions based on KSr2Nb5O15 modified electrode, *Procedia Chem.*, **1**, 293–296.

48. Collie, G. W. and Parkinson, G. N. (2011). The application of DNA and RNA G-quadruplexes to therapeutic medicines, *Chem. Soc. Rev.*, **40**, 5867–5892.

49. Kwok, C. K. and Merrick, C. J. (2017). G-quadruplexes: prediction, characterization and biological application, *Trends Biotechnol.*, **35**, 997–1013.

50. Tian, T., Xiao, H. and Zhou, X. (2015). A review: G-quadruplex's applications in biological target detection and drug delivery, *Curr. Top. Med. Chem.*, **15**, 1988–2001.

51. Ruttkay-Nedecky, B., Kudr, J., Nejdl, L., Maskova, D., Kizek, R. and Adam, V. (2013). G-quadruplexes as sensing probes, *Molecules*, **18**, 14760–14779.

52. Lu, M., Guo, Q. and Kallenbach, N. R. (1992). Structure and stability of sodium and potassium complexes of dT4G4 and dT4G4T, *Biochemisrty*, **31**, 2455–2459.

53. Willamson, J. R. (1994). G-quartet structures in telomeric DNA, *Annu. Rev. Biophys. Biomol. Struct.*, **23**, 703–730.

54. Hud, N. V., Smith, F. W., Anet, F. A. and Feigon, J. (1996). The selectivity for K+ versus Na+ in DNA quadruplexes is dominated by relative free energies of hydration: a thermodynamic analysis by 1H NMR, *Biochemistry*, **35**, 15383–15390.

55. Ueyama, H., Takagi, M. and Takenaka, S. (2002). A novel potassium sensing in aqueous media with a synthetic oligonucleotide derivative. fluorescence resonance energy transfer associated with guanine quartet-potassium ion complex formation, *J. Am. Chem. Soc.*, **124**, 14286–14287.

56. Bock, L. C., Griffin, L. C., Latham, J. A., Vermaas, E. H. and Toole, J. J. (1992). Selection of single-stranded DNA molecules that bind and inhibit human thrombin, *Nature*, **355**, 564–566.

57. Nagatoishi, S., Nojima, T., Juskowiak, B. and Takenaka, S. (2005). A pyrene-labeled G-quadruplex oligonucleotide as a fluorescent probe for potassium ion detection in biological applications, *Angew. Chem. Int. Ed.*, **44**, 5067–5070.

58. Yamane, T. and Davidson, N. (1961). On the complexing of desoxyribonucleic acid (DNA) by mercuric ion, *J. Am. Chem. Soc.*, **83**, 2599–2607.

59. Kosturko, L. D., Folzer, C. and Stewart, R. F. (1974). Crystal and molecular structure of a 2:1 complex of 1-methylthymine-mercury(II), *Biochemistry*, **13**, 3949–3952.

60. Liu, X., Tang, Y., Wang, L., Zhang, J., Song, S., Fan, C. and Wang, S. (2007). Optical detection of mercury(II) in aqueous solutions by using conjugated polymers and label-free oligonucleotides, *Adv. Mater.*, **19**, 1471–1474.

61. Ono, A. and Togashi, H. (2004). Highly selective oligonucleotide-based sensor for mercury(ii) in aqueous solutions, *Angew. Chem. Int. Ed.*, **43**, 4300–4302.

62. Wang, Z., Lee, J. H. and Lu, Y. (2008). Highly sensitive "turn-on" fluorescent sensor for Hg^{2+} in aqueous solution based on structure-switching DNA, *Chem. Commun.*, 6005–6007.

63. Li, T., Li, B., Wang, E. and Dong, S. (2009). G-quadruplex-based DNAzyme for sensitive mercury detection with the naked eye, *Chem. Commun.*, 3551–3553.

64. Ge, J., Li, X.-P., Jiang, J.-H. and Yu, R.-Q. (2014). A highly sensitive label-free sensor for mercury ion (Hg^{2+}) by inhibiting thioflavin T as DNA G-quadruplexes fluorescent inducer, *Talanta*, **122**, 85–90.

65. Ma, L., Wu, G., Li, Y., Qin, P., Meng, L., Liu, H., Li, Y. and Diao, A. (2015). A reversible metal ion fueled DNA three-way junction molecular device for "turn-on and -off" fluorescence detection of mercury ions (II) and biothiols respectively with high selectivity and sensitivity, *Nanoscale*, **7**, 18044–18048.

66. Liu, J. and Lu, Y. (2007). Rational design of "turn-on" allosteric DNAzyme catalytic beacons for aqueous mercury ions with ultrahigh sensitivity and selectivity, *Angew. Chem. Int. Ed.*, **46**, 7587–7590.

67. Lee, J.-S., Han, M. S. and Mirkin, C. A. (2007). Colorimetric detection of mercuric ion (Hg^{2+}) in aqueous media using DNA-functionalized gold nanoparticles, *Angew. Chem. Int. Ed.*, **46**, 4093–4096.

68. Li, T., Dong, S. and Wang, E. (2010). A lead(II)-driven DNA molecular device for turn-on fluorescence detection of lead(II) ion with high selectivity and sensitivity, *J. Am. Chem. Soc.*, **132**, 13156–13157.

69. Li, T., Wang, E. and Dong, S. (2009). Potassium–lead-switched G-quadruplexes: a new class of DNA logic gates, *J. Am. Chem. Soc.*, **131**, 15082–15083.

70. Hara-Chikuma, M., Yang, B., Sonawane, N. D., Sasaki, S., Uchida S. and Verkman, A. S. (2005). ClC-3 chloride channels facilitate endosomal acidification and chloride accumulation, *J. Biol. Chem.*, **280**, 1241–1247.

71. Gould, D. (1998). Human physiology. From cells to systems, 3rd ed. *J. Adv. Nurs.*, **28**, 680–682.

72. Russell, J. M. (2000). Sodium–potassium–chloride cotransport, *Physiol. Rev.*, **80**, 211–276.

73. Saha, S., Prakash, V., Halder, S., Chakraborty K. and Krishnan, Y. (2015). A pH-independent DNA nanodevice for quantifying chloride transport in organelles of living cells, *Nat. Nanotechnol.*, **10**, 645–651.

74. Modi, S., Swetha, M. G., Goswami, D., Gupta, G. D., Mayor, S. and Krishnan, Y. (2009). A DNA nanomachine that maps spatial and temporal pH changes inside living cells, *Nat. Nanotechnol.*, **4**, 325–330.

75. Giljohann, D. A. and Mirkin, C. A. (2009). Drivers of biodiagnostic development, *Nature*, **462**, 461–464.

76. Kelley, S. O. (2017). What are clinically relevant levels of cellular and biomolecular analytes?, *ACS Sens.*, **2**, 193–197.

77. Song, S., Wang, L., Li, J., Zhao, J. and Fan, C. (2008). Aptamer-based biosensors, *Trends Anal. Chem.*, **27**, 108–117.

78. Cho, E. J., Lee, J.-W. and Ellington, A. D. (2009). Applications of aptamers as sensors, *Annu. Rev. Anal. Chem.*, **2**, 241–264.

79. Stojanovic, M. N., de Prada, P. and Landry, D. W. (2001). Aptamer-based folding fluorescent sensor for cocaine, *J. Am. Chem. Soc.*, **123**, 4928–4931.

80. Stojanovic, M. N. and Kolpashchikov, D. M. (2004). Modular aptameric sensors, *J. Am. Chem. Soc.*, **126**, 9266–9270.

81. Kolpashchikov, D. M. (2005). Binary malachite green aptamer for fluorescent detection of nucleic acids, *J. Am. Chem. Soc.*, **127**, 12442–12443.

82. Yoshida, W., Sode, K. and Ikebukuro, K. (2006). Aptameric enzyme subunit for biosensing based on enzymatic activity measurement, *Anal. Chem.*, **78**, 3296–3303.

83. Huizenga, D. A. and Szostak, J. W. (1995). A DNA aptamer that binds adenosine and ATP, *Biochemistry*, **34**, 656–665.

84. Peng, P., Shi, L., Wang, H. and Li, T. (2017). A DNA nanoswitch-controlled reversible nanosensor, *Nucleic Acids Res.*, **45**, 541–546.

85. Tang, Z., Mallikaratchy, P., Yang, R., Kim, Y., Zhu, Z., Wang, H. and Tan, W. (2008). Aptamer switch probe based on intramolecular displacement, *J. Am. Chem. Soc.*, **130**, 11268–11269.

86. Xing, Y., Yang, Z. and Liu, D. (2011). A responsive hidden toehold to enable controllable DNA strand displacement reactions, *Angew. Chem. Int. Ed.*, **50**, 11934–11936.

87. Kong, L., Xu, J., Xu, Y., Xiang, Y., Yuan, R. and Chai, Y. (2013). A universal and label-free aptasensor for fluorescent detection of ATP and thrombin based on SYBR Green I dye, *Biosens. Bioelectron.*, **42**, 193–197.

88. Wang, H., Peng, P., Liu, S. and Li, T. (2016). Thioflavin T behaves as an efficient fluorescent ligand for label-free ATP aptasensor, *Anal. Bioanal. Chem.*, **408**, 7927–7934.

89. Lubrich, D., Lin, J. and Yan, J. (2008). A contractile DNA machine, *Angew. Chem. Int. Ed.*, **47**, 7026–7028.

90. Oha, S. S., Plakos, K., Lou, X., Xiao, Y. and Soh, H. T. (2010). In vitro selection of structure-switching, self-reporting aptamers, *Proc. Natl. Acad. Sci. U.S.A.*, **107**, 14053–14058.

91. Bock, L. C., Griffin, L. C., Latham, J. A., Vermaas, E. H. and Toole, J. J. (1992). Selection of single-stranded DNA molecules that bind and inhibit human thrombin, *Nature*, **355**, 564–566.

92. Kawakami, J., Imanaka, H., Yokota, Y. and Sugimoto, N. (2000). In vitro selection of aptamers that act with Zn^{2+}, *J. Inorg. Biochem.*, **82**, 197–206.

93. Minagawa, H., Onodera, K., Fujita, H., Sakamoto, T., Akitomi, J., Kaneko, N., Shiratori, I., Kuwahara, M., Horii, K. and Waga, I. (2017). Selection, characterization and application of artificial DNA aptamer containing appended bases with sub-nanomolar affinity for a salivary biomarker, *Sci. Rep.*, **7**, 42716.

94. Baker, B., Rebecca, R., Lai, Y., Wood, M. S., Doctor, E. H., Heeger, A. J. and Plaxco, K. W. (2006). An electronic, aptamer-based small-molecule sensor for the rapid, label-free detection of cocaine in adulterated samples and biological fluids, *J. Am. Chem. Soc.*, **128**, 3138–3139.

95. Liu, B., Zhang, B., Chen G. and Tang, D. (2014). An omega-like DNA nanostructure utilized for small molecule introduction to stimulate formation of DNAzyme–aptamer conjugates, *Chem. Commun.*, **50**, 1900–1902.

96. Tang, Y., Ge, B., Sen, D. and Y, H.-Z. (2014). Functional DNA switches: rational design and electrochemical signaling, *Chem. Soc. Rev.*, **43**, 518–529.

97. Buck, A. H., Campbell, C. J., Dickinson, P., Mountford, C. P., Stoquert, H. C., Terry, J. G., Evans, S. A. G., Keane, L. M., Su, T.-J., Mount, A. R., Walton, A. J., Beattie, J. S., Crain, J. and Ghazal, P. (2007). DNA nanoswitch as a biosensor, *Anal. Chem.*, **79**, 4724–4728.

98. Holliday, R. A. (1964). A mechanism for gene conversion in fungi, *Genet. Res.*, **5**, 282–304.

99. Chandrasekaran, A. R., Zavala, J. and Halvorsen, K. (2016). Programmable DNA nanoswitches for detection of nucleic acid sequences, *ACS Sens.*, **1**, 120–123.

100. Hansen, C. H., Yang, D., Koussa M. A. and Wong, W. P. (2017). Nanoswitch-linked immunosorbent assay (NLISA) for fast, sensitive, and specific protein detection, *Proc. Natl. Acad. Sci. U.S.A.*, **114**, 10367–10372.

101. Lequin, R. M. (2005). Enzyme immunoassay(EIA)/enzyme-linked immunosorbent assay (ELISA), *Clin. Chem.*, **51**, 2415–2418.

102. Porchetta, A., Ippodrino, R., Marini, B., Caruso, A., Caccuri, F. and Ricci, F. (2018). Programmable nucleic acid nanoswitches for the rapid, single-step detection of antibodies in bodily fluids, *J. Am. Chem. Soc.*, **140**, 947–953.

103. Cutts, J. C., Powell, R., Agius, P. A., Beeson, J. G., Simpson, J. A. and Fowkes, F. J. I. (2014). Immunological markers of plasmodium vivax exposure and immunity: a systematic review and meta-analysis, *BMC Med.*, **12**, 150–170.

104. Gubala, V., Harris, L. F., Ricco, A. J., Tan, M. X. and Williams, D. E. (2012). Point of care diagnostics: status and future, *Anal. Chem.*, **84**, 487–515.

105. Chao, J., Liu, H., Su, S., Wang, L., Huang, W. and Fan, C. (2014). Structural DNA nanotechnology for intelligent drug delivery, *Small*, **10**, 4626–4635.

106. Gehring, K., Leroy, J. L. and Gueron, M. (1993). A tetrameric DNA structure with protonated cytosine-cytosine base pairs, *Nature*, **363**, 561–565.

107. Liu, D. and Balasubramanian, S. S. (2003). A proton-fulled DNA nanomachine, *Angew. Chem. Int. Ed.*, **42**, 5734–5736.

108. Song, L., Ho, V. H. B., Chen, C., Yang, Z., Liu, D., Chen, R. and Zhou, D. (2013). Efficient, pH-triggered drug delivery using a pH-responsive DNA-conjugated gold nanoparticle, *Adv. Healthcare Mater.*, **2**, 275–280.

109. Zhang, W., Wang, F., Wang, Y., Wang, J., Yu, Y., Guo, S., Chen, R. and Zhou, D. (2016). pH and near-infrared light dual-stimuli responsive drug delivery using DNA-conjugated gold nanorods for effective treatment of multidrug resistant cancer cells, *J. Control. Release*, **232**, 9–19.

110. Liedl, T. and Simmel, F. C. (2005). Switching the conformation of a DNA molecule with a chemical oscillator, *Nano Lett.*, **5**, 1894–1898.

111. Motlagh, H. N., Wrabl, J. O., Li, J. and Hilser, V. J. (2014). The ensemble nature of allostery, *Nature*, **508**, 331–339.

112. Rossetti, M., Ranallo, S., Idili, A., Palleschi, G., Porchetta, A. and Ricci, F. (2017). Allosteric DNA nanoswitches for controlled release of a molecular cargo triggered by biological inputs, *Chem. Sci.*, **8**, 914–920.

113. Ranallo, S., Prévost-Tremblay, C., Idili, A., Vallée-Bélisle, A. and Ricci, F. (2017). Antibody-powered nucleic acid release using a DNA-based nanomachine, *Nat. Commun.*, **8**, 15150.

114. Davis, M. E., Chen, Z. and Shin, D. M. (2008). Nanoparticle therapeutics: an emerging treatment modality for cancer, *Nat. Rev. Drug Discov.*, **7**, 771–782.

115. Chen, Z. (2010). Small-molecule delivery by nanoparticles for anticancer therapy, *Cell*, **16**, 594–602.

116. Couvreur, P. (2013). Nanoparticles in drug delivery: past, present and future, *Adv. Drug Deliv. Rev.*, **65**, 21–23.

117. Chen, L., Di, J., Cao, C., Zhao, Y., Ma, Ying, Luo, J., Wen, Y., Song, W., Song, Y. and Jiang, L. (2011). A pH-driven DNA nanoswitch for responsive controlled release, *Chem. Commun.*, **47**, 2850–2852.

118. Liu, H. J. and Liu, D. S. (2009). DNA nanomachines and their functional evolution, *Chem. Commun.*, **19**, 2625–2635.

119. Hou, X., Guo, W., Xia, F., Nie, F. Q., Dong, H., Tian, Y., Wen, L. P., Cao, L., Yang, Y., Xue, J., Song, Y., Wang, Y., Liu, D. and Jiang, L. (2009). A biomimetic potassium responsive nanochannel: G-quadruplex DNA conformational switching in a synthetic nanopore, *J. Am. Chem. Soc.*, **131**, 7800–7805.

120. Nishikawa, M., Matono, M., Rattanakiat, S., Matsuoka N. and Takakura, Y. (2008). Enhanced immunostimulatory activity of oligodeoxynucleotides by Y-shape formation, *Immunology*, **124**, 247–255.

121. Ouyang, X., Li, J., Liu, H., Zhao, B., Yan, J., Ma, Y., Xiao, S., Song, S., Huang, Q., Chao, J. and Fan, C. (2013). Rolling circle amplification-based DNA origami nanostructrures for intracellular delivery of immunostimulatory drugs, *Small*, **9**, 3082–3087.

122. Zhu, G., Zheng, J., Song, E., Donovan, M., Zhang, K., Liu, C. and Tan, W. (2013). Self-assembled, aptamer-tethered DNA nanotrains for targeted transport of molecular drugs in cancer theranostics, *Proc. Natl. Acad. Sci. U.S.A.*, **110**, 7998–8003.

123. Hughes, B. (2010). Antibody-drug conjugates for cancer: poised to deliver?, *Nat. Rev. Drug Discov.*, **9**, 665–667.

124. Casi, G. and Neri, D. (2012). Antibody–drug conjugates: basic concepts, examples and future perspectives, *J. Control. Release*, **161**, 422–428.

125. Zhu, G., Ye, M., Donovan, M. J., Song, E., Zhao, Z. and Tan, W. (2012). Nucleic acid aptamers: an emerging frontier in cancer therapy, *Chem. Commun.*, **48**, 10472–10480.

126. Pharma, W. S. (2011). Interest surges in antibody drug conjugates, *Nat. Biotechnol.*, **29**, 297–298.

127. Bayraktar, S. and Rocha-Lima, C. M. (2010). Advanced or metastatic pancreatic cancer: molecular targeted therapies, *Mt. Sinai J. Med.*, **77**, 606–619.

128. Shieh, Y. A., Yang, S. J., Wei, M. F. and Shieh, M. J. (2010). Aptamer-based tumor-targeted drug delivery for photodynamic therapy, *ACS Nano*, **4**, 1433–1442.

129. Mongelard, F. and Bouvet, P. (2010). AS-1411, a guanosine-rich oligonucleotide aptamer targeting nucleolin for the potential treatment of cancer, including acute myeloid leukemia, *Curr. Opin. Mol. Ther.*, **12**, 107–114.

130. Trinh, T. L., Zhu, G., Xiao, X., Puszyk, W., Sefah, K., Wu, Q., Tan, W. and Liu, C. (2015). A synthetic aptamer-drug adduct for targeted liver cancer therapy, *PLoS One*, **10**, 1–17.

131. Bates, P. J., Kahlon, J. B., Thomas, S. D., Trent, J. O. and Miller, D. M. (1999). Antiproliferative activity of G-rich oligonucleotides correlates with protein binding, *J. Biol. Chem.*, **274**, 26369–26377.

132. Bates, P. J., Laber, D. A., Miller, D. M., Thomas, S. D. and Trent, J. O. (2009). Discovery and development of the G-rich oligonucleotide AS1411 as a novel treatment for cancer, *Exp. Mol. Pathol.*, **86**, 151–164.

133. Soundararajan, S., Wang, L., Sridharan, V., Chen, W., Courtenay-Luck, N., Jones, D., Spicer, E. K. and Fernandes, D. J. (2009). Plasma membrane nucleolin is a receptor for the anticancer aptamer AS1411 in MV4-11 leukemia cells, *Mol. Pharmacol.*, **76**, 984–991.

134. Qureshi, A., Gurbuz, Y. and Niazi, J. H. (2015). Label-free capacitance based aptasensor platform for the detection of HER2/ErbB2 cancer biomarker in serum, *Sens. Actuators, B*, **220**, 1145–1151.

135. Cardoso, A. R., Moreira, F. T. C., Fernandes, R. and Sales, M. G. F. (2016). Novel and simple electrochemical biosensor monitoring attomolar levels of miRNA-155 in breast cancer, *Biosens. Bioelectron.*, **80**, 621–630.

136. Kosaka, P. M., Pini, V., Ruz, J. J., da Silva, R. A., González, M. U., Ramos, D., Calleja, M. and Tamayo, J. (2014). Detection of cancer biomarkers in serum using a hybrid mechanical and optoplasmonic nanosensor, *Nat. Nanotechnol.*, **9**, 1047–1053.

137. Zhaoa, W., Cui, C. H., Bose, S., Guo, D., Shen, C., Wong, W. P., Halvorsen, K., Farokhzad, O. C., Teo, G. S. L., Phillips, J. A., Dorfmanj, D. M., Karnik, R. and Karp, J. M. (2012). Bioinspired multivalent DNA network for capture and release of cells, *Proc. Natl. Acad. Sci. U.S. A.*, **109**, 19626–19631.

Index